WAKING,
DREAMING,
BEING

WAKING, DREAMING, BEING

self and consciousness in neuroscience,
meditation, and philosophy

EVAN THOMPSON

COLUMBIA UNIVERSITY PRESS

NEW YORK

Columbia University Press
Publishers Since 1893
New York   Chichester, West Sussex
cup.columbia.edu
Copyright © 2015 Evan Thompson
All rights reserved

Library of Congress Cataloging-in-Publication Data
    Thompson, Evan.
  Waking, dreaming, being : self and consciousness in neuroscience,
meditation, and philosophy / Evan Thompson.
    pages   cm
  Includes bibliographical references and index.
  ISBN 978-0-231-13709-6 (cloth : alk. paper) —
  ISBN 978-0-231-53831-2 (electronic)
  1. Consciousness. 2. Self. I. Title.
  B808.9T456   2015
  128'.2—dc23

                                2014004887

Cover design: Alex Camlin
Cover image: The Buddha, c.1905 (pastel on paper), Redon, Odilon
(1840–1916) / Musée d'Orsay, Paris, France / Giraudon / Bridgeman Images
Book design: Lisa Hamm

References to websites (URLs) were accurate at the time of writing.
Neither the author nor Columbia University Press is responsible for URLs
that may have expired or changed since the manuscript was prepared.

For my mother and father:

Gail Gordon Thompson and William Irwin Thompson

Someone who dreams of drinking wine at a cheerful banquet may wake up crying the next morning. Someone who dreams of crying may go off the next morning to enjoy the sport of the hunt. When we are in the midst of a dream, we do not know it's a dream. Sometimes we may even try to interpret our dreams while we are dreaming, but then we awake and realize it was a dream. Only after one is greatly awakened does one realize that it was all a great dream, while the fool thinks that he is awake and presumptuously aware.

—Chuang Tzu (Zhuang Zi), trans. Victor H. Mair,
Wandering on the Way: Early Taoist Tales and Parables of Chuang Tzu

CONTENTS

FOREWORD

Were you to ask someone today what comes to mind on hearing the word "meditation," I suspect that they would immediately think of a person sitting cross-legged on a cushion with her eyes closed. If you then asked: "Do you meditate?" they would likely take this as a question about whether they regularly practice a spiritual exercise, probably derived from an Asian religion. Yet such usage of the term "meditation" in the West is relatively recent. A hundred years ago, the word would rarely have been used this way. To say you were "meditating" would have meant that you were "considering something thoughtfully" (according to Chambers), a definition that now strikes us as odd or archaic. Few today would find the seventeenth-century poet John Milton's line "And strictly meditate the thankless Muse" comprehensible at all.

This is but one example of how non-Western ideas have slowly infiltrated our ways of thinking and speaking. Such shifting use of language forms part of a complex historical and cultural process that began with the first translations of classical Asian texts at the end of the eighteenth century. Interest in Eastern ideas has waxed and waned over the decades since, but during the past fifty years or so—largely triggered by the countercultural movements of the 1960s—it seems to have increased exponentially. While it may have taken more than a century for "meditation" to have acquired its current usage, it seems to have taken only ten or fifteen years for the word "mindfulness" to have gained a similar currency in English.

When I lived as a Buddhist monk in the Tibetan community around the Dalai Lama in Dharamsala during the early 1970s, if someone had suggested that by 2013 mindfulness meditation would be freely available through the British National Health Service and that a U.S. congressman would have published a book called *A Mindful Nation*, I would have dismissed that person as a fantasist. Yet this is exactly what has happened. A term that was once used in its current sense only by a small number of Buddhist meditators has now spread virally into popular discourse.

The medical professionals who use mindfulness-based therapies have (or should have) no interest in promoting Buddhism. They have adopted the practice of mindfulness solely because clinical trials have shown that it is an effective treatment for certain pathological conditions such as relapse into depression. Yet this raises a crucial question: if one of its central practices can be shown to work in a nonreligious, secular setting, should we continue to regard Buddhism as a religion? According to the earliest canonical texts, Siddhattha Gotama presented his teachings as practical exercises that would lead to palpable and predictable results. Although the Buddha's teachings have given rise over the centuries to faith-based institutions and dogmas that we would unhesitatingly call "religious," might such developments be a digression from or even a betrayal of the founder's original intentions?

We need, therefore, to distinguish between Buddhism as a belief-based religion and Buddhism as a pragmatic philosophy with ethical and contemplative practices. Yet it is not so easy to know where to draw the line. Among contemporary Buddhists, opinions vary as to whether reincarnation, for example, should be treated as an empirical fact that will one day be confirmed by scientific evidence, or as merely an artifact of prescientific Indian cosmology. Some argue that Buddhism without reincarnation is unintelligible; others insist that the doctrine of reincarnation obscures the Buddha's primary concerns and should be discarded.

Evan Thompson refuses to take an a priori stand on such issues. He seeks instead to keep an open mind with regard to the metaphysical claims of traditional Buddhist, Hindu, and Taoist teachings, while at the same time subjecting them to a critique based on the findings of science. Likewise, he acknowledges the limitations of the scientific method in coming to terms with the felt reality of first-person experience, which involves what it is like to be a self endowed with consciousness. He thus envisages a synthesis of objective scientific rigor coupled with first-hand reports of contemplative experience. Moreover, although trained in the discipline of Western philosophy, he draws on the rich philosophical thought of India and China, thus building a way to a collaboration of ideas that is still resisted in the Eurocentric culture of the Western academy.

I first encountered the work of Evan Thompson in the early 1990s when I came across *The Embodied Mind: Cognitive Science and Human Experience*, which he coauthored with the neuroscientist Francisco Varela and the psychologist Eleanor Rosch. Reading this work from the perspective of a Buddhist scholar-practitioner who was neither a scientist nor a Western philosopher or psychologist, I found it a breath of fresh air to encounter authors (a scientist, a philosopher, and a psychologist) who were open to engaging respectfully and critically with non-Western traditions. *The Embodied Mind* is often spoken of as a pioneering work in developing the concept of "enactive" cognition, but it was equally pioneering as one of the first serious works of cross-cultural philosophy.

The Embodied Mind concludes with a question and a tentative answer:

When these two planetary forces, science and Buddhism, come genuinely together, what might not happen? At the very least, the journey of Buddhism to the West provides some of the resources we need to pursue consistently our own cultural and scientific premises to the point where we no longer need and desire foundations and so can take up the further tasks of building and dwelling in worlds without ground.

Evan Thompson's *Waking, Dreaming, Being* offers us a detailed and more fully articulated response to this same question.

In the twenty-two years that have elapsed since the publication of *The Embodied Mind,* the dialogue between science and Buddhism has developed in leaps and bounds. A key to this development has been the enthusiastic involvement of Tenzin Gyatso, the Dalai Lama of Tibet, in a series of discussions under the auspices of the Mind and Life Institute. This program began modestly in Dharamsala in 1987 as an "intellectual experiment" by the Dalai Lama, Francisco Varela, and the entrepreneur Adam Engle. In January 2013, it held its twenty-sixth conference, where the Dalai Lama and twenty leading scientists and philosophers addressed an audience of eight thousand Tibetan monks and nuns at Drepung Monastery in South India on the topic of "Mind, Brain and Matter."

Evan Thompson is uniquely placed to draw together the complex elements that constitute this ongoing dialogue among Buddhism, the natural sciences, and philosophy. Home educated in the countercultural Lindisfarne Association—founded in 1972 by his father, the writer, poet, and social critic William Irwin Thompson—he went on to earn a degree in Asian studies at Amherst College, followed by a doctorate in philosophy from the University of Toronto. He first met Francisco Varela while still a teenager, and later studied and collaborated with him in the Ecole Polytechnique in Paris in the late 1980s. As part of his unconventional upbringing, he was introduced to yoga and meditation practices as a child, and he continues to practice today.

Waking, Dreaming, Being is thus the culmination not only of a lifetime's concern with ideas but also of a life personally engaged with the core existential questions raised in this book: *Who am I? What is the self? How is it related to being conscious?* As an illustration of his concern for both first- and third-person perspectives, Thompson interweaves personal reflections with careful analyses of the latest research in neuroscience and illuminating interpretations of classical doctrinal and philosophical concepts from Asia. As one of those rare individuals who has sought to integrate these diverse approaches to human experience into his own way of being in the world, Thompson embodies a historical and cultural movement in which the anti-quated distinction between "East" and "West" is breaking down to be replaced, perhaps, with a more globally informed understanding of what it means to be fully human.

Stephen Batchelor
Aquitaine

PROLOGUE

The Dalai Lama's Conjecture

When I was eight years old my father gave me *Gautama Buddha: In Life and Legend* by Betty Kelen. I still have the copy, a 75-cent paperback, with my name in my own handwriting on the first page. I couldn't put the book down. I read it in the backseat of our old blue Volkswagen station wagon, as we drove along Highway 400 from York University in Toronto (where my father taught Humanities) to our home in Bradford, Ontario, about 40 miles north.

I asked my father why he sometimes marked sentences in books he was reading with a red pen. He told me they were important ones he wanted to remember and find again. Like father, like son: the red ink I marked on the paragraphs describing the aspiration to attain enlightenment and become a buddha hasn't faded against the yellow paper.

Something about this drama of enlightenment appealed to me. It was the 1970s, after all, and my father had already taught me breath-mantra yoga meditation and read to me at bedtime from Paramahansa Yogananda's *Autobiography of a Yogi.* The serenity and confidence of enlightenment enthralled me. Like most kids, I loved animals. So it's not surprising I also marked the passage telling the story of how the Buddha's wicked cousin, Devadatta, sent a rogue elephant to trample him, only to find the animal tamed by the Buddha's emanating love and calm composure.

Thirty-two years later I find myself, not exactly calm and composed, about to give a keynote speech to open a two-day conference at MIT called "Investigating the Mind: Exchanges Between Buddhism and the Biobehavioral Sciences on How the Mind Works."[1] I am to speak immediately after the opening remarks by the Dalai Lama. Walking onto the stage of Kresge Auditorium, I feel dizzy. At the center, to my left, sit the Dalai Lama and his English translator. Between us sits a row of prominent psychologists and brain scientists. To the Dalai Lama's right are two well-known Buddhist scholars and meditation teachers. Before us is an audience of over a thousand people. My legs shake midway through my speech, and I hold onto the lectern for support. And then the Dalai Lama sneezes—a loud sneeze that fills the auditorium. Smiling sheepishly, he breaks into a laugh, and his laughter proves contagious as everyone joins in. Startled at first, I immediately feel relieved, and my stage fright starts to dissipate.

I've been asked to give a speech honoring the memory of my close friend and mentor, Francisco Varela, and his vision of a new kind of science of the mind and brain.[2] Widely known for his work as a cognitive scientist and neurobiologist, Varela was also a practitioner of Buddhist meditation and a student of Buddhist philosophy. He cofounded the Mind and Life Institute, which organized this conference.

In the last decade of his life, a few years before this meeting, Varela made groundbreaking discoveries about how different areas of the brain coordinate their activities when we perceive and recognize something, such as a familiar face.[3] Building on this work, he proposed an approach to the scientific study of the mind called "neurophenomenology."[4] This approach combines the careful study of experience from within (phenomenology) with neuroscience or the scientific investigation of the brain.

As neuroscientists now appreciate, much of brain activity arises from within the brain rather than being determined by the outside world. Neuroscientists call this ongoing and spontaneous brain activity "intrinsic activity." But its relationship to our subjective experience remains unclear. Varela thought phenomenology could help address this issue.[5] He hypothesized that the flow of subjective mental events—what William James called the stream of consciousness—reflects intrinsic brain activity more than activity arising from external stimuli. The careful description of moment-to-moment experience could therefore help to reveal and make sense of hidden patterns of brain activity. But to play this role, phenomenology should be based on the kind of exacting exploration of the mind that Buddhists use in meditation.[6] Meditative insight provides a way to observe mental events as they happen from moment to moment; brain imaging and electrophysiology supply tools for recording brain activity during the flow of thought and feeling. Combining these perspectives would help scientists to relate the mind and the brain in meticulous ways.

As I come to the end of my speech, I propose that we need to see the dialogue between Buddhism and cognitive science from this neurophenomenological perspective. The full import of this dialogue isn't that we get a new Buddhist guinea pig to study. It's rather that we gain a new scientific partner whose firsthand expertise in mental life can play an investigative role directly within science. Working together, we can create a new and unprecedented kind of self-knowledge—a way of knowing ourselves that unites cognitive science and the contemplative view of the mind from within.

THE DALAI LAMA'S CONJECTURE

My speech honoring Varela took place in the morning, before the conference's first main session on attention and cognitive control. It's now the afternoon, and I'm happy to be sitting in the audience, listening to the panelists discussing mental imagery, the conference's second main topic. Matthieu Ricard, a French Tibetan Buddhist monk from Sechen Monastery in Nepal, has opened the session with a description of Buddhist visualization practices and their role in mental transformation. Steven Kosslyn, a psychologist at Harvard University, has followed with an overview of scientific studies of mental imagery. The discussion so far has focused on how the experience of mental imagery relates to the underlying processes in the brain.

Midway into this discussion, the Dalai Lama says something that takes me by surprise. It connects to a question about consciousness and the brain I've thought about for a long time. Varela and I debated the question just before he died, as we shared a quiet weekend in my father's apartment in Manhattan, writing what would be our last coauthored paper.[7] Is consciousness wholly dependent on the brain or does consciousness transcend the brain?

The Dalai Lama, addressing the Buddhists more than the scientists, wonders whether all conscious states—even the subtlest states of "luminous consciousness" or "pure awareness" without any mental images—require some sort of physical basis.[8] The thought is striking, given the traditional Indian and Tibetan Buddhist view that this sort of pure awareness isn't physical in nature (at least not in any ordinary sense). Sitting next to me, Richard Davidson, a psychologist, neuroscientist, and longtime participant in the science-Buddhism dialogue, whispers, "I've never heard him say anything like that before!"

The discussion returns to mental imagery, but I'm still thinking about what the Dalai Lama just said. He certainly can't mean that the basis for pure awareness is the brain. The traditional Buddhist view is that consciousness transcends the brain. For example, Buddhists believe in rebirth—that consciousness carries on from one life to the next—but the brain decays at death. So what exactly does the Dalai Lama have in mind? Does he mean something physical that affects or informs the brain, but isn't limited to that particular material structure? What sort of relation would there be between the physical basis and the consciousness it supports?

And what, I can't help thinking, if there were no convincing evidence for any type of consciousness without the brain, or for the continuation of any kind of consciousness after death? Could pure awareness be contingent on the brain? How far would the Dalai Lama be willing to go in considering these possibilities?

Four years later I was able to ask him these questions at a small meeting with leading neuroscientists, psychologists, and physicists at his refugee home in Dharamsala, India—a story I tell later in this book.

According to Indian and Tibetan Buddhist philosophy, the definition of consciousness is that which is luminous and knowing. Luminosity means the ability of consciousness to reveal or disclose. Knowing refers to the ability of consciousness to perceive or apprehend what appears. As the Dalai Lama explains in his book, *The Universe in a Single Atom: The Convergence of Science and Spirituality*, "As the primary feature of light is to illuminate, so consciousness is said to illuminate its objects. Just as in light there is no categorical distinction between the illumination and that which illuminates, so in consciousness there is no real difference between the process of knowing or cognition and that which knows or cognizes. In consciousness, as in light, there is a quality of illumination."[9]

Matthieu Ricard had already introduced this notion of luminous consciousness or pure awareness during his account of how certain types of Buddhist meditation rely on creating and contemplating visual mental images. Like the surface of a mirror, distinct from whatever image it happens to reflect, pure awareness is distinct from whatever transitory thought or mental image happens to arise. Luminosity is like the mirror's clean surface, and knowledge like its capacity to reflect an object. One can experience directly this luminous and knowing awareness, distinct from any passing thought or mental image, through certain types of meditation especially prized and cultivated in a number of Buddhist lineages.

Steven Kosslyn, at the end of his presentation, had expressed doubts about the notion of pure awareness. Introspection, the act of looking within one's own mind, he believes, can't occur without mental imagery, for what we become aware of when we introspect will always be some sensory or mental image. Anne Harrington, a historian of science at Harvard University, had then opened the discussion, claiming that the Western scientific tradition has no concept like that of pure awareness and hence wondering whether there's any way that brain scientists can begin to engage with Buddhists on this fundamental issue.

This is the context in which the Dalai Lama makes his remarks about the physical basis of consciousness. Speaking through his translator, Thupten Jinpa, he notes the tendency among Tibetan

Buddhists to think of these subtle states of pure awareness as if they lack embodiment or have no material basis. Yet he has come to think that even the subtlest "clear light state of mind," which manifests at the exact moment of death, must have some kind of physical base. He declares his view to be like the scientific standpoint that the brain is the basis for all mental events. And then, switching to his partial English, he explains that without the brain, the ordinary mind can't function. So, similarly, without some subtle physical base, there can't be subtle states of consciousness. "Whether there is something independent or not, I don't know."

I'm struck by the science-Buddhism encounter taking place within the Dalai Lama's own reflections. He has called attention to a profound type of consciousness supposedly revealed to highly trained contemplative insight, but currently missed by science. He has also forthrightly allowed that this type of consciousness might have a physical basis, maybe even a correlate in the brain, and hence that this correlate might be detectable by science (at least in principle). More striking still, he has ended his comments not with any statement of religious belief or dogma, but with a cheerful admission of his ignorance, along with his characteristic infectious chuckle.

FACING UP TO THE CHALLENGES

Is consciousness wholly dependent on the brain, or does consciousness transcend the brain? Given the cultural and philosophical differences between Western neuroscience and Tibetan Buddhism, how should we move forward in thinking about this issue?

Many Western scientists and philosophers would dismiss both the notion of pure awareness and the proposition that there may be types of consciousness that transcend the brain. Many Indian and Tibetan contemplative scholars and practitioners would be equally dismissive of the proposition that consciousness is a wholly biological process supported by the brain. Neither attitude attracts me as a philosopher committed to the deep importance of the dialogue beween Asian philosophical traditions and Western science.

Buddhism is a millennia-old tradition that has developed numerous forms of philosophical thought and religious practice across a huge range of cultural and historical settings. Western scientists and philosophers need to take seriously the knowledge this tradition has accumulated through its methods of training the mind and analyzing consciousness. We should study its ways of classifying mental functions and its philosophical accounts of the nature of awareness. We must also understand how this kind of knowledge grows out of an ethical concern with human suffering and liberation, as well as a deep existential understanding of the human fear of death.

Western psychology and neuroscience are still young—hardly more than a hundred years old. Yet in this short time we've acquired an impressive amount of knowledge about how the mind works, especially in relation to the evolution and development of the brain and living body as a whole. The Asian meditative and philosophical traditions have nothing comparable to this kind of knowledge. We must take it seriously, including its broad philosophical implications for how we should understand the mind and its place in nature.

A fair amount has been written in recent years about the ways Buddhism and cognitive science converge. But these traditions also challenge each other in forceful ways. Buddhism, along with other contemplative traditions, challenges the assumption that we can know everything there is to know about the mind without examining it carefully from within. Incorporating meditative insight into cognitive science signals a potentially profound transformation of science. The classical scientific ideal of a completely neutral perspective untainted by the observer—already shaken to its core by twentieth-century physics—confronts another kind of profound limit in the impossibility of understanding the mind without including the first-person exploration of how we experience our own consciousness.

At the same time, neuroscience and evolutionary biology challenge the view that the ultimate nature of consciousness is nonphysical, or more pointedly, nonbiological. Expanding cognitive science to include meditative insight need not require accepting all the traditional metaphysical beliefs and theories accompanying Indian and Tibetan contemplative knowledge. Many of those beliefs and theories involve ancient Indian conceptions of mind and matter very different from our modern scientific ones. Although cognitive science needs to give contemplative experience an active role to play in the scientific investigation of the mind, Buddhism and the other yogic traditions need to test their ancient Indian conceptions of mind, body, and rebirth against modern scientific knowledge.

If we're going to achieve a new kind of self-knowledge based on science and contemplative wisdom, as I believe we must, then both partners to this endeavor must put these challenges to each other openly and energetically. This is the spirit in which I've tried to write this book.

STAYING WITH THE OPEN QUESTION

Shortly before his death, Francisco Varela talked about the Tibetan Buddhist notion of "subtle consciousness" in an interview with Swiss filmmaker Franz Reichle.[10] Subtle consciousness isn't an individual consciousness; it's not an ordinary "me" or "I" consciousness. It's sheer luminous and knowing awareness beyond any sensory or mental content. It's rarely seen by the ordinary mind, except occasionally in special dreams, intense meditation, and at the very moment of death, when one's ordinary "I" or "me" consciousness falls apart. It's the foundation for every other type of consciousness, and it's believed to be independent of the brain. Neuroscience can't conceive of this possibility, while for Tibetan Buddhists it's unthinkable to dismiss their accumulated experience testifying to the reality of this primary consciousness.

Varela's position is to suspend judgment. Don't neglect the Buddhist observations and don't dismiss what we know from science. Instead of trying to seek a resolution or an answer, contemplate the question and let it sit there. Have the patience and forbearance to stay with the open question.

I try to do so in this book. For a philosopher, staying with the open question means turning it around and examining it from all sides, without trying to force any particular answer or conclusion. But it also means not being afraid to follow wherever the argument leads.

Where the argument has led me reflects my ongoing personal journey over many years. When my father taught me Rāja Yoga meditation when I was seven, he instilled in me a philosophical and spiritual worldview in which the mind and consciousness are the primary reality. Growing up as a teenager at the Lindisfarne Association—the community of scientists, artists, and contemplatives he founded in the 1970s—I experienced firsthand how the scientific and contemplative worldviews can enrich each other. This upbringing led me to study Asian religion and philosophy at Amherst College, and Western philosophy and cognitive science in graduate school at the University of Toronto. Yet, while pursuing my own research in philosophy of mind and cognitive science over the past twenty-five years, I've often found myself doubting whether consciousness—even in its most profound meditative forms—transcends the living body and the brain. At the same time, I think we need a much deeper understanding of the brain in order to do justice to the nature of consciousness and its embodiment. I also remain firmly committed to the partnership between the scientific and contemplative worldviews, which is crucial for giving meaning to human life in our time and culture.

To stay with the open question while following wherever the argument leads requires that we be resolutely empirical in our approach. By this I mean cleaving to experience and suspending judgment about speculative matters falling outside what's available to experience. Experience includes inward experience of the mind and body gained through meditation, and outward experience of the world gained through scientific observation and experimentation. In neither case can there be genuine knowledge without communal testing and agreement on what the valid findings are. Buddhism and science both share this critical and experiential stance.

Western science has so far explored one small corner of the mind—the one accessible to individuals with little contemplative insight into their own minds, reporting to equally inexperienced scientists. The encounter with contemplative traditions raises another prospect—individuals with a high degree of meditative expertise reporting to knowledgeable scientists. Varela's vision, in its boldest form, was that future cognitive scientists would be skilled in meditation and phenomenology, in addition to neuroscience, psychology, and mathematics. And contemplative adepts would be deeply knowledgeable in Western cognitive science.

My hope for this book is that it can help to foster this original kind of self-knowledge.

ACKNOWLEDGMENTS

Three people deserve special acknowledgment for everything they did to help me write this book.

The first person is Rebecca Todd. Rebecca gave her critical eye and ear to every word. From the earliest drafts she helped me to find the right voice, and she went over repeated drafts of every chapter in order to ensure the writing was as strong and clear as I could make it be. Besides being a first-rate cognitive neuroscientist and a former artist, as well as the love of my life, Rebecca grew up in the same 1970s countercultural milieu that I did—we first met when she was twelve and I was thirteen—so she knew exactly where I was coming from. I could not possibly have written this book without her.

The second person is my agent, Anna Ghosh. Anna understood right away what I was trying to do in this book and helped me to find a home for it. I am especially grateful to her for believing in the book when the proposal first crossed her desk.

The third person is my editor at Columbia University Press, Wendy Lochner. Wendy has been another enthusiastic believer in the book who has helped me improve it considerably along the way.

Several people read drafts of the chapters as I wrote them and gave helpful critical responses—Stephen Batchelor (to whom I'm also very grateful for the foreword), Jake Davis,

Georges Dreyfus, Matthew Mackenzie, Robert Sharf, William Irwin Thompson, and Jennifer Michelle Windt.

Gail Thompson, Gareth Todd Thompson, Hilary Thompson, and Maximilian Todd Williams deserve special thanks for their encouragement and support.

Many other people have helped me in the research and writing of this book. In particular, I wish to thank Lisa Adams, Miri Albahari, Bhikkhu Anālayo, James Austin, Michel Bitbol, Judson Brewer, Willoughby Britton, Heather Chapin, Kalina Christoff, Christian Coseru, Diego Cosmelli, Richard Davidson, Athena Demertzi, Ezequiel Di Paolo, Martin Dresler, Christine Dunbar, Katherine Duncanson, John Dunne, Tayler Eaton, Melissa Ellamil, Adam Engle, Norman Farb, Wolfgang Fasching, Owen Flanagan, Keiran Fox, Jonardon Ganeri, Shanti Ganesh, David Gardiner, Jay Garfield, Bruce Greyson, Joan Halifax, Aaron Henry, Jane Hirshfield, Amishi Jha, Thupten Jinpa, Al Kaszniak, Edward Kelly, Dr. Barry Kerzin, Dr. Alan Kindler, Belinda Kong, Lana Kuhle, Jean-Philippe Lachaux, Dorothée Legrand, Marc Lewis, Michael Lifschitz, Antoine Lutz, Alisa Mandrigin, Thomas Metzinger, Natasha Myers, Alva Noë, David Perlman, Luiz Pessoa, Joanna Polley, Adrienne Prettyman, Chakravarthi Ram-Prasad, Matthieu Ricard, Andreas Roepstorff, Harold Roth, Dorion Sagan, Sharon Salzberg, Clifford Saron, Eric Schwitzgebel, Mark Siderits, Heleen Slagter, Sean Michael Smith, Neil Theise, Amy Cohen Varela, John Vervaeke, Cassandra Vieten, B. Alan Wallace, Jeff Warren, Kristin Wilson, Dan Zahavi, Arthur Zajonc, and Philip David Zelazo.

A number of institutions have been crucial to this book. Were it not for the Mind and Life Institute and its cofounder and former chairman, Adam Engle, this book could not have been written. The Upaya Institute and Zen Center and its founder, Joan Halifax-Roshi, have provided essential support, especially in the form of our annual "Zen Brain" workshop and meditation retreat. Without the Lindisfarne Association and its founder, William Irwin Thompson, the necessary conditions for this book would never have been present. In addition, I wish to thank the Center for Subjectivity Research at the University of Copenhagen, the Columbia University Society for Comparative Philosophy, the Garrison Institute, the Insight Meditation

Society, Namgyal Monastery, the Sivananda Ashram Bahamas, and Sivananda Yoga Vedanta Toronto.

The writing of this book was supported by a Social Sciences and Humanities Research Council of Canada Standard Research Grant (2008–2012) and by a University of Toronto Jackman Humanities Institute Research Fellowship (2010–2011). I am grateful for their support.

Finally, I would like to thank my former colleagues in the Department of Philosophy at the University of Toronto for their support and encouragement, and my new colleagues in the Department of Philosophy at the University of British Columbia for welcoming me and this work.

INTRODUCTION

The central idea of this book is that the self is a process, not a thing or an entity. The self isn't something outside experience, hidden either in the brain or in some immaterial realm. It is an experiential process that is subject to constant change. We enact a self in the process of awareness, and this self comes and goes depending on how we are aware.

When we're awake and occupied with some manual task, we enact a bodily self geared to our immediate environment. Yet this bodily self recedes from our experience if our task becomes an absorbing mental one. If our mind wanders, the mentally imagined self of the past or future overtakes the self of the present moment.

As we start to fall asleep, the sense of self slackens. Images float by, and our awareness becomes progressively absorbed in them. The impression of being a bounded individual distinct from the world dissolves. In this so-called hypnagogic state, the borders between self and not-self seem to fall away.

The feeling of being a distinct self immersed in the world comes back in the dream state. We experience the dream from the perspective of the self within it, or the dream ego. Although the entire dream world exists only as a content of our awareness, we identify our self with only a portion of it—the dream ego that centers our experience of the dream world and presents itself as the locus of our awareness.

At times, however, something else happens. We realize we're dreaming, but instead of waking up, we keep right on dreaming with the knowledge that we're dreaming. We enter what's called a lucid dream. Here we experience a different kind of awareness, one that witnesses the dream state. No matter what dream contents come and go, including the forms the dream ego takes, we can tell they're not the same as our awareness of being in the dream state. We no longer identify only with our dream ego—the "I" as dreamed—for our sense of self now includes our dreaming self—the "I" as dreamer.

Similarly, while meditating in the waking state, we can simply witness being conscious and watch whatever sensory or mental events occur within the field of our awareness. We can also watch how we may identify with some of them as "Me" or appropriate some of them as "Mine."

We usually lose touch with this ability to be mindful when we fall asleep. We regain it in a vivid way when we have a lucid dream. Some Indian and Tibetan traditions of philosophy and meditation claim we can recover this mindfulness or witnessing awareness even during deep and dreamless sleep. If this is true, then there must be more to consciousness than just the contents of our waking and dreaming minds.

According to the Indian yogic traditions, which broadly construed include Buddhism, we can distinguish three aspects of consciousness.[1] The first aspect is awareness, which is often likened to a light that reveals whatever it shines upon. The second aspect is whatever the light illuminates, that is, whatever we happen to be aware of from moment to moment. The third aspect is how we experience some of these contents of awareness as "I" or "Me" or "Mine." To understand how we enact a self, therefore, we need to understand three things—the nature of awareness as distinct from its sensory and mental contents, the mind-body processes that produce these contents, and how some of these contents come to be experienced as the self.

In the following chapters, I take this threefold framework of awareness, contents of awareness, and self-experience—or what the

Indian tradition calls "I-making"—and put it to work in cognitive science. Whereas the Indian thinkers mapped consciousness and I-making in philosophical and phenomenological terms, I show how their insights can also help to advance the neuroscience of consciousness, by weaving together neuroscience and Indian philosophy in an exploration of wakefulness, falling asleep, dreaming, lucid dreaming, out-of-body experiences, deep and dreamless sleep, forms of meditative awareness, and the process of dying.

The organizing principle for this book comes from the Indian tradition. The ancient Indian texts called the *Upanishads* contain the world's first recorded map of consciousness.[2] The earliest texts—dating from the sixth or seventh century B.C.E.—delineate three principal states of the self—the waking state, the dream state, and the state of deep and dreamless sleep. Later texts add a fourth state—the state of pure awareness. Waking consciousness relates to the outer world and apprehends the physical body as the self. Dream consciousness relates to mental images constructed from memories and apprehends the dream body as the self. In deep and dreamless sleep, consciousness rests in a dormant state not differentiated into subject and object. Pure awareness witnesses these changing states of waking, dreaming, and dreamless sleep without identifying with them or with the self that appears in them. I use this fourfold structure to organize my exploration of consciousness and the sense of self across the waking, dreaming, and deep-sleep states, as well as meditative states of heightened awareness and concentration.

In the yogic traditions, meditation trains both the ability to sustain attention on a single object and the ability to be openly aware of the entire field of experience without selecting or suppressing anything that arises. In both modes of meditation—focused attention (or one-pointed concentration) and open awareness—one learns to monitor specific qualities of experience, such as moment-to-moment fluctuations of attention and emotion, that are difficult for the restless mind to see.[3] One of the guiding ideas of this book is that individuals who can move flexibly and reliably between these different modes of awareness and attention, and who can describe in precise terms how their experience feels from moment to moment, offer a new source of information about the self and consciousness for neuroscience and the philosophy of mind.[4]

Let me now give a brief overview of the book's main ideas. In the chapters themselves I present these ideas using neuroscience, philosophy, literature, and stories from my own experience.

Chapter 1 explains the formative Indian image of light or luminosity as the basic nature of consciousness. Indian philosophers often define consciousness as that which is luminous and knowing. "Luminous" means having the power to reveal; "knowing" means being able to apprehend whatever appears. In the waking state, consciousness reveals and apprehends the outer world through the senses; in the dream state, consciousness reveals and apprehends the inner world of mental images. This chapter also introduces the ancient Indian map of consciousness, which comprises the four states of wakefulness, dreaming, deep and dreamless sleep, and pure awareness.

Chapter 2 focuses on attention and perception in the waking state. I compare theories and findings from cognitive neuroscience with Indian Buddhist theories of attention and perception. According to both perspectives, although the stream of consciousness may seem to flow continuously, it's really made up of discrete moments of awareness that depend on how attention shifts from one thing to another. I review evidence from neuroscience showing that focused attention and open awareness forms of meditation have measurable effects on how attention structures the stream of consciousness into discrete moments of awareness. I conclude by using both Buddhist philosophy and cognitive neuroscience to argue that in addition to these discrete moments, we also need to recognize a more slowly changing background awareness that includes the sense of self and that shifts across waking, dreaming, and dreamless sleep.

Chapter 3 takes up the question, raised in the prologue, of whether the basic nature of consciousness as pure awareness is dependent on the brain or transcends the brain. I describe a dialogue on this question with the Dalai Lama at his refugee home in Dharamsala, India, and explain the basis in Buddhist philosophy for the Dalai Lama's view that consciousness transcends the brain. I argue, however, that there's no scientific evidence to support this view. All the evidence available to us indicates that consciousness, including pure awareness, is contingent on the brain. Nevertheless, my viewpoint isn't a materialist one, for two reasons. First, consciousness has a cognitive primacy that materialism fails to see. There's no way to step outside consciousness and measure it against something else. Science always moves within the field of what consciousness reveals; it can enlarge this field and open up new vistas, but it can never get beyond the horizon set by consciousness. Second, since consciousness has this kind of primacy, it makes no sense to try to reductively explain consciousness in terms of something that's conceived to be essentially nonexperiential, like fundamental physical phenomena. Rather, understanding how consciousness is a natural phenomenon is going to require rethinking our scientific concepts of nature and physical being.

Chapters 4, 5, and 6 concern falling asleep, dreaming, and lucid dreaming. I begin with the state leading into sleep, the hypnagogic state, in which strange images make their way before our eyes and we hear sounds or what seem like conversations going on around us or inside us. Whereas normal waking consciousness is ego-structured—we experience ourselves as bounded beings distinct from the outside world—this structure dissolves in the hypnagogic state. There's no ego in the sense of an "I" who acts as a participant in a larger world, and there's no larger world in which we feel immersed. Instead, there's a play of images and sounds that holds consciousness spellbound. In short, two key features mark the hypnagogic state—a dissolution of ego boundaries and an attention drawn to what consciousness spontaneously imagines.

The ego structure of consciousness returns in the dream state. In the hypnagogic state we look at images and they absorb us; in the dream state we experience being in the dream world. Sometimes we experience it from an inside or first-person perspective; sometimes we see ourselves in it from an outside or third-person perspective. These two perspectives also occur in memory, where they're known as "field memory" and "observer memory." Yet even in the case of the observer perspective in a dream, we experience ourselves as a subject situated in relation to the dream world. At the same time, the spell-bound attention that arises in the hypnagogic state also characterizes the dream state, so it too is a kind of captivated consciousness.

All this changes in a lucid dream. The defining feature of a lucid dream is being able to direct attention to the dreamlike quality of the state so that one can think about it as a dream. When this happens, the sense of self shifts, for one becomes aware of the self both as dreamer—"I'm dreaming"—and as dreamed—"I'm flying in my dream."

In these three chapters I review findings from sleep science that show that each state—the hypnagogic state, dreaming, and lucid dreaming—is associated with its own distinct kind of brain activity.

Brain-imaging studies of lucid dreaming offer a fascinating way to investigate what neuroscientists call the "neural correlates of consciousness." Lucid dreamers can use eye movements to signal when they become lucid, and scientists can monitor what's going on in the brain at the same time. In Tibetan Buddhism, "dream yoga" includes learning how to have lucid dreams in order to practice meditation in the dream state. This kind of meditation is thought to be especially powerful for learning to transform negative emotions into positive emotions, such as anger into equanimity, and for learning to recognize the basic nature of consciousness as pure awareness. By combining these ancient yoga practices with modern methods from sleep science, we can envision a new kind of dream science that integrates dream psychology, neuroscience, and dream yoga.

I end my discussion of dreaming by criticizing the standard neuroscience conception of the dream state as a form of delusional hallucination. Instead, I argue that dreaming is a kind of spontaneous imagination. I also argue that the dreaming mind isn't a passive epiphenomenon of the sleeping brain, for intentional mental activity in dreaming, especially in lucid dreaming, actively affects the sleeping brain.

Chapter 7 examines out-of-body experiences. In an out-of-body experience, you feel as if you're located outside your body, often at an elevated vantage point. Yet far from showing the separability of the self from the body, out-of-body experiences reinforce the strong connection between the body and the sense of self. These aren't experiences of disembodiment; they're experiences of altered embodiment. You see your body as an object at a place that doesn't coincide with the felt location of your visual and vestibular awareness. In this way, there's a dissociation between your body as an object of perception and your body as a perceptual subject and attentional agent. Out-of-body experiences reveal something crucial about the sense of self: you locate yourself as an experiential subject wherever your attentional perspective feels located, regardless of whether this happens to be the place you see your body as occupying.

Out-of-body experiences provide no evidence that one can have an experience without one's biological body, for the body remains present throughout. Furthermore, experiences with many of the features of out-of-body experiences can be brought about by direct electrical stimulation of certain brain regions and by virtual reality devices. So it's reasonable to assume that out-of-body experiences depend on activity at specific regions of the brain and therefore as a general rule are contingent on the living body.

Chapter 8 asks whether consciousness is present in deep and dreamless sleep. Most neuroscientists and philosophers of mind today think of dreamless sleep as a blackout state in which consciousness fades or disappears completely. In contrast, the Indian philosophical schools of Yoga and Vedānta, as well as Indian and Tibetan Buddhism, maintain that a subtle form of consciousness continues. I present the Indian philosophical case for deep sleep being a mode of consciousness and show that none of the behavioral or physiological evidence from sleep science suffices to rule out there being a mode of consciousness in dreamless sleep. Hence the standard neuroscience way of defining consciousness as that which disappears in dreamless sleep needs to be revised. Yoga, Vedānta, and Buddhism assert that the subliminal consciousness present in dreamless sleep can become cognitively accessible through meditative mental training. I present some preliminary evidence from sleep science in support of this idea. I end the chapter by proposing that we need to enlarge sleep science to include contemplative ways of training the mind in sleep. This project will require sleep scientists, sleep yogis, and contemplative scholars of the Indian and Tibetan traditions to work together to map the sleeping mind. In short, we need a new, contemplative kind of sleep science.

Chapter 9 investigates what happens to the self and consciousness when we die. Neuroscience and biomedicine talk about death as if it were essentially an objective and impersonal event instead of a subjective and personal one. From a purely biomedical perspective, death consists in the breakdown of the functions of the living body along with the disappearance of all outer signs of consciousness. Missing from this perspective is the subjective experience of this breakdown and the existential significance of the inevitable fact of one's own death. In contrast, Tibetan Buddhism presents a vivid account of the progressive breakdown of consciousness and the dissolution of the sense of self during the dying process. It also describes how to face this process in a meditative way. According to Tibetan Buddhism—as well as Yoga and Vedānta—great contemplatives can disengage from the sense of self as ego as they die. Resting in an experience of pure awareness, they can watch the dissolution of their everyday "I-Me-Mine" consciousness and witness their own dying with equanimity.

The Tibetan Buddhist tradition also claims that sometimes the bodies of great yogis don't die in the usual way. After their hearts stop beating and their breathing ceases, these individuals are said not to decay for days or even weeks. I discuss a number of recent reports of such phenomena and how they're viewed from Western scientific and medical perspectives. Scientific studies have only just begun. One reason this kind of investigation has value is that it can help science to see that a full understanding of death—even in biomedical terms—requires understanding how the mind meets death and may affect the dying process.

Near-death experiences during cardiac arrest provide another important case for investigating how the mind meets death and the relationship between consciousness and the body. Although these experiences are often presented as challenging the view that consciousness is contingent on the brain, I argue that none of the evidence brought forward to support this position is convincing. Instead, all the evidence to date, when examined carefully, supports the view that these experiences are contingent on the brain.

At the same time, we should avoid the trap of thinking that the reports of near-death experience after resuscitation from cardiac arrest must be either literally true or literally false. This way of thinking remains caught in the grip of a purely third-person view of death. Dying and death must also be understood from the first-person perspective. We need to stop using accounts of these experiences to justify either neuroreductionist or spiritualist agendas and instead take them seriously for what they are—narratives of first-person experience arising from circumstances that we will all in some way face.

Chapter 10 targets the view widespread in neuroscience and "neurophilosophy" that the self is nothing but an illusion created by the brain. I call this view "neuro-nihilism." I argue that although the self is a construction—or rather a process that's under constant construction—it isn't an illusion. A self is an ongoing process that enacts an "I" and in which the "I" is no different from the process itself, rather like the way dancing is a process that enacts a dance and in which the dance is no different from the dancing. I call this the "enactive" view of the self. This chapter presents a systematic statement of the enactive view and shows how I-making happens at multiple biological, psychological, and social levels. The discussion combines elements from Buddhist philosophy (specifically from the "Middle Way" or Madhyamaka school), biology, cognitive science, and the neuroscience of meditation.

Although these chapters are meant to be read in sequence, I've written them so they can be read on their own in any order. So, for example, if you're interested in the issue about pure awareness and the brain, you can jump to chapter 3 and read it straightaway. Or if you want to know how contemplative approaches are crucial for thinking about death in our modern biomedical culture, you can go to chapter 9 (which is also published separately by Columbia University Press as a short e-book). Throughout, no specialized knowledge of cognitive science or Western or Indian philosophy is presupposed; everything is explained along the way.

Although cognitive science and the Indian yogic traditions of philosophy and meditation form the core of this book, I also draw from a wide range of other sources—poetry and fiction, Western philosophy, Chinese Daoism, and personal experience. By weaving together these diverse sources, I hope to demonstrate a new way to relate science and what many people like to call spirituality. Instead of being either opposed or indifferent to each other, cognitive science and the world's great contemplative traditions can work together on a common project—understanding the mind and giving meaning to human life. Two extreme and regressive tendencies mark our era—the resurgence of religious extremism and outmoded belief systems, and the entrenchment of scientific materialism and reductionism. Neither mindset realizes the value of meditation and the contemplative way of life as a source of wisdom and firsthand knowledge essential to a mature cognitive science that can do justice to our entire way of being—to our spirit, to use an older idiom.[5] This book upholds a different vision. By enriching science with contemplative knowledge and contemplative knowledge with cognitive science, we can work to create a new scientific and spiritual appreciation of human life, one that no longer requires or needs to be contained within either a religious or an antireligious framework.

WAKING,
DREAMING,
BEING

1

SEEING

What Is Consciousness?

What exactly is consciousness? The oldest answer to this question comes from India, almost three thousand years ago.

Long before Socrates interrogated his fellow Athenians and Plato wrote his *Dialogues*, a great debate is said to have taken place in the land of Videha in what is now northeastern India. Staged before the throne of the learned and mighty King Janaka, the debate pitted the great sage Yājñavalkya against the other renowned Brahmins of the kingdom. The king set the prize at a thousand cows with ten gold pieces attached to each one's horns, and he declared that whoever was the most learned would win the animals. Apparently Yājñavalkya's sagacity did not entail modesty, for while all the other priests kept silent, not daring to step forward, Yājñavalkya called out to his student to take possession of the cows. Challenged by eight great Brahmins, one by one, Yājñavalkya demonstrated his superior knowledge. As a favor to the king, he allowed him to ask any question he wanted. In the ensuing dialogue, told in the "Great Forest Teaching" (*Bṛhadāraṇyaka Upaniṣad*)—a text dating from the seventh century B.C.E. and the oldest of the ancient Indian scriptures called the *Upanishads*—Yājñavalkya gave the first recorded account of the nature of consciousness and its main modes or states.[1]

AN ILLUMINATING QUESTION

The dialogue begins with the king, knowing exactly where he wants to lead the sage, asking a simple question: "What light does a person have?" Or, as it can also be translated: "What is the source of light for a person here?"

"The sun," replies the sage. "By the light of the sun, a person sits, goes about, does his work, and returns."

"And when the sun sets," asks the king, "then what light does he have?"

"He has the moon as his light," comes the reply.

"And when the sun has set and the moon has set, then what light does a person have?"

"Fire," answers the sage.

Persisting, the king asks what light a person has when the fire goes out, and he gets in reply the clever answer, "Speech." Yājñavalkya explains: "Even when one cannot see one's own hand, when speech is uttered, one goes toward it." In pitch-black darkness, a voice can light your way.

The king, however, still isn't satisfied and demands to know what light there is when speech has fallen silent. In the absence of sun, moon, fire, and speech, what source of light does a person have?

"The *self (ātman)*," Yājñavalkya answers. "It is by the light of the self that he sits, goes about, does his work, and returns."

This answer makes plain that the dialogue has been moving backward, from the distant, outer, and visible to the close, inner, and invisible. Nothing is brighter than the sun, or the moon at night, but they reside far away, at an unbridgeable distance. Fire lies closer to hand; it can be tended and cultivated. Speech, however, is produced by the mind. Darkness can't negate the peculiar luminosity of language, the power of words to light up things and to close the distance between you and another. Yet speech is still external in its being as physical sound. The sun, moon, fire, and speech—we know each one by means of outer perception. The self, however, can't be known through outer perception, because it resides at the source of perception. It isn't the perceived, but that which lies behind the perceiving.

The self dwells closest, at the maximum point of nearness. It's never *there*, but always *here*. How could we possibly find our way around without it? How could outer sources of light reveal anything to us, if they weren't themselves lit up by the self? And yet, precisely because the self is so intimate, it seems impossible to have any clear view of it and to know what it is.

Finally, the king is able to ask the question he has all along been aiming toward: "*What is the self?*"

A REVEALING ANSWER

Yājñavalkya answers that the self (*ātman*) is the inner light that is the person (*puruṣa*). This light, which consists of knowledge, resides within the heart, surrounded by the vital breath. In the waking state, the person travels this world; in sleep, the person goes beyond this world. The person is his own light and is self-luminous.

As this answer unfolds, it becomes clear that the "light" Yājñavalkya is talking about is what we would call "consciousness." Consciousness is like a light; it illuminates or reveals things so they can be known. In the waking state, consciousness illuminates the outer world; in dreams, it illuminates the dream world.

It's here, in Yājñavalkya's answer to the king's question about the self, that we find the first map of consciousness in written history.

WAKING, DREAMING, SLEEPING

Yājñavalkya explains to the king that a person has two dwellings—this world and the world beyond. Between them lies the borderland of dreams where the two worlds meet. When we rest in the intermediate state of dreams, we see both worlds. The dream state serves as an entryway to the other world, and as we move through it we see both bad things and joyful things.

In the waking state, we see the outer world lit up by the sun. Yet we also see things when we dream. Where do they come from, and what makes them visible? What is the source of the light illuminating things in the dream state?

Yājñavalkya explains that in the dream state we take materials from the entire world—this world and the other one—break them down, and put them back together again. Although the dream state lies between the two worlds, it's a state of our own making. The person creates everything for himself in dreams and illuminates it all with his own radiance:

> When he falls asleep, he takes with him the material of this all-containing world, himself breaks it up, himself re-makes it. He sleeps by his own radiance, his own light. Here the person becomes lit by his own light.
>
> There are no chariots, nor chariot-horses, nor roads there, but he creates chariots, chariot-horses and roads. There are no pleasures, no enjoyments, nor delights there, but he creates pleasures, enjoyments and delights. There are no ponds, nor lotus-pools, nor rivers there, but he creates ponds, lotus-pools and rivers. For he is a maker.[2]

Like a great fish swimming back and forth between the banks of a wide river, the person alternates between waking and dreaming. Yet the self never attaches fully to either state, as the fish never touches the riverbanks when it swims between them.

There's also a third state, the state of deep and dreamless sleep. Here the person rests quietly with no desires:

> As a hawk or eagle, tired after flying around in the sky, folds its wings and is carried to its roosting-place, even so the person runs to the state where he desires no desire and dreams no dream. . . .
>
> As a man closely embraced by a beloved wife knows nothing outside, nothing inside, so the person, closely embraced by the self of wisdom, knows nothing outside, nothing inside. That is the form of him in which his desires are fulfilled, with the self as his desire, free from desire, beyond sorrow.[3]

These images present deep and dreamless sleep as a sought-for state of peace and bliss. Conventional characteristics and burdens drop away: "Here a father is not a father, a mother is not a mother . . . a thief is not a thief, a murderer not a murderer . . . a monk not a monk, an ascetic not an ascetic."[4] Instead, we rest in the embrace and wisdom of the cosmic or universal self (*ātman*), which is free from desire and without fear.

If deep sleep is peaceful and blissful, does this mean we're somehow conscious in deep sleep? Is awareness present, or is deep sleep the oblivion of awareness? Put another way, is deep sleep a state of consciousness, like waking and dreaming, or is it a state where consciousness is absent, as most neuroscientists think today?

Yājñavalkya's description of deep and dreamless sleep—and many later Indian interpretations of what he meant—implies that consciousness pervades deep sleep. Consider the following rich but enigmatic passage: "Though then he does not see, yet seeing he does not see. There is no cutting off of the seeing of the seer, because it is imperishable. But there is no second, no other, separate from himself, that he might see."[5]

This passage seems to be saying that although there are no longer any dream images to be seen ("he does not see"), there remains a kind of awareness in dreamless sleep ("yet seeing he does not see"). As the sun cannot stop shining, so the self cannot lose all consciousness; specifically, it cannot lose the basic luminosity of awareness ("there is no cutting off of the seeing of the seer"). In deep sleep, however, this awareness doesn't witness any object separate from itself—no waking world of perceptible things and no dream world of images ("there is no second, no other, separate from himself, that he might see"). So the awareness here must be of a subtle and subliminal kind, devoid of images and desires, while peaceful and at ease.

In later texts of the *Upanishads*, as well as other Indian philosophical works, dreamless sleep is described as lacking the obvious or gross subject/object duality that's present in the waking and dreaming states. In the waking state, the subject appears as the body and the object appears as what we perceive. In dreams, the subject appears as the dream ego or self-within-the-dream and the object as the dream world. In deep sleep, consciousness doesn't differentiate this way between subject and object, knower and known. Instead, it rests as one quiescent "mass." Consciousness withdraws into itself while its function of being directed toward external objects lies dormant. Yet this dormancy isn't a total loss or oblivion of awareness; it's a peaceful absorption that offers a foretaste of the lucid bliss belonging to the self-realized consciousness liberated from illusion.

Later philosophers belonging to the Yoga and Vedānta schools would also offer the following argument in support of the idea that consciousness continues in deep sleep: if there were no awareness at all in deep and dreamless sleep, then you couldn't have the memory, "I slept well," immediately upon waking up. Memory is the recollection of past experience; when you remember something, you recall an earlier experience and you recall it as your own. In remembering you slept peacefully, you recall something from deep sleep, so that state must have been a subtly conscious one. We'll examine this argument in light of the neuroscience of consciousness in chapter 8.

Yājñavalkya's progression from the waking state to the dreaming state to the deep-sleep state recapitulates his earlier progression from the sun, moon, fire, and speech to the self. Both narratives move increasingly away from what is outer and obvious to what is inner and subtle. Both trace the visibility of something in the waking and dreaming states back to its source in the basic luminosity of consciousness.

SUBTLE CONSCIOUSNESS

We've now uncovered an important difference between Western cognitive science and the Indian yogic philosophies. Cognitive science focuses on the contrast between the *presence* and the *absence of consciousness*—for example, between being awake and being under anesthesia, or between being able to report seeing a stimulus, such as the image of a face, and not being able to report seeing it, even though you show some other kind of behavioral or brain response to its presence. The Indian yogic traditions, however, focus on the contrast between *coarse* or *gross consciousness* and *subtle consciousness*— for example, between waking perception of outer material objects and subliminal awareness in deep sleep.

From a meditative perspective, consciousness comprises a continuum of levels of awareness, ranging from gross to subtle. Gross consciousness is waking sense perception, which tells you about things outside you, like the words you're reading now, and gives you the feeling of your body from within. Dreaming is subtler because you withdraw from the outside world and create what you see and feel on the basis of memory and imagination. Deep sleep is subtler still because it's consciousness without mental images. Subtle aspects of consciousness are also said to manifest in certain states of deep meditation where all overt thinking and perceiving cease, as well as at the time of death. These subtler or deeper aspects of consciousness aren't apparent to the ordinary untrained mind; they take a high degree of meditative awareness to discern.

In Western philosophy of mind, it's common to distinguish between two meanings of the word "conscious." On the one hand, we can say you're conscious of something when it appears to you some way in your experience. Feeling a pain or having a visual experience of the color red are two standard examples philosophers give of a conscious experience. As they say, there is "something it's like" for you to see color or to feel pain. In this sense, a mental or bodily state is conscious when there is something it's like for the subject to be in that state. Philosophers call this concept of consciousness "phenomenal consciousness" ("phenomenal" here means how things seem or appear in experience). On the other hand, we can also say you're conscious of something when you can report or describe it, or reason about it, or use it to guide how you act or behave. Philosophers call this concept "access consciousness" ("access" here means available for use in thought and action). In short, "consciousness" can mean awareness in the sense of *subjective experience* or awareness in the sense of *cognitive access*.

One reason philosophers make this distinction is to point out that explaining consciousness in the sense of cognitive access doesn't necessarily explain consciousness in the sense of subjective experience. As philosopher David Chalmers puts it, "even when we have explained the performance of all the cognitive and behavioral functions in the vicinity of experience—perceptual discrimination, categorization, internal access, verbal report—there may still remain a further unanswered question: *Why is the performance of these functions accompanied by experience?* A simple explanation of the functions leaves this question open."[6]

Another reason for making the distinction between phenomenal consciousness and access consciousness is to allow for the possibility that you could be subliminally or implicitly conscious of something without being able to report and describe your experience, at least not fully or explicitly. In other words, you could be phenomenally aware of something while lacking full cognitive access to that awareness. Perhaps you experienced the image on the screen, but it went by so fast you weren't able to form the kind of memory needed for a verbal report of exactly what it was. This is one way your phenomenal consciousness might outstrip the cognitive capacities or resources you have for accessing your experience.

One way to think about the Indian yogic idea of subtle consciousness is to see it as pointing to deeper levels of phenomenal consciousness to which we don't ordinarily have cognitive access, especially if our minds are restless and untrained in meditation. According to this way of thinking, as we'll see throughout this book, much of what Western science and philosophy would describe as unconscious might qualify as conscious, in the sense of involving subtle levels of phenomenal awareness that could be made accessible through meditative mental training.

There's another important difference between the Indian yogic and modern Western views. According to the standard cognitive science view, waking sense experience is the basis for all consciousness. According to many Indian (and Tibetan) views, however, gross or sensory consciousness depends on subtle consciousness. We'll look closely at these contrasting viewpoints when we examine pure awareness (chapter 3) and dreamless sleep (chapter 8).

THE FOURTH

In the dialogue between Yājñavalkya and King Janaka, we can see two ways of thinking about the self side by side—an older way, with roots in the Vedic thought of ancient India (ca. 1500 B.C.E.), and a newly emerging one, proper to the *Upanishads* (ca. 700–400 B.C.E.). For the older way of thinking, waking, dreaming, and deep sleep are *places* or *locations* to which the "inner person" (*puruṣa*) travels. When you fall asleep, you go to the place of dreams, which lies between this world and the other world beyond it. When you tire of dreaming, you go to the place where there are no dreams, the blissful realm of dreamless sleep. As Indian thought evolved, these places gradually transformed into *states* or *modes* of consciousness. This conceptual transformation culminated in the *Māṇḍūkya Upaniṣad*, a somewhat later text (ca. first century B.C.E. to the first or second century C.E.) that presents in a few short verses the famous doctrine of the four states of consciousness—waking, dreaming, deep and dreamless sleep, and "the fourth," or pure awareness.[7]

The *Māṇḍūkya Upaniṣad* describes these states as the four "feet" or "quarters" of the self (*ātman*). The first quarter is the waking state. Here consciousness turns outward and experiences the physical body as the self. Waking consciousness takes enjoyment in the "gross" objects of sense perception, yet no object holds its interest for long, because attention, motivated by desire, constantly flits from one thing to another. Consciousness in the waking state is restless, dissatisfied, and constantly on the move.

The second quarter of the self is the dream state. Here consciousness turns inward and experiences the dream ego as the self. Dreaming consciousness takes enjoyment in the dream images fabricated from "subtle" mental impressions caused by past experience and belonging to memory. Like waking, dreaming is a restless state, for desire and attention constantly jump from one thing to another.

The third quarter of the self is the state of deep and dreamless sleep. Here desire disappears, the "whirling" of the mind subsides, and consciousness becomes quiescent. The self rests in a "single mass" of absorbed and peaceful consciousness.

The fourth quarter of the self is the pure awareness beneath or behind waking, dreaming, and deep sleep, not conditioned by these changing states. The *Māṇḍūkya Upaniṣad* describes it first negatively and then positively:

> Not with consciousness turned inward [dreaming], not with con-
> sciousness turned outward [waking], not with consciousness turned
> both ways, not a mass of consciousness [deep sleep], not conscious,
> not unconscious—folk consider the fourth to be unseen, inviolable,
> unseizable, signless, unthinkable, unnameable, its essence resting
> in the one self, the stilling of proliferation, peaceful, gracious (*śiva*),
> without duality (*advaita*). That is the self: so it should be understood.[8]

Called simply "the fourth" (*turiya*), this mode of consciousness is pure nondual awareness.[9] Unlike waking, dreaming, and deep sleep, pure awareness isn't a state in the sense of a transient and discrete condition; hence it isn't a mere "quarter" like the other three. Instead, it's the constant, underlying source for these changing states, as well as a stage of meditative realization. As the underlying source for waking, dreaming, and deep sleep, "the fourth" is sheer awareness, defined by its quality of luminosity. As a stage of meditative realization, it's the deeper, background awareness that can witness these changing states without mistakenly identifying with them as the self. To borrow an image from Andrew Fort, an American scholar of Indian religion, pure awareness is like pervasive radio waves, which are obscured by the constant static of mental activity—thoughts, mental images, emotions, and memories.[10] We take the static to be real and to be who we really are, but it's only superimposed on the wave. When we remove or see through the superimposition, then the wave's true nature stands revealed. Or, to change analogies, removing the superimposition is like waking up from a dream, seeing through it, like knowing that you're dreaming. "The fourth" is the supreme wakefulness that reveals the true self as the witnessing awareness behind waking, dreaming, and deep sleep. This higher wakefulness is said to bring true freedom, serenity, and bliss.

We've been talking about waking, dreaming, and deep sleep as discrete states, but there's another way to think about them that will be especially important in this book. Rather than being discrete, these states are contained within one another; they interpenetrate. For example, when you have a dream and know you're dreaming, you're awake within the dream state; you're having a so-called lucid dream. And when you daydream, you're dreaming in the waking state.

The Indian philosopher Śaṅkara, who lived from 788 to 820 C.E., called attention to this crucial point in his commentary on the *Māṇḍūkya Upaniṣad*.[11] He said we experience all three states of waking, dreaming, and deep sleep within the waking state. Waking is perception, dreaming is remembering, and deep sleep is total self-absorption. In perception, we experience "gross" material objects. In memory, as in dreams, we experience "subtle" mental images and impressions. In total self-absorption, like deep sleep, active perceiving and remembering cease. In these and other ways, such as lucid dreaming, these states aren't opposed but flow into and out of one another.

I have yet to mention the most memorable feature of the *Māṇḍūkya Upaniṣad*—the way it links the four quarters of the self to the sacred Vedic syllable or mantra OM (or AUM). The text begins by identifying OM with all there is, with the past, present, and future, and with anything beyond these three times. OM is the sound of *brahman*, the nondual source and basis of the phenomenal universe that's also identical to the transcendent self, *ātman*. As one syllable, OM is the self; its constituent phonemes are the states of consciousness, from gross to subtle. "A" expresses the waking state; it's a rough sound produced by the mouth wide open. "U" expresses the dreaming state; it's a subtle sound produced with the aid of the lips. "M" expresses the deep-sleep state; still more refined, it's voiced with closed lips. Reverberating inside the throat, "Mmm" vocalizes the blissful and dreamless consciousness of deep sleep. "The fourth," however, being unique and incomparable, has no constituent phonemes. We can think of it as the silence from which all sounds emerge, or as the unity of the three phonemes in the one syllable OM, expressing the unity of the three states in the one nondual awareness.

DEATH

Let's return to the dialogue between the sage Yājñavalkya and King Janaka. Pleased with what he's heard, but knowing Yājñavalkya can impart still more, the king rewards him with a thousand cows but demands that he continue. The sage, realizing now that the king aims to extract all his secret knowledge, explains the nature of death and liberation.

As we return from dreaming to waking, so at death we pass from this life to another one. When the body becomes weak from sickness or old age, the self departs, leaving this body lifeless: "As he goes, the breath follows; as the breath goes, the senses (*prāṇa*) follow. He becomes a being of consciousness; he follows consciousness. His knowledge and action take hold of him, as does his former experience."[12]

In dying, consciousness withdraws from the senses, turns inward, and departs the body. But consciousness doesn't perish; it takes on a new body and begins another life conditioned by the knowledge and actions, or habits and dispositions, accumulated over the course of the previous life. Dying is like a caterpillar getting ready to jump; death and rebirth are like the caterpillar jumping from one blade of grass to another: "As a caterpillar, reaching the end of a blade of grass and taking the next step, draws itself together, so the self, dropping the body, letting go of ignorance and taking the next step, draws itself together."[13]

How things turn out in the new life depends on actions done in the life before. The doer of good is reborn into happy and favorable circumstances, the doer of evil into painful and difficult circumstances. Each life offers new opportunities to do good or ill, and thus to be reborn again into favorable or unfavorable conditions. No god or divine judge decrees this result; it happens because of the cause-and-effect relation between desire and its results: "They say, 'As one desires, so does one become, for the person is made of desire.' As he desires, so does his will become; as his will is, so is the action he does; as is the action he does, so is what he gets back."[14]

For a person made of desire, death offers no release, because desire binds one to the cycle of life, death, rebirth, and redeath. This cycle carries on incessantly, like the cycle of waking, dreaming, deep sleep, and waking up.

Someone free of desire, however, suffers no rebirth. Having fully recognized the true self (*ātman*), he no longer identifies with his body and ordinary mind in waking and dreaming. Desire—the fuel that powers the wheel of rebirth and redeath—is all used up. At death such a person casts off his body, like a snake shedding its skin, and becomes one with *brahman,* the infinite ground of all being.

Yājñavalkya ends his speech with the famous declaration of the *Upanishads,* "The self is 'not this, not this.'"[15] Ultimately, the self, understood now as universal pure awareness, can't be described in positive terms, only indicated by a series of negations—ungraspable, indestructible, unbound, not affected by what has been done or not done. Whoever knows this truth becomes calm, composed, patient, and at peace. He sees the self in himself and sees the self as all.

"This is the world of *brahman,*" Yajñavalkya tells King Janaka, "and I have taken you to it."

"I offer you my kingdom," replies the king, "and myself to be your servant."

Yajñavalkya declines the king's offer. Choosing instead to renounce the world, he becomes a wandering ascetic, thereby showing that he has indeed let go of desire.

WHAT IS CONSCIOUSNESS?

We're now ready to go back to our guiding question. Given the four-fold framework of waking, dreaming, deep sleep, and pure awareness, can we say precisely what makes each of these a form of consciousness?

According to the yogic traditions of Indian philosophy, consciousness is that which is *luminous* and has the capacity for *knowing.* A lot is packed into these words, so let's go slowly.

"Luminous" means having the power to reveal, like a light. Without the sun, our world would be veiled in darkness, but without consciousness, nothing could appear. Consciousness is fundamentally that which reveals or makes manifest because it is the crucial precondition for appearance. Nothing, strictly speaking, *appears* unless it appears *to* some consciousness. Without consciousness, the world can't appear to perception, the past can't appear to memory, and the future can't appear to hope or anticipation. The point extends to science: without consciousness, there's no appearance of the microscopic world through electron microscopes, no appearance of distant stars through telescopes, and no appearance of the brain through magnetic resonance imaging (MRI) scanners. Simply put, without consciousness there's no observation, and without observation there are no data.

"Knowing" means having the ability to apprehend whatever appears. When you're conscious of something, you grasp or apprehend it in a certain way. A certain pattern of light and color appears to you, and you grasp or apprehend it as a sunset. Western philosophers call this capacity of the mind "intentionality." When we use this word, it is not to mean being able to do something on purpose. We use the word in a special sense to mean "aboutness," or being mentally directed toward something in perception or thought. When you see the sunset, your visual perception is about the sunset; it's the object of your seeing. When you remember the sunset, you think about that past event; it's the object of your memory. In such cases, not only does something *appear* to your consciousness, but also you *apprehend* it in a certain way, depending on your senses and cognitive capacities.

Let's now apply this way of thinking about consciousness to waking, dreaming, and deep sleep.

In the waking state, the world shows up for us in perception, in certain ways and not others, depending on our senses and cognitive capacities. For example, we see the world as having certain colors, but some animals see little in the way of color while other animals, such as birds, see colors we can't see.[16] To say the world shows up or appears to us in these ways is to say that our consciousness reveals the world and apprehends it in these ways. The primary means or instruments by which consciousness accomplishes this are sense perception and conceptualization (we can't apprehend what we see as a sunset unless we can conceptualize it as a sunset). Thus, the waking state is that state in which consciousness apprehends the outer world through sense perception and conceptualization.

In the dream state, what show up or appear are mental images. In an ordinary dream, we don't recognize or apprehend them as dream images; we take them to be real things outside us. In this way, our apprehension is a misapprehension, our knowing a misknowing. In a lucid dream—a dream where we know we're dreaming—the images still appear, but now we apprehend them as dream images, and we can conceptualize or think about our state as a dream state. In both kinds of dream state, consciousness relies heavily on memory and other conceptual and imagery-related processes. Thus, the dream state is that state in which consciousness apprehends the inner world of mental images.

What about deep sleep? If deep sleep is a state of consciousness, then what appears and what's apprehended in this state?

Unlike contemporary Western philosophy of mind, where there is almost no discussion of deep sleep, only an occasional passing mention of it as an obvious case of the complete absence of consciousness, Indian philosophy contains many rich examinations of this state as well as debates about whether and in what sense it qualifies as a state of consciousness.[17] Some of these discussions are related to certain meditation practices, notably the yoga of sleep, known today as *yoga nidrā*. We'll look closely at this in chapter 8. For now, I'll mention only Śaṅkara's influential view, which he aligned to Yājñavalkya's view from a thousand years before: in deep sleep, consciousness enters into an unknowing but blissful state, due to the absence of images, desires, and activities—in short, due to the absence of what in Yoga is called the usual "whirling" of the mind.[18] In this way, deep sleep offers a foretaste of the lucid and knowing blissfulness of pure awareness.

To summarize, in this book I'll use the word "consciousness" to mean experience in all its forms across waking, dreaming, deep sleep, and meditative states of awareness. In all these modes, consciousness is that which makes something manifest and apprehends it in some way. In order to describe how consciousness functions to reveal and apprehend phenomena, I'll distinguish among three aspects—awareness, the contents of awareness (what we're aware of from moment to moment), and ways of experiencing certain contents of awareness as being or belonging to the self (our sense of self or "I-Me-Mine").

SELF-ILLUMINATION VERSUS OTHER-ILLUMINATION

One more question remains. If consciousness is luminous and has the capacity for knowing, if it's what reveals and apprehends, then what reveals consciousness? To put the question another way, what reveals your conscious experiences to you?

Indian philosophy contains numerous intricate discussions of this issue.[19] Broadly speaking, there are two opposed camps. "Self-illumination" theories say that every conscious experience is revealed to itself. "Other illumination" theories say that for a conscious experience to be revealed, there needs to be a second, higher-level cognition of that experience.

Western philosophy of mind has debates parallel to the Indian ones between self-illumination versus other-illumination theories.[20] Without going into the details of the many Indian and Western versions of these theories, let me explain why I think the self-illumination viewpoint is the better one.

According to the other-illumination viewpoint, in order for your seeing the sunset to be a conscious seeing, it needs to be revealed to you as your seeing, and in order for that to happen, you need to have some kind of higher-level, inner cognition of your seeing. In other words, you need to have some kind of inner mental perception of your seeing or some kind of thought along the lines of "I'm seeing the sunset" or "this seeing is my experience," and your seeing needs to cause this higher-level, inner cognition to occur.

The question is whether that higher-level cognition is itself conscious or not. If it is conscious, then it needs to be revealed to you as your cognition, and this requires that there be a third-level cognition of the second-level cognition. And if this third-level cognition is also conscious, then there needs to be a fourth-level cognition of this third-level one, and so on, leading to a vicious infinite regress.

So let's suppose that the higher-level cognition isn't conscious and that it's not revealed to you as your cognition. Instead, it occurs nonconsciously. But how could a nonconscious cognition make your seeing into a conscious seeing? How could it reveal the seeing to you as yours? Since the higher-level cognition is nonconscious, it has no "luminosity," so how could it "light up" so that your seeing doesn't happen in the dark but feels some way to you? More generally, how could two states that are supposedly nonconscious in themselves — the second-level cognition and the first-level perception — come together and make one of them into a conscious state? Why should an interaction between two otherwise unconscious states cause one of them to become conscious?

The basic problem with the other-illumination viewpoint is that it projects the subject-object structure of ordinary perception onto consciousness itself. At the most basic level, however, your consciousness isn't revealed to you as an object. When you see the sunset, your seeing isn't present to you as another object of awareness like the sunset. Neither is your seeing simply absent to you. Rather, your seeing reveals itself in the sunset's appearing to you visually. To use a grammatical metaphor, your awareness of the sunset is a transitive or object-directed awareness, but your seeing experience is intransitive and reflexive. In this way, your seeing is self-aware.

This kind of self-awareness isn't a higher-level, introspective, or reflective self-awareness. It's not a second-level awareness whose object is the first-level awareness. Rather, it's contained within or belongs to the first-level awareness. Western phenomenologists call this "prereflective self-awareness," because it already belongs to the first-level awareness before any reflection or introspection happens.

According to the self-illumination viewpoint, consciousness is self-luminous or self-revealing. The traditional analogy is that of a light, which shows itself while illuminating the other things around it. A light illuminating other things doesn't require another light to be seen. So, consciousness, in revealing other things, doesn't need another consciousness to be revealed. Put another way, in its witnessing of outer objects of perception and inner mental images and thoughts, consciousness also witnesses itself. This kind of self-witnessing, however, isn't like seeing your image in a mirror; it doesn't involve any kind of doubling or subject-object structure. What the analogy of light aims to convey is that the luminosity of consciousness is also essentially a self-luminosity: consciousness, in its nature, is self-manifesting or self-revealing. Western philosophers call this feature of consciousness "reflexivity."

Here, then, is a fuller answer to our question, what is consciousness? Consciousness is that which is luminous, knowing, and reflexive. Consciousness is that which makes manifest appearances, is able to apprehend them in one way or another, and in so doing is self-appearing and prereflectively self-aware.

A MEDITATIVE MAP OF CONSCIOUSNESS

In the ancient Indian dialogues of the *Upanishads*, whose oldest portions precede the first Greek philosophers by at least a hundred years, we see philosophy in its earliest written expression. To my mind, the *Upanishads* stands as humanity's first truly philosophical work. Its sages scorn blind adherence to the sacrificial ritual and priest craft codified in the *Rig Veda* (India's oldest text), and instead uphold direct experience through meditation as the path to knowledge and liberation. Woven into the poems, stories, and dialogues of the *Upanishads* is a distinctly philosophical form of thought: consciousness is the necessary precondition of all knowing; nothing can be known unless consciousness reveals it; it's impossible to think consciousness away or imagine its absence; consciousness most fundamentally isn't the object seen, the act of seeing, or the person who sees; consciousness most fundamentally is the luminous witnessing awareness. In these ways, the *Upanishads* privileges consciousness above all else.

Yet the *Upanishads* offers no mere speculative philosophy but instead a philosophy steeped in meditative experience—in the exploration of consciousness from within using heightened attention, concentration, and awareness. Relying on meditation, the sages of the *Upanishads* devised humanity's oldest recorded map of consciousness. Refined over the millennia, this map guided especially the Yoga and Vedānta systems of Hindu thought, but also deeply informed Buddhism, which rejected the authority of the *Upanishads* and traveled out of India to the rest of the Asian world. To quote my father, William Irwin Thompson:

> The *Upanishads* is a watershed in the evolution of consciousness. Instead of being ethnocentric and dividing all global history between B.C. and A.D., we should really divide it between before *Upanishads* and after *Upanishads*—B.U. and A.U.—because the sophisticated psychology of consciousness in the *Upanishads* represents a quantum leap forward in human development. In the *Rig Veda* there is a complex and obscure symbolic code that elliptically refers to states of experience for those who have already had the experiences, but in the *Upanishads* there are radical yogi psychologists who insist that one can be a Brahmanical priest and throw butter on the fire until the cows come home but still never become truly enlightened.[21]

This radical yogi map of consciousness will guide us in this book. The following chapters will explore the waking state, dreaming, lucid dreaming, deep sleep, dying and death, pure awareness, and the nature of the self, drawing not only from the *Upanishads* and the philosophical systems of Yoga and Vedānta but also from Indian and Tibetan Buddhism.

This exploration, however, won't be limited to the meditative view of consciousness from within, because we also want to know how these modes of consciousness are related to the brain and the rest of the living body. Today we know that distinct electrochemical and metabolic patterns of brain activity occur during waking, dreaming, lucid dreaming, and deep sleep. We also know that different brain systems are associated with different ways we can experience having or being a self. Given this knowledge, it's natural to ask how meditative experiences of "pure awareness" are related to the brain or living body as a whole. We also need to ask what happens to consciousness during the dying process, and whether there's any evidence for the survival of consciousness or for rebirth.

Our exploration begins in the next chapter with attention and perception in the waking state.

2

WAKING

How Do We Perceive?

'm sitting in the Cognitive Neuroscience and Brain Imaging Laboratory of the Salpêtrière Hospital in Paris, where Diego Cosmelli, one of Francisco Varela's last doctoral students, is showing me and fellow philosopher Alva Noë what happens when each eye sees a different image at the same time. The two images I'm viewing are a photograph of a woman's face and an expanding ring with a checkerboard pattern. A special setup keeps the two images separate by projecting the face to my right eye and the expanding checkerboard ring to my left eye. As I stare at the screen, the checkerboard ring starts to give way and change into the woman's face. The face breaks through in disconnected patches, which merge and take over the whole screen. A few seconds later, the face falls apart and the expanding checkerboard rings return, removing the face completely. But this too lasts only several seconds, as the face eventually reasserts its presence. Watching these two images vie with each other, I can sense how their alternation somehow takes place within me. After a while, I begin to feel I can intentionally affect how these images behave. By paying careful attention to a fixed spot at the center of the screen, I can keep one image there a little longer. I can mentally hold onto the image as it starts to fade and sometimes even bring it back. But it's hard to be sure, because there seems to be no regular pattern to the changing images, just a spontaneous and unpredictable alternation.

Psychologists and vision scientists call what I'm experiencing "binocular rivalry." Binocular rivalry happens when two images are presented simultaneously, one to each eye. Instead of the images fusing to form a third stable image, they appear to alternate unpredictably, so that the whole perception is "bistable."

You don't need a fancy lab, however, to experience binocular rivalry. Take a cardboard tube or a rolled-up piece of paper and look through it with one eye while placing your hand about ten inches in front of the other eye. Angle the tube so that it points just behind the hand the other eye is viewing. Your hand should now seem to have a hole in it, through which you can see what's at the other end of the tube. The hole appears because each eye has a different view, and the two views compete for perceptual dominance. When you see the hole, the view through the tube dominates, while the other eye's view of that part of your hand is suppressed. If you keep looking, the hole will eventually disappear; now the other eye's view of the hand dominates. Over time, the hole will return and then disappear again as the view from each eye dominates in turn.

Scientists have used binocular rivalry to try to track down the brain activity associated specifically with conscious perception. Although the stimulus itself doesn't change, your visual awareness does, so we know the change is happening in you and not in it. First you see the checkerboard rings, then you see the face, and so on. So we can ask the following question: What's the difference in your brain between the times when you say you see just one image and the times when you say you don't see that image anymore (even though it's still there, affecting your eyes and visual system)? What changes in your brain? As we'll see later, the answers to these questions from the neuroscience of consciousness evoke classical Indian Buddhist ideas about the nature of perception.

A PARADIGM SHIFT

Asking how a moment of visual awareness is contingent on the brain seems far removed from the early Indian Upanishadic view of consciousness presented in the first chapter. For one thing, the philosopher-sages of the *Upanishads* didn't know much about the brain. For another thing, they saw consciousness as the primary reality. Consciousness, they maintained, is the infinite ground of all being (*brahman*) and the true self (*ātman*). The absolute primacy of consciousness is evident in the Upanishadic theory of creation, according to which the undifferentiated and primordial unity of *ātman/ brahman* differentiates into "name-and-form" (*nāmarūpa*), or what we apprehend on the basis of thoughts and concepts ("name") and what we apprehend through the senses ("form").[1]

The earliest Indian Buddhist texts, however, present the Buddha as breaking sharply with this viewpoint.[2] In the *Suttas* or recorded sayings of the Buddha, preserved in the so-called Pāli Canon, the Buddha repeatedly states that consciousness is contingent or dependent on conditions, and he forcefully rejects the Upanishadic view that one and the same consciousness lies behind the changing mental states and changing bodily states that make up a person. This is especially clear in his conversation with the monk Sāti, who mistakenly attributes this view to the Buddha:

[Sāti:] "As I understand the Dhamma taught by the Blessed One, it is this same consciousness that runs and wanders through the round of rebirths, not another."

[Buddha:] "What is that consciousness, Sāti?"

[Sāti:] "Venerable sir, it is that which speaks and feels and experiences here and there the result of good and bad actions."

[Buddha:] "Misguided man, to whom have you ever known me to teach the Dhamma in that way? Misguided man, have I not stated in many ways consciousness to be dependently arisen, since without a condition there is no origination of consciousness? . . .

"Consciousness is reckoned by the particular condition dependent upon which it arises. When consciousness arises dependent on the eye and forms, it is reckoned as eye-consciousness; when consciousness arises dependent on the ear and sounds, it is reckoned as ear-consciousness; when consciousness arises dependent on the nose and odours, it is reckoned as nose-consciousness; when consciousness arises dependent on the tongue and flavours, it is reckoned as tongue-consciousness; when consciousness arises dependent on the body and tangibles, it is reckoned as body-consciousness; when consciousness arises dependent on the mind and mind-objects, it is reckoned as mind-consciousness. Just as fire is reckoned by the particular condition on which it burns . . . as a log-fire . . . as a grass-fire . . . as a cow-dung fire . . . so too, consciousness is reckoned by the particular condition dependent on which it arises."[3]

As the image of the fire indicates, consciousness arises dependent on conditions, but it also has its own causal influence on things. Indeed, the Buddha often states both that consciousness is contingent upon "name-and-form" and that consciousness in turn conditions "name-and-form":

"Then, monks, it occurred to me: . . . 'By what is name-and-form conditioned?' Then, monks, through careful attention, there took place in me a breakthrough by wisdom: 'When there is consciousness, name-and-form comes to be; name-and-form has consciousness as its condition.'

"Then, monks, it occurred to me: 'When what exists does consciousness come to be? By what is consciousness conditioned?' Then, monks, through careful attention, there took place in me a breakthrough by wisdom: 'When there is name-and-form, consciousness comes to be; consciousness has name-and-form as its condition."[4]

In the early Indian Buddhist usage, "name-and-form" covers the whole range of what is available to consciousness. "Form" (*rūpa*) stands for the "four great elements" of earth, water, fire, and wind, which are understood as the four basic material qualities of solidity, cohesion, temperature, and motion. These elements are what impinge upon the five senses, and they constitute our experience of matter. "Name" (*nāma*) stands for the five basic mental processes that are responsible for identifying or recognizing something based on apprehending it through a concept. These five mental processes are "contact" between a sense organ and its object (such as the eye and a visual form), the sensory "feeling" and the "perception" arising with this contact, and the factors of "attention" and "intention" (or "volition") that co-occur with the feeling and perception. In the words of Buddhist scholar-monk Anālayo: "Contact and attention provide the first input of a previously unknown object. This object is then felt and perceived, and eventually something will be done with it. The whole complex of mental operations that in this way takes place finds its conjunction in the 'name' under which the hitherto unknown object will be remembered and conceptualized."[5] In cognitive science language, whereas "form" provides the input of experienced materiality, "name" provides the sensory registration, feeling, attention, perceptual identification, and intention required for an object to be cognized and acted upon.

To say that name-and-form depends on consciousness means that consciousness is required in order for the physically and mentally presented aspects of an object (its "form" and "name") to be experienced, whereas to say that consciousness depends on name-and-form means that name-and-form is required in order to provide the content of what is experienced through any of the six types of consciousness (eye consciousness, ear consciousness, nose consciousness, tongue consciousness, body consciousness, and mind consciousness).[6]

In a famous simile the Buddha describes name-and-form and consciousness as supporting each other, like two bundles of reeds propping each other up: "Well, then, friend, I will make up a simile for you. . . . Just as two sheaves of reeds might stand leaning against each other, so too, with name-and-form as condition, consciousness [comes to be]; with consciousness as condition, name-and-form [comes to be]."[7] In other words, consciousness and the makeup of an individual being—its living body, environment, and perceptual and cognitive systems taken as a whole complex—are mutually supporting; one is not found without the other. In Anālayo's words, "It is the interplay of these two aspects—consciousness on the one side and name-and-form on the other—that makes up the 'world' of experience."[8]

Let's return now to our question about the brain basis for a moment of visual awareness, such as seeing one of the two images in binocular rivalry. What we're asking about, to use the early Indian Buddhist terminology, is the contingency of visual consciousness on name-and-form: specifically, how a moment of visual awareness depends on the physical and psychological makeup of the perceiver. As we'll see, investigating binocular rivalry has proven especially fruitful for addressing this question.

NOW YOU SEE IT, NOW YOU DON'T

One of the main goals of the neuroscience of consciousness is to discover the so-called "neural correlates of consciousness"—the neural processes that correlate directly with conscious experience, also known as the NCC.[9] Neuroscientists distinguish between two main kinds of NCC. On the one hand, there are the brain activities that correlate directly with the level of consciousness—awake, asleep, dreaming, alert, drowsy, and so on. On the other hand, there are the brain activities that correlate directly with specific conscious experiences, such as the visual experience of the color red. Binocular rivalry is one of the main experimental paradigms scientists have used to hunt for neural correlates of consciousness in the second sense, in this case the neural correlates of specific visual experiences.

Binocular rivalry seems ideal for this quest because it seems to offer a way to differentiate between conscious and unconscious visual contents—"now you see it, now you don't"—at the level of the brain. In other words, it seems to offer a way to dissociate the neural activity corresponding directly to conscious visual perception from the neural activity associated with unconscious processing of the stimulus. Although the stimulus stays constant, the conscious perception changes dramatically every few seconds. What neuroscientists seek to uncover are the neural processes that correlate directly with one image being perceptually dominant (now you see it) versus the neural processes responsive to the same image when it's suppressed (now you don't).

Neuroscientist Nikos Logothetis, at the Max Planck Institute for Biological Cybernetics in Tübingen, Germany, pioneered this approach in the 1990s in a series of experiments with macaque monkeys.[10] These animals have a visual system similar to ours, they experience binocular rivalry, and they can be trained to pull a lever when they see one image or the other. Logothetis and his colleagues recorded the firing of individual neurons in a number of visual brain areas while the monkey viewed a binocular rivalry display. These areas included the primary visual cortex (V1)—whose neurons are tuned to one eye or the other and respond to basic features of the visual world, such as contrast, orientation, motion, direction, and speed—and higher visual areas that respond to object categories, such as faces or houses. These scientists found that neural activity at early stages of the visual pathway closer to the retina was better correlated with the stimulus, regardless of what the animal indicated it was perceiving, whereas the proportion of neurons whose activity correlated with the animal's perception increased at later visual stages. Neural activity in the earliest visual cortical area, the primary visual cortex (V1), correlated almost entirely with the stimulus, independent of perception. Meanwhile, almost all the recorded neurons in the inferotemporal cortex (IT)—the last stage in the ventral visual pathway and an area crucial for object recognition—responded only to the dominant perceptual image. For example, neurons in IT sensitive to butterfly images but not sunburst images responded only when the animal indicated that it saw the butterfly and not the sunburst image presented to the other eye.

Other studies in humans have also shown that activity in later object-recognition areas of the ventral visual pathway reflects the reported perception in binocular rivalry.[11] These human studies, however, also found strong correlations between the reported perception and activity in early visual areas, including V1 and the lateral geniculate nucleus, or LGN (a part of the thalamus that receives signals from the retina, sends connections to V1, and receives strong feedback connections from the cortex).

What these monkey and human studies indicate is that there's no single site in the brain correlated with reportable conscious visual perception. Put another way, the NCC for a specific visual experience of an object seems to consist of brain activities distributed over multiple areas, including both early visual areas sensitive to basic sensory qualities and higher visual areas sensitive to object categories, as well as frontal and parietal areas involved in voluntary attention. As Logothetis wrote in 1999, summarizing the results of his studies: "the findings to date already strongly suggest that visual awareness cannot be thought of as the end product of such a hierarchical series of processing stages. Instead it involves the entire visual pathway as well as frontal parietal areas, which are involved in higher cognitive processing. The activity of a significant minority of neurons reflects what is consciously seen even in the lowest levels we looked at, V1 and V2; it is only the proportion of active neurons that increases at higher levels in the pathway."[12]

Given these findings, we can ask the following questions. If there's no single locus for the brain activity associated with conscious perception, is there something that links or holds together the widespread and distributed neural activities that occur during a moment of conscious perception? Does the back-and-forth shift from "now you see it" to "now you don't" correspond to shifting relations between multiple brain areas? Is it possible to track how these areas coordinate their activities when a "now you see it" moment occurs?

WAVES OF CONSCIOUSNESS

The binocular rivalry demonstration Diego Cosmelli showed me in Paris was part of an experiment he had designed to address these questions.[13] To find out what happens in the brain during the flow of alternating images, Cosmelli used the approach Francisco Varela called "neurophenomenology." Neurophenomenology combines the careful study of experience from within with investigations of the brain and behavior from without. It uses descriptions of direct experience to guide the study of the brain processes relevant to consciousness.

Cosmelli asked the participants in his experiment to explore and describe their experience with the binocular rivalry stimulus. Their descriptions indicated that their experience had an inherent ebb and flow like the rise and fall of ocean waves. Dominance periods for each image recurred in time, but the transitions between them happened in variable and shifting ways. Sometimes the transition from one image to the other began in the center of the field and moved outward; sometimes it began on one side, or from the top or the bottom, and then progressively took over the other image. Most participants claimed it was hard to give a stable description of how these transitions happened, because they arose in a different way each time. Yet everyone said they saw one image for a while and then the other image, no matter what they did or how much they tried to prevent it.

Cosmelli's question was whether we could follow these waves of consciousness in the brain. Does the brain activity during perceptual rivalry show corresponding patterns that recur in time but with variable transitions between them?

Taking our lead from the way experience flows through time means we need more refined methods to analyze brain activity than the ones neuroscientists often use. Scientists often simplify the binocular rivalry experience for experimental purposes by treating it as a clear-cut alternation between two otherwise static visual states—the perception of one image and then the other. The experimenter instructs the participant to indicate with a button press the moment when the perceptual switch happens; this report provides a fixed reference point in time for defining the average brain state when the participant sees one image and the average brain state when she no longer sees it. The experimenter then contrasts these average measures to see what's specific to the conscious perception of the image. This approach has provided a wealth of important findings about the basic properties of binocular rivalry and their basis in the brain.

If we don't simplify the binocular rivalry experience this way, but instead appreciate it as a flowing experience consisting of variable and recurrent visual fluctuations, then we can't assume there's such a thing as an "average transition" from one image to the other. Average measures wipe out the variable aspects of a phenomenon. So assuming there's an average transition in binocular rivalry would prevent us from being able to take account of variations in how the whole experience unfolds in time and how these variations are reflected in brain activity.

Cosmelli and his colleagues in Varela's lab therefore developed a new statistical framework that considered as significant any neural activity that recurs in time, regardless of when it happens. This enabled them to follow the spontaneous flow of brain activity during binocular rivalry while they tracked neural patterns that occurred again and again, but never in precisely the same way.

These scientists also took advantage of the fact that stimuli like the expanding checkerboard rings—which moved outward from center to periphery at a rate of five times a second—will produce corresponding neural responses at the same frequency in the brain. The neural signals will go through an up-and-down cycle with peaks and troughs, like a pendulum swinging back and forth in what physics calls an oscillation, and the frequency of this so-called neural oscillation will match the expansion of the checkerboard rings. By looking at how these oscillating neural responses evolve, one can determine what distinctive features they acquire during conscious perception of the checkerboard image, compared to when that image is suppressed. To track these brain responses, Cosmelli and his colleagues used a neuroimaging technique called magnetoencephalography (MEG), which measures the magnetic fields produced by electrical activity in the brain.

What these scientists found was that neural responses oscillating with the same frequency as the checkerboard rings occurred at several distinct brain regions throughout the entire viewing period, but only during conscious perception of the checkerboard image did they become precisely synchronized. To visualize these "synchronous neural oscillations," think of the "wave" at a sporting event, where successive small groups of spectators stand and raise their arms at the same time and then sit down again. Each complete up-and-down cycle is like an oscillation, and the phases of these oscillations—the points on these cycles where some people are on their way up and other people are on their way down—have to be precisely coordinated. Similarly, Cosmelli and his colleagues found that during conscious perception, the neural responses located in different brain regions were oscillating in synchrony. More precisely, the phase of each neural response—where exactly it was in its oscillatory cycle— was in sync with the phases of the other neural responses. Neuroscientists call this phenomenon "neural phase synchrony." Thus, the NCC for the perception of the expanding checkerboard rings consisted of a large-scale pattern of neural phase synchrony occurring within and between many regions of the brain.

Other experimental results by Varela and his colleagues support this idea of a moment of reportable conscious perception being correlated with large-scale synchronous oscillations in the brain. Earlier in Varela's lab, before Cosmelli's experiment, Eugenio Rodriguez had shown that a large-scale pattern of neural synchrony correlates with the "now you see it" or "Aha!" moment when you suddenly see a face in what initially seemed to be meaningless shapes.[14] Rodriguez used the electroencephalogram (EEG) to measure electrical activity at multiple sites on the scalp. Such activity reflects the underlying electrical activity in the cortex and shows complex patterns in different frequency ranges. In addition, the dominant frequency component changes according to the level of consciousness. For example, in the waking state, the so-called alpha rhythm (8–12 hertz or cycles per second) is the most prominent feature of the EEG, but fast rhythms in the beta (12–30 Hz) and gamma (30–80 Hz) frequency ranges are also strongly present, whereas in deep sleep, slow-frequency delta waves (0.5–4 Hz) are dominant. Rodriguez and Varela found that gamma oscillations, which are known to be associated with perceptual recognition and attention, occurred regardless of whether the ambiguous stimuli were seen as meaningless shapes or as faces, but at the "Aha!" moment when the stimuli were seen as faces, the gamma oscillations became synchronized at parieto-occipital and fronto-temporal regions. In other words, the NCC for the "Aha!" moment consisted in the phase synchrony of the ongoing gamma oscillations. Building on this work, Lucia Melloni and Rodriguez later showed that a moment of reportable conscious perception (compared to unconscious processing of the same stimuli) correlates with a large-scale pattern of synchronous oscillations in the gamma band.[15] Other studies have also confirmed that the neural correlates of reportable conscious perception consist of widespread patterns of neural synchrony.[16]

But what about the flow or continuum of such moments of conscious perception? The novelty of Cosmelli's study was that it addressed this question.

Cosmelli found that the back-and-forth rhythm from "now you see it" to "now you don't" corresponds to the formation and dissolution of a large-scale pattern of neural synchrony. Each time the perceptual transition from the face to the checkerboard rings started to happen, a new synchronous neural network was being formed, like the "wave" starting up at a hockey game. As a small group of fans starts to stand up and swing their arms, so, at the beginning of the perceptual transition, only a few cortical responses were synchronized. As the perceptual dominance of the checkerboard rings developed, the synchrony wave grew in scale, increasing both within occipital visual areas and between these and frontal cognitive areas. This large-scale synchrony pattern persisted for several seconds and coincided with full perceptual dominance of the checkerboard rings. As suppression of the image began, the wave in the brain subsided as the long-range synchronous activity fell apart, leaving isolated patches of local synchrony in occipital visual regions. During full suppression, few areas of synchrony remained and the brain pattern returned to the pretransition situation. Then the whole pattern of synchrony buildup began again, but in a slightly different way, a new wave in the stadium of the brain.

In summary, this study suggests that the alternation between the two images in binocular rivalry corresponds to waves of synchronous oscillations in the brain. The way conscious perception varies in time corresponds to the way neural synchrony varies in time. The arising of a new perceptual moment corresponds to the formation of a new pattern of neural synchrony, and the subsiding of this perceptual moment and its replacement by another one correspond to the dissolution of the synchrony pattern and its replacement by another one. In both the perceptual experience and the brain activity, there's a rhythm of distinct moments that alternate repeatedly but never exactly in the same way. Waves of conscious perception correspond to waves of synchronous oscillations in the brain.

One more study, by Sam Doesburg and Lawrence Ward at the University of British Columbia, adds another level to these ideas.[17] Inspired by Cosmelli's study, these scientists used EEG to investigate the electrical brain rhythms occurring during binocular rivalry. They used the image of a butterfly and the image of a group of maple leaves. Doesburg and Ward found that the perceptual switch from one image to the other corresponded to the formation of a new pattern of gamma synchrony between cortical areas. But they also found that these fast synchronous oscillations (38–42 Hz) were connected to slower oscillations in the theta range (5–7 Hz). Roughly speaking, the fast gamma waves were superimposed on the slower theta waves, so that one slow wave carried many fast waves. The slow waves also affected the fast ones by varying their amplitude (height) and the way they synchronized. In these ways, the slower theta rhythms played a role in shaping the faster gamma synchrony patterns.

Doesburg and Ward propose that this coupling between fast and slow electrical brain rhythms supports discrete and successive moments of perceptual experience. On the one hand, the fast gamma synchrony integrates neural activities occurring at different brain regions and thus supports the binding together of sensory features (such as black and white shapes) into a coherent perception (such as seeing a face in profile). On the other hand, the slow theta rhythms define discrete and successive "frames" or moments of perception. According to this view, these fast and slow electrical brain rhythms structure the flow of conscious perception so that it consists of discrete and successive moments whose content can either stay the same—you continue to see the butterfly—or change every few hundred milliseconds—you see the butterfly change into the maple leaves. In short, the slow rhythms divide the sensory stream into discrete temporal units or moments of perception, while the fast rhythms bind the features discriminated within a given moment into a coherent perception. In this way, the slow rhythms define the temporal context of perception (what you perceive as happening "now"), while the fast rhythms set the content (now you see a butterfly, now you see a bunch of maple leaves).

But is the flow of conscious perception really discrete, like the frames of a movie or a series of snapshots? Maybe this way of thinking about perception comes from relying too much on the unusual situation of binocular rivalry. Doesn't ordinary perception seem to be a continuous flow? How does this appearance of continuity come about, if perception is really discontinuous? As we'll see now, these questions were addressed long ago in Indian Buddhist philosophy in ways that parallel and cast new light on the current scientific discussions.

MIND MOMENTS

"Consciousness, then, does not appear to itself chopped up in bits. Such words as 'chain' or 'train' do not describe it fitly as it presents itself in the first instance. It is nothing jointed; it flows. A 'river' or a 'stream' are the metaphors by which it is most naturally described. *In talking of it hereafter, let us call it the stream of thought, of consciousness, or of subjective life.*"[18]

So William James wrote in 1890, introducing the metaphor of the "stream of consciousness" into Western psychology. Over a thousand years before, the same image figured prominently in the Buddhist philosophical tradition known as the Abhidharma (Abhidhamma in Pāli). There the Buddha is portrayed as saying: "The river never stops: there is no moment, no minute, no hour when the river stops: in the same way, the flux of thought."[19]

For both James and the Abhidharma, mental states don't arise in isolation from each other; rather, each state arises in dependence on preceding states and gives rise to succeeding ones, thus forming a mental stream or continuum. However, James and the Abhidharma have different views about the nature of the mental stream.

According to James, although the mental stream is always changing, we experience these changes as smooth and continuous, even across gaps or breaks. The gaps and changes in quality that we do feel or notice—for example, when we wake up from a deep sleep—don't undermine the feeling that our consciousness is continuous and whole. And the gaps and changes in quality that we don't notice aren't felt as interruptions because we're not aware of them.

The Abhidharma philosophers agree that the mental stream is always changing, but they argue that it appears to flow continuously only to the untrained observer. A deeper examination indicates that the stream of consciousness is made up of discontinuous and discrete moments of awareness. Whether the Abhidharma philosophers arrived at this view through inner observation or through logical analysis premised on an atomistic view of the mind, or some combination of both, is a matter of scholarly debate. In any case, they believed that discrete moments of awareness or "mind moments" can be identified, described, and catalogued; moreover, their duration is said to be measurable. As we'll see, the measures given in certain Abhidharma texts of how long these moments last bear comparison with modern scientific estimates of the duration of a moment of perception.

We first need some familiarity with the Abhidharma view of how the mind works.[20] The Abhidharma builds on the basic Buddhist insight that each moment of awareness arises contingent upon a host of physical and mental processes, and in turn conditions the arising of the next moment of awareness. What we call "the mind" is a stream of momentary mental events, each of which can be analyzed into a number of basic constituents. Every mental event consists of a "primary awareness" together with various constituent "mental factors." The primary awareness belongs to one of the six types of awareness the Buddha delineates in the passage quoted above—visual awareness, auditory awareness, olfactory awareness, gustatory awareness, tactile awareness, and mental awareness (the awareness of a thought, emotion, memory, mental image, and so on). Vasubandhu, the fourth-century C.E. author of the classic work *Treasury of Abhidharma*, defines awareness as the impression or bare apprehension of something.[21] We never experience bare apprehension, however, because every moment of consciousness always arises in conjunction with a number of constituent mental factors. These factors qualify awareness by making it pleasant or unpleasant, focused or unfocused, calm or agitated, ethically wholesome or unwholesome, and so on.

The key insight is that every moment of consciousness not only apprehends a particular object, in the sense of a particular sensory or mental appearance, but also characterizes that object in various ways. More precisely, there's no way for consciousness to apprehend an object—for something to appear to consciousness—without that object or appearance being characterized by consciousness in some way or other. A moment of gustatory consciousness, for example, is never just an awareness of a particular taste and texture; it's an awareness that's attentive or distracted, that experiences the taste as pleasant or unpleasant, that categorizes the taste as the flavor of a mango, that discriminates the mango as ripe or unripe, and so on. In these and many other ways, every moment of consciousness is "about" or "directed toward" an object of experience.

The Abhidharma philosophers thus agree with Western phenomenologists, notably Edmund Husserl (1859–1938), that all consciousness is consciousness *of* something in one way or another. Phenomenologists call this feature of consciousness "intentionality." Western phenomenology and the Abhidharma agree that intentionality, being directed toward an object, belongs to the nature or being of consciousness; it isn't something that gets added to consciousness from the outside.

What's unique to the Abhidharma, however, is the way it analyzes the directedness of consciousness into the basic structure of primary awareness and constituent mental factors. Without the mental factors, consciousness couldn't grasp an object. In the words of a twentieth-century Tibetan teacher, Geshe Rabten: "A primary mind [primary awareness] is like a hand whereas the mental factors are like the individual fingers, the palm, and so forth. The character of a primary mind is thus determined by its constituent mental factors."[22] According to this image, the occurrence of a moment of consciousness consists in the grasping of a particular object of awareness by means of the mental factors, and the flow of consciousness consists in the picking up and putting down of successive objects by means of successive sets of mental factors.

The Abhidharma maps of the mind list over fifty distinct mental factors, specify their functions, and group the factors into various categories. These maps express the Abhidharma tendency to codify crucial phenomenological insights into elaborate theoretical systems. The core insight is that how we're aware deeply conditions what we're aware of, and that how we're aware can be ethically wholesome or unwholesome. The overall scheme is ethical and is meant to support Buddhist practice; to this end, the mental factors conditioning awareness are categorized according to whether they are positive (reduce suffering and increase well-being), negative (increase suffering), or neutral (neither positive nor negative in themselves). The lists and categories differ from one Abhidharma school to another, but the schools generally agree that there are at least five ethically neutral "ever-present" mental factors that are always functioning in every moment of consciousness: "contact," "feeling," "perception," "intention," and "attention." They perform the most rudimentary and essential cognitive functions, without which consciousness of an object would be impossible.

The first mental factor, "contact," consists in a three-way relationship between a sensory or mental object, the corresponding sensory or mental faculty, and the consciousness dependent upon these two elements. For example, a moment of visual consciousness arises in dependence on the faculty of vision and the presence of something visible; contact is the concurrence of these three elements. As the contemporary Buddhist scholar Bhikkhu Bodhi explains: "[Contact] is the mental factor by which consciousness mentally 'touches' the object that has appeared, thereby initiating the entire cognitive event."[23]

A moment of consciousness, however, never simply "touches" its object in some affectless way. On the contrary, it feels the object as pleasant, unpleasant, or neutral. ("Neutral" doesn't mean the absence of feeling; it means being neither pleasant nor unpleasant.) "Feeling" is the mental factor that performs this affective function. With sensory or mental contact, there occurs a basic affective quality or feeling tone, on the basis of which consciousness evaluates its object.

In addition to feeling its object, consciousness also discerns it. "Perception" is the mental factor that plays this role; it discriminates, discerns, or identifies the object by distinguishing it from other objects. Discernment is the basis for recognition, for being able to re-identify the object over time.

According to the Abhidharma, every moment of consciousness is also goal-directed, in the sense that it approaches its object with an intention or motivation. "Intention" (also translated as "volition") is the mental factor responsible for this goal-directed function. Normally we think and feel so rapidly and habitually that we don't see intention at work, but the Abhidharma maintains that it's always there, operating subliminally in each moment of consciousness. This factor is also what determines the ethical quality of consciousness, that is, whether the consciousness is wholesome (lessens suffering) or unwholesome (increases suffering).

Finally, "attention" is the mental factor that enables consciousness to orient toward its object, and to target and refer to it. It guides or binds all the other mental factors to the object of the primary awareness. In Bhikkhu Bodhi's words: "Attention is the mental factor responsible for the mind's advertence to the object, by virtue of which the object is made present to consciousness. Its characteristic is the conducting . . . of the associated mental states [factors] towards the object. Its function is to yoke the associated states [factors] to the object."[24] Thus, according to the Abhidharma, there is no consciousness of a sensory or mental object without the mental orientation and reference that attention supplies.

The Abhidharma distinguishes attention from other factors that also focus consciousness on its object but that aren't present in every mental state. Two such factors in the Tibetan Abhidharma are "concentration" and "mindfulness." Concentration is the ability of the mind to focus exclusively or single-pointedly on the object; mindfulness is the ability to keep the object in focus without forgetting or floating away from it. Concentration differs from attention because it involves not just attending to an object but also sustaining that attention over time. Similarly, mindfulness involves more than attention because it retains the object in awareness from moment to moment, repeatedly bringing it back to mind and preventing it from slipping away in forgetfulness. Concentration and mindfulness belong to the category of so-called "object-ascertaining" factors. These factors don't establish the basic attentional orientation to an object; they presuppose it. They function to make the object more determinate for consciousness, given that attention has targeted it. The object-ascertaining factors of concentration and mindfulness are vital to Buddhist meditation and are present only when the object is apprehended with some degree of clarity and sustained focus.

We can now appreciate that each discrete moment of consciousness is a structured cognitive event, involving at least a minimal level of feeling, discernment, attention, and intention. According to the Abhidharma, each of these momentary cognitive events arises and passes away in rapid succession. Our waking cognition of the world is thus discrete instead of a seamless flow. Contemporary visual science offers an analogous idea: although it seems as if we're seeing many things at one time, our eyes are actually darting about quickly from one thing to another and back again. Our impression of a seamless visual world doesn't come from taking in everything all at once or in a smooth progression; it comes from the rapid way our eyes sample the scene and from our knowing we can look anywhere we need to in order to get more information.[25] Similarly, from the Abhidharma perspective, although the flow of consciousness seems continuous, this appearance is like our impression of continuity while watching a movie; in reality, the arising and passing away of each cognitive event happens rapidly, like the high-speed sequence of movie frames.

Given this viewpoint, it's natural to wonder how much time it takes for a moment of consciousness to arise and subside. How long does a moment of consciousness last? Are mind moments measurable?

Although the Abhidharma texts do address these questions, the answers they give aren't straightforwardly observational, for they combine what look like empirical observations with what we would regard as metaphysical considerations.[26] Several texts estimate the length or duration of a moment of consciousness in ways that sound observational. For example, the *Treasury of Abhidharma* says, "there are sixty-five instants in the time it takes a healthy man to snap his fingers," a measure that works out roughly to 1/65th of a second, or 15 milliseconds, for a mind moment.[27] We also find statements reflecting views of the mind that seem far removed from direct experience. For example, the following statement occurs in the Pāli Theravada Abhidhamma: "in the time it takes for lightning to flash or the eyes to blink, billions of mind-moments can elapse."[28] Furthermore, metaphysical problems arise for Buddhist philosophy from supposing that mental events have any kind of duration, no matter how short. If something endures unchanged for even a millisecond, then it seems to violate the cardinal Buddhist principle of impermanence—that everything is constantly changing, that nothing is static for any period of time, no matter how brief. In classical Indian Buddhist philosophy, this thought led from a conception of mind moments as first arising, then briefly performing their function, and then finally passing away (the position of the Sarvāstivāda school) to a conception of mind moments as instantaneous, flashing into and out of existence at one dimensionless instant of time (the position of the Sautrāntika school).

If the classical Abhidharma philosophers were alive today, they might wonder whether experimental psychology and neuroscience would have anything to say about these questions. If we set aside abstract metaphysical issues about the nature of time—whether time is discrete and momentary, or continuous—is there any scientific evidence for measurable discrete moments of experience in the stream of consciousness?

MIND THE GAP

In 1979, when I was sixteen, I took part in an experiment Francisco Varela devised to investigate whether perception is continuous or discrete. I'd never been to a neuroscience lab, and the prospect of seeing my own brain waves was enticing. Francisco and I set off from Sixth Avenue and 20th Street, where we lived at the Lindisfarne Association, to the New York University Brain Research Laboratories at 550 First Avenue. I sat in a dark room with electrodes fixed to my scalp and watched two small lights flash on and off. My task was to say whether the lights were simultaneous or sequential, or whether there was one light moving from left to right.

It's well known in experimental psychology that there is a certain minimal window of time within which two successive events will be consistently perceived as happening at the same time. For example, if you're shown two successive lights with less than about 100 milliseconds between them, you'll see the lights as simultaneous. If the interval is slightly increased, you'll see one light in rapid motion. If the interval is further increased, you'll see the lights as sequential. These phenomena of "apparent simultaneity" and "apparent motion" have sometimes been interpreted as supporting the idea of a discrete "perceptual frame," according to which stimuli are grouped together and experienced as one event when they fall within a period of approximately 100 milliseconds.

If perception is discrete—if it unfolds as a succession of perceptual frames with a gap between each frame and the next one—then we can make the following prediction: whether two distinct events will be judged as simultaneous or sequential depends not just on the time interval between them but also on the relation between the timing of each event and the way perception falls into discrete and successive frames, that is, the ongoing process of perceptual framing. In particular, two events with the same time interval between them can be perceived as simultaneous on one occasion and as sequential on another occasion, depending on their temporal relationship to perceptual framing: if they fall within the same perceptual frame, they're experienced as simultaneous, but if they fall in different perceptual frames, they're experienced as sequential. In short, what you perceive as one event happening "now" depends not just on the objective time of things but on how you perceptually frame them.

It was precisely this idea that Francisco wanted to test. Already in his early years as a young scientist, Francisco's research was strongly motivated by a vision of the brain as a self-organizing system with its own complex internal rhythms. (Although popular today, this vision was a small minority view in the 1970s, when most scientists thought of the brain as a sequential-processing computer.) These rhythms, he believed, bring forth meaningful moments of perception in a fluctuating and periodic way. Francisco was also intrigued by the parallels between the Abhidharma notion of mind moments and the neuroscience view of discrete perceptual frames created by the brain's self-generated rhythms. A month or so before my visit to the NYU Brain Research Lab, Francisco and I had talked about the Buddhist idea of mind moments and the gaps between them as we walked to the old Paragon Book Gallery on East 38th Street, where he bought a hard-to-find copy of Louis de la Vallée Poussin's classic French translation of Vasubandhu's *Treasury of Abhidharma*. It was only after the experiment that Francisco told me what he really wanted to do was measure a mind moment.

In the experiment, Francisco recorded the brain's ongoing EEG alpha rhythm and used it to trigger when the two lights flashed on and off. The hypothesis was that seeing the lights as either simultaneous or in apparent motion would depend on when they occurred in relation to the phase of the ongoing alpha rhythm. Like a surfer catching a wave, if the lights arrived at a certain point of the repeating alpha cycle, they would be seen as simultaneous, but if they missed the wave, they'd be seen as in apparent motion. In other words, presenting two flashes of light always with the same time interval between them, but at different phases of the alpha rhythm, would result in noticeably different perceptions.

The results supported the hypothesis: when the lights were presented at the positive peak of the alpha rhythm, they were almost always seen as in apparent motion, but when they were presented at the negative peak (the opposite phase), they were seen as simultaneous.[29] In the published study, a figure showed my visual performance along with that of two other participants (see the bar labeled "ET" in figure 4 of the original study). At an interval of 47 milliseconds between flashes, my discrimination between simultaneity and apparent motion was at a chance level, but there was a change in the probability of my perceiving the lights as simultaneous when they were presented at either the positive or the negative peak of my ongoing alpha rhythm.

Here's how Francisco and I described what we took to be the significance of this study in our 1991 book, *The Embodied Mind*:

> Experiments such as these suggest that there is a natural parsing in the visual frame and that such framing is at least partially and locally related to the rhythm of one's brain in the range of duration of about 0.1–0.2 milliseconds at its minimum. Roughly stated, if the lights are presented at the beginning of the frame, the likelihood of seeing them occur simultaneously is much greater than if they are presented toward the end of the frame: when they are presented toward the end of the visual frame, the second light can fall, as it were, in the next frame. Everything that falls within a frame will be treated by the subject as if it were within one time span, one "now."[30]

Unfortunately, these promising results have proved difficult to replicate, both by Francisco in a follow-up study and by other scientists today.[31] Nevertheless, the experiment is widely cited as precisely the kind of experiment that would be needed to demonstrate definitively the discrete nature of perception; furthermore, new and more sophisticated studies are extending and deepening this line of research into the relationship between electrical brain rhythms and the discrete character of perceptual awareness.[32] For example, recent experiments show that whether a visual stimulus is consciously detected or not depends on when it arrives in relation to the phases of the brain's ongoing alpha (8–12 Hz) and theta (5–7 Hz) rhythms.[33] You're more likely to miss the stimulus when it occurs during the trough of an alpha wave; as the alpha wave crests, you're more likely to detect it.

Such findings support the idea that perception happens through successive periodic cycles instead of as one continuous process. Like a miniature version of the wake-sleep cycle, neural systems alternate from moment to moment between phases of optimal excitability, when they're most "awake" and responsive to incoming stimuli, and phases of strong inhibition, when they're "asleep" and least responsive. Moments of perception correspond to excitatory or "up" phases; moments of nonperception to inhibitory or "down" phases. A gap occurs between each "up" or "awake" moment of perception and the next one, so that what seems to be a continuous stream of consciousness may actually be composed of discrete and episodic pulses of awareness.

This picture of a pulsing and gappy stream of consciousness may hold even when you're paying constant attention to something. It's well known that different aspects of attention, such as "alerting" (maintaining an alert state), "orienting" (turning attention toward a target), and "executive control" (monitoring and resolving conflicts in the presence of competing targets), are associated with different electrical brain rhythms.[34] It's also well known that "sustained attention"—attending to an area and keeping your attention focused there—enhances your ability to detect targets presented in that area. Scientists often use the metaphor of the "spotlight" to describe how your attention can move around your field of vision and focus selectively on certain areas, so that you detect more efficiently what falls within the spotlight. Although this metaphor suggests that sustaining your attention at a location is like continuously shining a light there, recent studies have found that the way sustained attention enhances perception is discrete and periodic, as if the spotlight blinks on and off every 100–150 milliseconds like a strobe light.[35]

One study used targets presented at the level of light intensity at which individuals detected the target only 50 percent of the time (the so-called "individual luminance threshold").[36] When the targets were presented at attended versus nonattended locations—that is, when they were presented within the "spotlight" of attention—detection performance fluctuated over time along with the phase of the ongoing EEG theta oscillations occurring just before the stimulus presentation. In other words, when the ongoing theta rhythm reached a certain phase, a stimulus presented just afterward was more likely to catch the wave and be detected.

If attention samples information periodically, as such studies suggest, and if attention is trainable, as the scientific studies of meditation to be discussed later in this chapter indicate, then we can hypothesize that long-term meditation practice may increase the sampling rate of attention by tuning certain brain rhythms, specifically the theta oscillations associated with attention. I will come back to this idea later.

The recent studies of attention and perception I've been discussing reinforce another point Francisco liked to emphasize—that the brain activity preceding an event is crucial for determining the significance of that event for perception and action. We can't understand brain activity just by looking at the neural response after the stimulus arrives; we need to examine the ongoing activity prior to the stimulus, which sets up how the stimulus will be received. Put another way, the brain meets the stimulus on its own terms, so to comprehend any neural response we need to see it as emerging from the context of the brain's ongoing activity.

This insight motivated one of Francisco's last published studies.[37] With his Ph.D. student, Antoine Lutz, he used neurophenomenology to investigate the brain's ongoing activity in relation to conscious perception. They found that experientially distinct kinds of attentional states occurring prior to a moment of perception—states they called "steady readiness," "fragmented readiness," and "unreadiness"— were associated with distinct gamma frequency phase-synchrony patterns, and that these EEG patterns predicted both the subsequent brain activity during the perception and how subjectively stable and vivid the perception was reported as being.

The experimental findings I've been presenting are consistent with the Abhidharma view that the stream of consciousness consists of discrete and successive moments of awareness. The neuroscience findings also indicate that the immediately preceding moments of the stream of consciousness strongly influence the characteristics of the following ones.

But what about the duration of these moments of awareness? How do the Abhidharma measures and their scientific counterparts compare?

MEASURING AWARENESS

As mentioned earlier, the Abhidharma measures that seem observational work out to around 10–20 milliseconds as the time it takes for a minimal moment of awareness to occur. This estimate might seem remarkable from the perspective of cognitive science. It's significantly less than the 100–250 millisecond time periods usually given for a moment of reportable perceptual awareness, and it suggests that individuals may be able to discern events as fast as 10–20 milliseconds occurring in their own stream of consciousness. Is there any scientific evidence that speaks to these issues?

The short answer is yes. A recent study showed that some people can be aware of a target stimulus that's presented for only 17 milliseconds and followed immediately by another stimulus that masks it.[38] These individuals were "high achievers" compared to the others in the study who weren't reliably aware of such rapidly presented stimuli. In addition, recent scientific studies have shown that certain types of meditation improve attention in precisely measurable ways on a millisecond time scale, so you could start out as an "average achiever" and become a "high achiever" as a result of meditative mental training. Taken together, these findings suggest that being able to identify and describe discrete moments of awareness happening as fast as 10–20 milliseconds is by no means beyond the human mind, especially the mind trained in meditation. Thus the Abhidharma philosophers' estimates of roughly 15 milliseconds for an inwardly observable mind moment do not seem unreasonable.

There's a long answer, however, that's also worth giving. We need to take account of two key points that have emerged from recent scientific studies of perceptual awareness. First, determining whether someone is aware of a stimulus is a difficult matter. There's no single, definitive way of defining and assessing awareness; rather, there are multiple criteria and methods we can use. Second, awareness isn't the same for everyone; people differ in their awareness of target stimuli.

One obvious way of finding out whether you're aware of something is to ask you. According to this method, you're unaware when you sincerely report not having seen the target stimulus. Here the decisive factor is what you, the individual subject, report about your awareness, so the criterion of awareness is a "subjective criterion." This approach respects your first-person perspective and the unique access you have to your experiences. But if your mental stream includes subliminal moments of awareness you're not able to report, then this kind of subjective criterion will miss them.

A different approach to finding out whether you're aware of something is to give you a forced-choice task, where you have to say either "yes" or "no" every time you are asked whether the stimulus was present. If you perform better than chance, then, according to this "objective criterion" of awareness, you're aware of the stimulus, even if you deny being able to see it. This approach can reveal moments of awareness you may not be able to report, but it provides very little, if any, information about what your experience is like for you.

Recent experiments using various subjective and objective criteria of awareness have shown that individuals differ in their awareness of target stimuli.[39] For example, in the study I just mentioned, neuroscientist Luiz Pessoa and his postdoctoral researcher Remigiusz Szczepanowski used both objective and subjective criteria to investigate the visual awareness of rapidly presented emotional stimuli.[40] The target stimulus was a photograph of a face with a fearful expression; it was presented for 17, 25, 33, or 41 milliseconds, and then immediately followed by another stimulus that served to "mask" the target. (This technique is called "backward masking.") The target plus the mask lasted 100 milliseconds. The participants had to indicate "fear" or "no fear." Calculating their performance—how many times they got it right and how many times they got it wrong—provided a measure of their ability to detect the target stimulus, according to an objective criterion. The participants also had to rate their confidence in their answers on a 1–6 scale. The relationship between their confidence and their accuracy provided another measure of their awareness, according to a subjective criterion. The reasoning was that if the participants could not only detect the target better than chance but also make use of or rely on this ability, then higher confidence ratings should be linked with correct responses more often than with incorrect ones, and likewise, lower confidence ratings should be linked with incorrect responses more often than with correct ones. Szczepanowski and Pessoa found that all the participants could reliably tell when their responses were

correct for the 41- and 33-millisecond targets, and that some participants could also reliably tell when their responses were correct for
the 25-millisecond targets. In addition, they found that nearly all
the participants could reliably detect the 17-millisecond targets, but
only a few participants—the "high achievers"—could also reliably
tell when their responses were correct for these targets. In other
words, for most participants, but not the "high achievers," there
was a "dissociation" between the objective and subjective measures
of awareness for the rapid targets. This suggests that although these
participants could detect the targets, they had no inner access to or
knowledge of their sensitivity.

An Abhidharma thinker could describe these findings as follows.
When the participants detect the 17-millisecond targets, there's a primary visual awareness with the target as its focus, together with the
"ever-present" mental factors of sensory "contact" with the target,
a "feeling" tone in response to the fearful face, a "discernment" of
the face as fearful (versus neutral or happy, the control conditions),
an "attentional" orientation toward the target, and an "intention" or
motivation to detect the target. In addition, for the "high achievers"
who can reliably tell when their responses are correct for the 17-milli-
second targets, further "object-ascertaining" mental factors are present, and these account for the cognitive access the participants have
to their visual sensitivity. Such factors might include "determined
attention," which achieves stability of focus, and "mindfulness,"
which retains or holds the target long enough in working memory so
that the participants can evaluate the correctness of their responses.

At this point, we need to recall the distinction, introduced in chapter 1, between "phenomenal consciousness" and "access consciousness," that is, between consciousness in the sense of felt experience
and consciousness in the sense of being accessible to thought, action
guidance, and verbal report. One way to describe the "high achievers" is to say that they have some degree of cognitive access to their
visual sensitivity to the 17-millisecond targets, whereas the other
participants don't.

But now comes a difficult question: Should we think of the visual
sensitivity to the 17-millisecond targets as a nonconscious process

that becomes phenomenally conscious only when it's cognitively accessible? Or should we think of the visual sensitivity itself as a phenomenally conscious process that's cognitively inaccessible to everyone except the "high achievers"?

According to the first way of thinking, only the "high achievers" are phenomenally conscious of the fast targets. According to the second way of thinking, nearly everyone has phenomenal consciousness of the fast targets, but only the "high achievers" are able to access and rely on it in their thinking, action guidance, and verbal reports.

What would the Abhidharma thinkers say? On the one hand, the distinction between "phenomenal consciousness" and "access consciousness" comes from recent Western philosophy of mind and shouldn't be projected onto Buddhist philosophy. On the other hand, according to the Abhidharma, we do need to differentiate between a basic moment of awareness and a person's ability to discern that awareness and its qualities, thereby enabling him to adjust his thought, speech, and action accordingly. So, from the Abhidharma perspective, it makes sense to say that any participant detecting the 17-millisecond target has at least a low-level phenomenal awareness of the target, but only some of the participants—the "high achievers"—have cognitive access to this awareness.

Now comes the crucial point of this long answer to our question about how the Abhidharma and cognitive science measures for mental events compare. If meditation has measurable effects on attention and awareness, including the electrical brain rhythms associated with them, then we have good reason to believe that the "dissociation zone" between objective and subjective measures of awareness could change as a result of meditation. In other words, we have good reason to believe that meditation could increase both one's sensitivity to the moment-to-moment flow of events (measured according to objective criteria of awareness) and one's inner cognitive access to that sensitivity, including one's ability to report and describe it (so that one also shows increased sensitivity according to subjective criteria of awareness). The supposition, in short, is that meditation can refine awareness measured according to both objective and subjective criteria, and that some of the accounts we read in the Abhidharma texts of the rapid moment-to-moment flow of consciousness may have drawn upon this kind of refined awareness.

What, then, is the scientific evidence for meditation having these effects? And what does this evidence tell us about attention and perceptual awareness in the waking state?

DON'T BLINK

"Millions of items of the outward order are present to my senses which never properly enter into my experience. Why? Because they have no *interest* for me. *My experience is what I agree to attend to.* Only those items which I *notice* shape my mind—without selective interest, experience is an utter chaos."[41] Later in his *Principles of Psychology*, William James also adds: "For the moment, what we attend to is reality."[42]

If what we experience is what we attend to, and what we attend to is reality for us, then what we ignore or fail to notice will have no reality for us, even though it may affect us in all sorts of ways.

A compelling example from experimental psychology is the "attentional blink." When your task is to identify two visual targets presented within less than 500 milliseconds of each other in a rapid sequence of other visual stimuli, you'll often miss the second target even though you notice the first one. It's as if your attention blinks after you notice the first target, and the second one goes by in that instant. For the moment, what you attend to—the first target—is reality for you, and what you fail to notice immediately afterward— the second target—has no reality for you, even though it's there, affecting your visual system.

If meditation trains attention and awareness, does it affect the attentional blink? Can we use the attentional blink task as a way to assess the effects of meditation practice on attention and awareness?

The answer to both questions is yes, as neuroscientist Heleen Slagter showed in a series of experiments she conducted as a postdoctoral researcher in Richard Davidson's lab at the University of

Wisconsin-Madison. They investigated whether meditation practitioners would show improved performance on the attentional blink task after a three-month intensive retreat in Theravada Buddhist Vipassanā or "insight" meditation. These retreats are conducted in silence, and individuals meditate for ten to twelve hours a day. Slagter and Davidson compared the performance of the practitioners on the attentional blink task before and after the retreat, and they also compared the performance of the practitioners with that of a control group of novices who were interested in meditation, took a one-hour Vipassanā meditation class, and were asked to practice for twenty minutes each day for a week before the experiment. After the three-month retreat, the attentional blink of the practitioners was significantly reduced, that is, the practitioners showed significantly improved detection of the second target (compared to the novice group, who also showed improvement). This improvement was also correlated in the practitioners (but not in the novices) with EEG measures showing more efficient brain responses to the first target.[43]

To appreciate these findings, we need to know more about Theravada Vipassanā meditation. This type of meditation involves practicing both "focused attention" and "open monitoring." These terms, although derived from traditional Buddhist meditative vocabularies, were recently coined by scientists and contemplative scholars in order to delineate the specific kinds of mental processes involved in various Buddhist and non-Buddhist meditation practices, ranging from Vipassanā to Yoga to Zen.[44]

In focused attention or concentration meditation, you direct your attention to a chosen object, such as the sensation of the breath entering and leaving your nostrils, and you keep your attention focused on that object from moment to moment. Inevitably your mind wanders as distracting thoughts and feelings arise. At some point, you notice your attention is no longer focused on the object. You're instructed simply to recognize that your mind has wandered, to release the distraction, and to return your attention to the object.

Repeated practice of focused attention meditation is said to develop a number of attentional skills. The first is a kind of watchfulness or vigilance that stays alert to distracting thoughts and feelings but without losing attentional focus. The second is the ability to disengage from distractions without getting caught up in them. The third is the ability to redirect the attention to its chosen focus. Developing these skills leads to attentional flexibility and an acute ability to catch your mind as it starts to wander. Eventually, focused attention practice leads to "one-pointed concentration"—the ability to sustain your attention effortlessly on the object for longer and longer periods of time.

In open monitoring meditation—or "open awareness" meditation, as I prefer to call it—you cultivate an "objectless" awareness, which doesn't focus on any explicit object but remains open and attentive to whatever arises in experience from moment to moment. One way to do this is to relax the focus on an explicit object in focused attention meditation and to emphasize instead the watchful awareness that notices thoughts and feelings as they arise from moment to moment. Eventually, you learn to let go of the object of attention and to rest simply in open awareness without any explicit attentional selection.

Open awareness meditation trains awareness of awareness, or what psychologists call meta-awareness. In open awareness meditation, meta-awareness takes the form of witnessing thoughts, emotions, and sensations as they arise from moment to moment, and observing their qualities. This style of practice leads to an acute sensitivity to implicit aspects of experience, such as the degree of vividness in awareness from moment to moment or the way that transitory thoughts and feelings typically capture attention and provoke more thoughts and habitual emotional reactions. One learns to see how habits of identifying with sensations, thoughts, emotions, and memories—in other words, with specific contents of awareness—create the sense of self.

Theravada Vipassanā meditation training usually begins with focused attention on the sensation of breathing; the practice then shifts to open awareness once some degree of attentional stability has been attained. In general terms, Vipassanā meditation cultivates a moment-to-moment awareness and clear comprehension of whatever arises in the field of experience. Such "mindfulness" is "nonreactive" in the sense that ideally one simply observes or witnesses the coming and going of sensory and mental events without getting caught up in cognitive and emotional reactions to them. Put another way, Vipassanā cultivates an awareness that's "nongrasping" or "nonclinging" because one discerns whatever arises without holding onto it.

We can see these features of Vipassanā meditation reflected in the results of the attentional blink study. After three months of intensive Vipassanā practice, there was less mental "clinging" or "sticking" to the first target, so that attention was open and ready for the second target, making it easier to detect. This reduction in mental clinging or stickiness was reflected in the EEG brain waves, which showed that fewer attentional processes were devoted to the first target after intensive meditation training, making more attentional processes available for the second target. Furthermore, the individuals who showed the largest decrease over time in the neural activity they required to detect the first target also showed the greatest improvement in detecting the second target. Thus, a more efficient neural response to the first target seems to facilitate detecting the second one.

Recall the Abhidharma image of awareness as like a hand and the associated mental factors as like the fingers, so that the flow of consciousness consists in the picking up and putting down of successive objects by means of successive sets of mental factors. The attentional blink study suggests that Vipassanā meditation makes you better at quickly picking up a sensory object and then quickly letting go of it, so that you're ready for the next one.

But there's more. Earlier in this chapter I described how electrical brain rhythms in the theta frequency range (5–7 Hz) may define discrete and successive moments of perception and attentional sampling. Slagter and Davidson found that intensive Vipassanā meditation practice affected these theta rhythms in ways that were linked to improved performance on the attentional blink task.[45]

First, for both the practitioners and the novices, the neural oscillations in the theta frequency range "phase-locked" to the targets when the targets were consciously perceived. If you think of the incoming stimuli and the ongoing brain activity as making up a partner dance, then the brain stays in step with its stimulus partners by matching its activity at a certain frequency and phase to their arrival. Slagter and Davidson determined that whenever the targets were consciously seen, the brain had stayed in step with them by matching the phase of its theta oscillations to their occurrence.

Second, Slagter and Davidson found that in the practitioners, but not in the novices, the theta phase-locking to the second target increased following intensive meditation. The brain got better at staying in step with the second target. More precisely, there was a reduction in the variability of theta phase-locking from trial to trial, which is to say that the brain's matching of its theta waves to the second target became more precise and consistent. Furthermore, the individuals who showed the largest decrease in the neural processes required to detect the first target also showed the greatest increase in theta phase-locking to the second target. In this way, more efficient neural responses to the first target were linked to greater neural attunement to the second target. That is, individuals who required fewer resources to stay in step with the first target were also better at staying in step with the second one.

In these experiments, the participants were instructed not to meditate during the attentional blink task. So the results show that meditation can improve performance on attention-demanding tasks that don't require meditation. But what if the participants were asked to meditate while performing an attention-demanding task? How would treating the task as a meditation practice affect performance?

In another study, Slagter, Davidson, and Antoine Lutz asked the same meditation practitioners from the attentional blink study to perform an attention-demanding task as a focused attention meditation.[46] This was a standard selective attention auditory task called a "dichotic listening task." Two different auditory stimuli were presented simultaneously, one to each ear, and the participants were asked to attend to the tones presented to one ear, to press a button each time they detected an intermittent nonstandard tone in that ear, and to ignore the tones presented to the other ear. Performing this task as a focused attention meditation meant sustaining attention from moment to moment on the attended side in order to detect the target tones, while also constantly monitoring the quality of this attention.

This study found that three months of intensive Vipassanā meditation increased the ability to sustain attention from moment to moment on the chosen object. Three measures indicated this increase in attentional stability. First, there was an increase in the phase-locking of theta oscillations to the target stimuli. Once again, the brain got better at staying in step with the targets, so that the consistency of its response increased from trial to trial. Second, there was a reduction in the amount of time it took to make a button press in response to the deviant tone, as well as a reduction in the variability of this reaction time, which is to say that the time it took to make the button press also became more consistent from trial to trial. Third, individuals who showed the greatest increase in theta phase-locking to the target also showed the largest decrease in their reaction time variability.

These studies using the attentional blink task and the dichotic listening task indicate that intensive Theravada Vipassanā meditation improves attention and affects brain processes related to attention. In the past five years, many other studies using other tasks and a wide range of meditation styles have shown that meditation improves perceptual sensitivity and strengthens the ability to sustain attention on a chosen object from moment to moment.[47]

One striking example used binocular rivalry to examine focused attention meditation. Olivia Carter and Jack Pettigrew found that Tibetan Buddhist monks were able to change the perceptual switching rate of the two competing images when they practiced focused attention meditation.[48] Half of the twenty-three monks reported that the amount of time one image remained dominant increased considerably while they practiced focused attention meditation as well as immediately after meditation. Three monks reported that the image remained completely stable with no switching for an entire five-minute period of focused attention meditation. In some cases, one of the two images was completely dominant; in other cases, the non-dominant image remained faintly or partly visible behind the dominant one, so that the conscious perception was of two superimposed images. As Carter and Pettigrew observe, "These results contrast sharply with the reported observations of over 1000 meditation-naïve individuals tested previously."[49]

Carter and Pettigrew conclude their report by "highlighting the synergistic potential for further exchange between practitioners of meditation and neuroscience in the common goal of understanding consciousness."[50] All of the studies I've been discussing should be seen in this light. Although many meditation studies emphasize mental training and its effects on the brain, these in particular also advance the neuroscience of consciousness.

Seen in this light, the finding that meditation in expert practitioners alters binocular rivalry points to the strong role that voluntary attention can play in affecting basic visual processes and thereby determining what we see. Earlier we saw that the neural correlates of conscious perception in binocular rivalry involve synchronous oscillations in the fast gamma band coupled to slower theta oscillations. Given Carter and Pettigrew's findings, it's reasonable to assume that focused attention meditation strongly affects these neural rhythms.

Similarly, the findings from Davidson's lab suggest that intensive Vipassanā meditation affects the ongoing brain activity that's specifically related to conscious perception. In both the attentional blink task and the dichotic listening task, Vipassanā meditation may increase the sampling rate of attention by fine-tuning the theta oscillations that shape the stream of sensory events into discrete moments of conscious perception.

Using these results, we can respond to an objection that might arise about the Abhidharma notion of mind moments. According to the Abhidharma philosophers, what appears as a uniform stream of consciousness to the untrained observer is really an articulated sequence of discrete moments of awareness. You might object, however, that if Vipassanā meditation changes experience and how the brain operates, then we have no right to assume that ordinary, premeditative consciousness is discrete and not uniform. Maybe premeditative consciousness is uniform and Vipassanā meditation makes it discrete. Given this possibility, it's unwarranted to project onto premeditative experience how experience seems after meditative training.

This objection is important. As a general policy, we must avoid the fallacy of projecting qualities from later trained experience onto earlier untrained experience. In the present case, however, we have independent data from psychology and neuroscience that ordinary perception and attention may be discrete, at least in certain respects or under certain conditions. We also have data about the electrical brain rhythms linked to these discrete functions. Given these findings, as well as the findings from the Vipassanā meditation studies on how meditation affects the same cognitive functions and electrical brain rhythms, it seems legitimate to conclude that meditation can sensitize you to discrete and gappy features of awareness you ordinarily overlook.

In summary, these neuroscience findings complement the Abhidharma view. Intensive Vipassanā meditation seems to refine the way the brain rhythmically organizes the sensory flow into discrete moments of perception, lessening the brain's tendency to get stuck on one momentary perception, thereby enabling the brain to be present to whatever arises in the next moment.

BEYOND THE GAPS

If consciousness includes a gappy sequence of moments of awareness, what happens during the gaps? Does consciousness stop in each gap and then somehow start up again in the next moment? How do we account for continuity across the gaps? And what about more slowly changing background aspects of consciousness, such as being awake and alert versus daydreaming, falling asleep, dreaming, or being deeply asleep? How do these global levels of consciousness relate to the faster sequence of moments of perceptual awareness?

To deal with these questions from either the perspective of Buddhist philosophy or the perspective of neuroscience, we need to enlarge and enrich our conception of consciousness. Our last task in this chapter will be to trace this expansion in order to pave the way for the chapters to come.

First, we need to be clear about what we're asking. We're not asking why we don't notice the gaps between moments of awareness. As philosopher Daniel Dennett writes, "The discontinuity of consciousness is striking because of the *apparent* continuity of consciousness. . . . Consciousness may in general be a gappy phenomenon, and as long as the temporal edges of the gaps are not positively perceived, there will be no sense of the gappiness of the 'stream' of consciousness."[51] The Abhidharma philosophers would agree, but they would add that we can come to notice the temporal edges of the gaps when we've trained our minds in meditation, and in this way we can sense the gappiness in the stream of consciousness.

The questions asked above point to a different issue: how consciousness manages to function coherently, given that it's gappy. If consciousness is strictly momentary, in the sense that there's no consciousness whatsoever that persists during the gaps, then what accounts for its coherent functioning, not only from moment to moment but also across longer stretches of time? For example, what accounts for longer-lasting traits of consciousness, such as the attentional stability arising from meditation practice? Why don't the gaps between moments of awareness disrupt these continuities?

These questions arose early in the history of Indian Buddhist philosophy and eventually became pressing for the Abhidharma thinkers. Abhidharma theory dissects experience into discrete and momentary components, such as the six kinds of primary awareness and the various associated mental factors. The practical purpose of this dissection is to determine which unwholesome mental factors are present in a given moment of awareness, so that one can counteract them by cultivating the appropriate wholesome antidote factors in the next moment. For example, generosity is an antidote to greed; loving-kindness is an antidote to hatred; and intelligence is an antidote to delusion. For this practice to work, however, there must be some underlying continuity of mind, so that it makes sense to speak of latent unwholesome tendencies or dispositions and of transforming them into wholesome tendencies or dispositions. Hence, the relentless commitment to analyzing the stream of consciousness into a gappy sequence of discrete moments of awareness threatens to undermine the existential purpose of the whole Abhidharma framework. Buddhist scholar William Waldron calls this dilemma the "Abhidharma Problematic."[52]

One way Indian Buddhist philosophy dealt with this problem was by enlarging and enriching the Abhidharma notion of consciousness beyond the six modes of primary awareness. For example, the Theravada school distinguished between cognitively active forms of consciousness, such as waking perception, and a passive or inactive form of consciousness, which occurs in deep and dreamless sleep as well as in the gaps between moments of active consciousness. This passive consciousness is said to persist from the moment of conception until death, and thereby also to serve as the basis for the continuity of the individual. The Theravada school called this the "life-continuum" or "factor of existence" (*bhavaṅga*) consciousness. Bhikkhu Bodhi describes how it functions:

> When an object impinges on a sense door, the bhavanga is arrested and an active cognitive process ensues for the purpose of cognizing the object. Immediately after the cognitive process is completed, again the bhavanga supervenes and continues until the next cognitive process arises. Arising and perishing at every moment during this passive phase of consciousness, the bhavanga flows on like a stream, without remaining static for two consecutive moments.[53]

A traditional image for this dynamic alternation between active and passive modes of consciousness is that of a spider resting at the center of its web and then moving quickly outward to catch its prey. As Buddhist scholar Rupert Gethin explains:

> The web extends out in different directions and when one of the threads of the web is struck by an insect the spider in the middle stirs, and then runs out along the thread and bites into the insect to drink its juice. Similarly, when one of the senses is stimulated, the mind, like the spider, wakes up and adverts to the "door" of the particular sense in question. Like a spider running out along the thread, the mind is then said in due order to perceive the object, receive it, investigate it,

and establish its nature. Finally, again like our spider, the mind enjoys and savours the object.[54]

This image suggests that the passive life consciousness can accumulate tendencies or store elements from past experience by "digesting" what the active and cognitive modes of consciousness provide. Nevertheless, according to the Theravada theory, the life consciousness doesn't exist at exactly the same time as the cognitive awareness; it exists only in the gaps between active moments, when the active mode of awareness ceases. Active cognitive awareness and passive life consciousness alternate, turning on and off from moment to moment; in addition, the passive life consciousness exists from moment to moment in states where cognitive awareness is entirely absent, such as deep sleep. In these ways, there's a *jutxaposition* of active and passive modes of consciousness, but there's no *superposition* of the two modes at the same time. For this reason, the passive life consciousness seems to be a kind of "stopgap" consciousness, not a subliminal or base consciousness that supports the sensory and mental modes of awareness and provides for deeper and longer-lasting continuities.[55]

Another Indian Buddhist school, however, the Yogācāra school, argued that there is an underlying and more continuous base consciousness.[56] The Yogācāra philosophers called this the "store consciousness." Although the store consciousness too is momentary (because time in Buddhist metaphysics is momentary), it exists at all times, not just when the mind is cognitively active. As a base consciousness, it is always functioning subliminally. It serves as the support for the active forms of cognitive awareness in waking life and connects them to the more passive forms of awareness in dreams and deep sleep. Its material support is the energetic body, not any particular sense organ. As a "store," it contains "seeds" or latent dispositions that eventually "ripen" and manifest in the stream of active waking consciousness and in dreams. In this way, the store consciousness contains all the basic habits that are built up or accumulated throughout life (and from one lifetime to the next, in the Buddhist view).

The Yogācāra philosophers also added one more type of mental consciousness to their scheme—the preattentive "mind," also described as the "afflicted-mind consciousness." It can be thought of as an "ego consciousness" because it provides a preattentive sense of "I" or "Me." What makes this consciousness afflicted is that it habitually projects this "I" feeling onto the store consciousness, which it mistakenly takes to be a separate and independent self. In reality, however, the store consciousness isn't a separate and independent ego that's present in each experience and that functions as the owner of experience. Although it exists at every moment, it's constantly changing, like a flowing river, so nothing in it is wholly present from one moment to the next, and hence nothing in it could function as the owner of awareness. Thus the feeling that consciousness at its deepest level is somehow "I" or "mine" is based on a profound illusion. Because this illusion causes so much distress, the habitual ego consciousness is an afflicted consciousness.

Let me give an example to illustrate the Yogācāra view that our ego consciousness is fundamentally distorted. Take a moment of visual awareness such as seeing the blue sky on a crisp fall day. The ego consciousness makes this visual awareness feel as if it's "my" awareness and makes the blue sky seem the separate and independent object of "my" awareness. In this way, the ego consciousness projects a subject-object structure onto awareness. According to the Yogācāra philosophers, however, the blue sky isn't really a separate and independent object that's cognized by a separate and independent subject. Rather, there's one "impression" or "manifestation"[57] that has two sides or aspects—the outer-seeming aspect of the blue sky and the inner-seeming aspect of the visual awareness. What the ego consciousness does is to reify these two interdependent aspects into a separate subject and a separate object, but this is a cognitive distortion that falsifies the authentic character of the impression or manifestation as a phenomenal event.

Now take a memory of the blue sky, occurring a few months later on a gray winter day. The ego consciousness makes the memory appear as "mine" and thereby makes the present memory and the past perception seem to belong to one and the same "I" as their owner. According to the Yogācāra view, however, there is no separate "I" that owns these experiences; there's only a stream of mind moments, with an earlier perception moment giving rise to a later memory moment (by way of the intervening active mind moments and the deeper passive continuity of the store consciousness). Moreover, the feeling of "I" will have undergone all sorts of changes between the time of the perception and the time of the memory, so the impression that one and the same "I" is wholly present from one experience to the next is an illusion.[58]

As we now can see, the notion of consciousness is multifaceted and has a variety of meanings we need to clarify.

On the one hand, "consciousness" can mean perceptual or cognitive awareness, that is, the kind of awareness that targets a specific object of perception, thought, or emotion, like a spider targeting its prey. This kind of consciousness is transitive—it takes an object, as when we talk about "seeing the blue sky," "tasting the coconut," "remembering the dream," "fearing the snake," and so on.

Western philosophers of mind call this "state consciousness." When we say someone is conscious of the face and not the checkerboard rings, or someone is conscious of the first visual target but not the second one, we're referring to particular states of transitive or object-directed consciousness that we specify in terms of their phenomenal contents (how things seem to the subject who's in the conscious state).

On the other hand, "consciousness" can mean life consciousness or sentience. This kind of consciousness is intransitive, as when we talk about being conscious versus being unconscious, or being sentient versus being insentient.

Western philosophers of mind call this "creature consciousness."[59] Creature consciousness pertains to a whole subject of experience, not to the individual states of that subject. When we talk about waking consciousness versus dreaming consciousness, or about being awake and conscious versus being in a coma and unconscious, we're talking about global conditions or levels of awareness that pertain to the whole creature or sentient being as a subject of experience.

In addition, "consciousness" can mean self-consciousness. Self-consciousness comes in different forms, but what's important now is what we can call minimal self-consciousness, namely, the feeling that your consciousness belongs to you, that you are the subject of your awareness.

Notice, however, that this aspect of consciousness has to do with the *sense* of self—the *feeling* that *this awareness here is yours*. We can't assume that there really is a self that exists separately as the owner of this awareness. Whether there is a self is a further question not decided simply by the fact that we feel that there's a self. After all, as we've seen, according to the Yogācāra school, this feeling is a deep-seated cognitive and emotional distortion created by the ego consciousness (the afflicted-mind consciousness). We'll come back to this issue in the book's last chapter.

In this chapter, we've been focusing on consciousness in the sense of the perceptual awareness of an object. Our examples have been alternating images in binocular rivalry, two quickly flashing lights, rapidly presented visual targets in the attentional blink task, and standard and deviant tones in the dichotic listening task. What we're now in position to see, however, is that accounting for this kind of consciousness still leaves us needing to account for global conditions of awareness (waking, dreaming, and so on), the sentience or life consciousness they modulate, and the sense of self.

To my mind, the Yogācāra view represents an important innovation in Indian Buddhist thought because it recognizes the need to distinguish these global and more slowly changing background aspects of consciousness from more rapidly changing episodes of sensory and cognitive awareness. It also recognizes the need to account for how the sense of self arises and how it conditions the entire field of sensory and mental awareness. In these ways, the Yogācāra view (as well as other Indian yogic views to be discussed later) enables us to distinguish among the following aspects of consciousness: awareness and its global modulations across waking, dreaming, and deep sleep; the particular transitory contents of awareness (what we're aware of from moment to moment); and ways of experiencing certain contents of awareness as "I-Me-Mine."

Interestingly, we can trace an analogous line of thought in the neuroscience of consciousness. Binocular rivalry seems to give us a way to dissociate the neural activity corresponding directly to conscious vision from the neural activity associated with unconscious processing of the same stimulus. One of the working assumptions in the neuroscience of consciousness is that if we could find the neural correlates of consciousness for a particular sensory experience, such as the visual perception of a face, this finding might generalize to other sorts of conscious experiences. Put another way, if we could determine what the brain does when it makes an unconscious visual content into a conscious one—as it does in binocular rivalry—then we might be able to determine what makes a content conscious for any sensory modality.

But there's a limitation to this way of thinking. Although binocular rivalry consists in the alternation between one and the same visual content being consciously seen and not seen, this alternation always takes place within the total field of a subject's conscious awareness. In other words, binocular rivalry doesn't give us a contrast between the presence of consciousness and the absence of consciousness; it gives us a contrast between the presence of a particular visual content within consciousness and the absence of that content from consciousness. When you're in a binocular rivalry setup, you're awake with a coherent field of awareness, and you report the coming and going of a particular content within that field. Your consciousness as such, however, never disappears; on the contrary, the experimental setup depends precisely on your being conscious the entire time and being able to report the changing contents of your awareness.

You might think that neuroscience could build up to an explanation of the total field of awareness by putting together all the neural correlates of all the contents of consciousness at a particular time. The thought is that there are neural correlates for the sights you see, the sounds you hear, the odors you smell, the body you feel as yours, and so on, and that binding them all together in the right way would compose your total field of awareness. The philosopher John Searle calls this the "building block model" of consciousness.[60] The assumption is that consciousness is made up of a whole bunch of individual experiences that are somehow bound together from moment to moment. Searle points out the problem with this: "Given that a subject is conscious, his consciousness will be modified by having a visual experience, but it does not follow that the consciousness is made up of various building blocks of which the visual experience is just one."[61]

Searle contrasts the building block model with what he calls the "unified field model." According to this model, the neural correlates of individual conscious states aren't sufficient for those states, because those states presuppose that the subject is already conscious with a field of awareness. Any given conscious experience—such as seeing one of two images in binocular rivalry, or detecting the first target and then the second one in the attentional blink task—is a modulation of the already present field of awareness. In Searle's words: "Conscious experiences come in unified fields. In order to have a visual experience, a subject has to be conscious already, and the experience is a modification of the field."[62] Thus, instead of trying to find the neural correlates for individual conscious experiences, the unified field approach investigates the neural basis for the whole field of awareness, including what happens in the brain and body as that field changes within and across such global states as waking, dreaming, and deep sleep.

We thus arrive by another route at essentially the same distinction discussed earlier—between being conscious in the sense of having transitory moments of object-directed awareness and being conscious in the sense of being a conscious creature with a persisting field of awareness that changes across waking and sleeping and is ordinarily permeated with a sense of self.

In the chapters to come, we'll focus mainly on consciousness as the total field of awareness and explore how the field of awareness and the sense of self change as we move from waking into dreaming and deep sleep. Our next task, however, is to take up the issue of pure awareness and its relationship to the brain.

3

BEING

What Is Pure Awareness?

'm sitting in the audience at the "Investigating the Mind" conference at MIT, listening to the Dalai Lama, neuroscientists, psychologists, and Buddhist scholars talk about mental imagery. Matthieu Ricard, a French Tibetan Buddhist monk, has been talking about the experience of pure awareness, the source from which mental images arise. The Dalai Lama enters the discussion and makes this comment:

> The discussion of pure awareness triggered a thought for me that is probably more appropriate for a discussion among Buddhists, but I have thought about it for a long time. There are references in Buddhism to phenomena like pure awareness, which is sometimes described as "the clear-light state of mind" or "most subtle state of mind" that becomes manifest at the very moment of death and carries on into the intermediate state after this life. I feel that even in such extremely subtle states of consciousness, the mental state must have some physical base, however subtle it may be. Sometimes there is a tendency among Buddhists to think of these very subtle states of consciousness as if there were no embodiment or material basis for them.
>
> The spirit of my thinking on this is very similar to the basic scientific standpoint: that the brain is the basis for all cognitive events. Without the brain, there could be no function

of the mind. So I don't know whether that subtle consciousness could exist independently without a physical base. I don't know. (*laughing*)[1]

The conversation moves on, but the Dalai Lama's remark leaves me wanting to hear more. What exactly did he mean when he said pure awareness must have some physical base? What did he mean by "physical"? What are his thoughts about how mental phenomena and physical phenomena are related to each other? What's his position on what Western philosophers call the mind-body problem?

Three years later I asked the Dalai Lama these questions at another meeting we had with scientists and Buddhist scholars at his refugee home in Dharamsala, India. Before I tell that story, however, I need to tell you how the Dalai Lama came to be talking about pure awareness with neuroscientists and psychologists in front of an audience of over a thousand people at MIT. I also need to explain the Dalai Lama's key role in helping to create the new field of "contemplative neuroscience," and why a group of us traveled to Dharamsala to talk with him about his book on Buddhism and science, *The Universe in a Single Atom: The Convergence of Science and Spirituality.*[2]

THE MIND AND LIFE DIALOGUES

In September 1983 my father took me to Alpbach, Austria, to the International Symposium on Consciousness, where he had been invited to speak. I had just graduated from Amherst College, where I had majored in Asian studies and studied Buddhist philosophy with Robert Thurman, so the conference, which featured the Dalai Lama, quantum physicist David Bohm, and neuroscientist Francisco Varela, was the perfect graduation present. It was at this conference that Varela and the Dalai Lama first met, and the friendship that sparked between them helped to create a new kind of collaborative dialogue between Buddhism and the science of the mind.

The Dalai Lama already had a strong interest in science and had been learning about physics from David Bohm since they had met in

1979.[3] But he hadn't had much opportunity to learn about the cognitive and brain sciences, let alone to learn from a neuroscientist like Varela who was also a practicing Tibetan Buddhist. Eventually, through the combined efforts of Varela, lawyer and entrepreneur Adam Engle, and anthropologist Joan Halifax, a weeklong conversation on mind science with the Dalai Lama and six scientists took place in Dharamsala in October 1987.[4] After the meeting, Engle asked the Dalai Lama whether he would like to hold future private meetings on Buddhism and the sciences of mind and life; the Dalai Lama said yes, and the "Mind and Life" dialogue series was born.

After the third Mind and Life dialogue in 1990, the Mind and Life Institute was created. In 1998, the institute expanded its efforts to create a collaborative research program for investigating the mind with scientists, Buddhist contemplatives, and contemplative scholars. Two neuroscientists, Francisco Varela in Paris and Richard Davidson at the University of Wisconsin-Madison, agreed to begin pilot studies in their labs with long-term meditation practitioners. It was also decided that the topics for the future Mind and Life dialogues would be ones that could stimulate this colloborative research program for investigating the mind.

The first of these new dialogues took place in April 2000 in Dharamsala and focused on the topic of "Destructive Emotions."[5] At this meeting Varela presented to the Dalai Lama his groundbreaking EEG studies of visual perception, which he planned to extend with long-term meditators serving as expert first-person witnesses on the nature of moment-to-moment conscious experience.[6] The Dalai Lama, encouraged by the scientists' interest in how meditative mental training might influence the brain and behavior, asked them to investigate how contemplative practices for training attention, awareness, and emotional balance might be beneficial according to Western scientific criteria, and, if they were beneficial, to find nonsectarian and secular ways to teach them so more people could benefit.

Varela, however, was unable to participate in this research for long, for he died of liver cancer a year later, on May 28, 2001, at his home in Paris. Only a few days before, on May 21–22, Richard

Davidson had hosted the ninth Mind and Life dialogue at the University of Wisconsin-Madison on the theme, "Transformations of Mind, Brain, and Emotion." Varela had been planning to present his EEG and MEG studies of meditative states, but his Ph.D. student, Antoine Lutz, presented their work in his stead. A live webcam was set up so Varela could watch the discussions from his home. On the final day of the meeting, the Dalai Lama looked into the video camera and said: "Good morning, my dear friend, and in some sense I also consider you as a spiritual brother. I didn't realize what a strong feeling [it is] missing you here. I heard you were watching our meeting yesterday, and also today, so I wanted to express my deep feeling to you as a human brother and to your contribution. In science, especially in neurology, you made great contributions, and also in our work on the dialogue between the science of the mind and Buddhist thought. So we'll never forget that. Until my death I will remember you."

Varela had also been working with several scientists on another project—a public Mind and Life conference that would bring together neuroscientists and psychologists with the Dalai Lama and other Buddhist contemplative scholars. In 2003 the conference convened at MIT, with the title "Investigating the Mind: Exchanges Between Buddhism and the Bio-Behavioral Sciences on How the Mind Works." Over two days the conference explored the topics of attention and cognitive control, mental imagery, and emotion. Besides the scientists who spoke on stage with the Dalai Lama, many leading scientists and clinical researchers from around the world were in the audience. The conference proved to be a turning point, a critical moment in the emergence of a new field of collaborative research on the effects of contemplative practices on the brain and behavior—the field now called "contemplative neuroscience."

CONTEMPLATIVE NEUROSCIENCE

On the first afternoon of the "Investigating the Mind" conference, a group of scientists and Buddhist scholars and meditation teachers got together after the main session for a meeting at the conference hotel. Richard Davidson and Varela's former student Antoine Lutz, now a postdoctoral researcher in Davidson's lab, were going to present some of the first results from their EEG and brain-imaging studies of meditation in long-term Tibetan Buddhist monks. Here, for the first time, we would see contemplative neuroscience in action.

A year later, in November 2004, the *Proceedings of the National Academy of Sciences* (*PNAS*) published some of their findings.[7] The first thing that struck me when I read the advance online publication was the unusual list of institutional affiliations for the authors—a major research university in the United States and a Tibetan Buddhist monastery in Nepal. Davidson, the senior scientist responsible for the study, directs the Laboratory for Affective Neuroscience at the University of Wisconsin-Madison (he now also directs the Center for Investigating Healthy Minds, founded in 2009). Matthieu Ricard, a French Tibetan Buddhist monk who holds a Ph.D. in biology from the Pasteur Institute in France, lives and works at Shechen Monastery in Nepal. And Lutz, the first author and lead investigator, after ten years at the Waisman Laboratory for Brain Imaging and Behavior at the University of Wisconsin-Madison, is now a tenured professor at the University of Lyon Neuroscience Research Center in France.

The title of the *PNAS* paper, "Long-Term Meditators Self-Induce High-Amplitude Gamma Synchrony During Mental Practice," tells the basic finding. Eight long-term Tibetan Buddhist meditators, when they practiced a particular kind of meditation, generated striking EEG brain waves. The same kind of gamma frequency pattern that we've seen to be closely associated with reportable conscious experience occurred when the monks practiced their meditation, but didn't occur in the novice meditators, who served as the experimental control subjects. Most striking, however, the gamma frequency pattern in the monks was especially strong and well organized. Specifically, the size of the gamma brain waves (the amplitudes of the oscillations) was greater than any others previously reported in healthy individuals, and the phases of these fast oscillations were precisely synchronized. In short, the meditative practice for the monks, but not for the novices, was correlated with an exceptionally strong and large-scale pattern of gamma frequency phase synchrony—the same kind of pattern closely associated with alert and clear conscious awareness.

Tibetan Buddhists call the kind of meditation the monks were practicing in this study "pure compassion" meditation. Compassion is the wish that others be free from suffering. Usually one directs this wish toward a specific person or group, but pure compassion is described as a state of being in which an unconditional feeling of benevolence pervades the mind. Pure compassion meditation aims to suffuse awareness with what psychologists would call the "affect" of compassion—the strong emotional feeling that motivates selfless action to relieve someone's suffering. Tibetan Buddhists believe this kind of meditation lessens fixation on the ego, counteracts afflictive states of mind such as hatred and jealousy, creates a general sense of well-being and an unrestricted mental readiness to help others, and prevents mental dullness in meditation practice.

Pure compassion meditation belongs to the style of practice that scientists and contemplative scholars call "open monitoring" meditation, or as I prefer to call it, open awareness meditation (see chapter 2). One cultivates an "objectless" awareness, a mode of awareness that remains open and attentive to whatever arises in the field of consciousness without selecting or preferentially focusing on any particular thing. This style of practice is said to lead to an acute sensitivity to implicit aspects of consciousness, such as the degree of vividness in awareness from moment to moment, or the way transitory thoughts and feelings typically capture attention and provoke more thoughts and habitual emotional reactions. Through this practice, one comes to appreciate the distinction between awareness itself— the luminous or open and unobstructed quality of the mind—and the changing contents of awareness—the particular thoughts, emotions, and sensations that come and go from moment to moment. In the pure compassion version, one generates the affect of compassion within the open awareness state.

When the monks practiced pure compassion meditation in Lutz and Davidson's study, they alternated between sixty-second meditation periods and thirty-second rest periods. The ability to enter and exit a vivid meditative state in a short period of time reflects a high level of meditative expertise. The strong gamma synchrony occured during the meditation period but not in the rest period. This EEG pattern indicates that the monks' brains entered into a highly coherent mode of activity during pure compassion meditation. The high amplitude of the gamma oscillations—the large size of the brain waves—most likely reflects the size and number of the neural populations firing at this frequency, as well as how precisely the individual neurons in each population are firing in rhythm. The synchrony of the oscillations—the way that the EEG waves are in sync with one another across distant areas of the scalp—reflects the large-scale coordination of the neural populations into a larger but temporary functional network (like many people doing the "wave" at a soccer game). More simply put, during the meditation practice, numerous "neural assemblies"—populations of neurons that fire together—rapidly established communication and thereby formed a massive interconnected network.

Varela liked to use the term "brainweb" to describe these kinds of shifting large-scale neural networks.[8] In the World Wide Web or Internet, geographically distant computers transfer data to one another through changing networks that are supported by a more permanent system of hard-wired connections. In the brainweb, distant brain regions establish temporary networks by exchanging signals that pull their neural populations into coordinated firing rhythms. Phase synchrony in the EEG at the scalp is thought to reflect the formation of these large-scale networks in the cortex.

Can long-term meditation practice create new brainweb sites and links? Can practicing meditation affect the architecture of the brainweb? Three additional findings from the 2004 *PNAS* study suggest it can.

First, even before the meditation began, the ratio of the faster gamma oscillations to the slower alpha and theta oscillations in the resting state or baseline EEG was higher for the monks than for the novice meditators. Thus these two groups already showed different EEG patterns even before the meditation periods began.

Second, again for the monks but not for the novices, the ratio of the fast to slow rhythms increased sharply during the meditation period and stayed higher afterward. As an analogy, imagine you're a concert pianist and you play with flow and concentration for sixty seconds, then rest for thirty seconds, then play again, then rest again, and so on. The way your mind and body feel in the third resting period is going to be different from how they felt in the first resting period before you started playing, because each successive resting state will reflect the accumulating flow and concentration of your playing. Similarly, the brains of the monks, but not those of the inexperienced meditators, didn't return to the initial premeditation state during the resting periods, but instead entered a new and qualitatively distinct resting state, with elevated gamma activity reflecting the previous meditation state and the cumulative time spent in meditation. In short, whereas the baseline resting state stayed the same for the inexperienced meditators, for the monks the baseline was flexible and changeable as a function of their meditation practice.

Finally, there was a positive correlation between the amount of premeditation gamma activity and the length of meditation training throughout life. Monks with more hours of formal sitting practice had higher amounts of premeditation gamma activity in their resting-state EEG than did monks with fewer hours of practice.

All these findings together suggest that meditation is a unique kind of mental skill and that long-term meditation practice can bring about long-lasting changes in the brain.

Of course, these findings can't tell us definitively that meditation brings about these brain changes, since the EEG patterns could reflect individual differences that existed prior to learning meditation (and that might have led these monks to become serious meditators). To find out whether meditation causes long-lasting changes in the brain and behavior, we need studies that examine directly the effects of meditation practice, such as the studies of Theravada Vipassanā meditation presented in chapter 2. These show that focused attention and open awareness forms of meditation enhance attentional processes in the brain and improve performance on attention-demanding perceptual tasks. The studies provide direct evidence that meditation affects the brain.

CLARITY OF CONSCIOUSNESS

In the previous chapter we saw that reportable conscious awareness correlates with large-scale gamma phase synchrony in the EEG. I mentioned two examples from the study of visual perception—the "Aha" moment in perceptual recognition, and the perceptual dominance of one image versus another in binocular rivalry. A few other EEG studies besides the Lutz and Davidson study have found pronounced gamma frequency activity in advanced meditators, but not in beginners, for a variety of meditation practices, including Rāja Yoga and Theravada Vipassanā meditation.[9] So there seems to be some relation between meditative expertise—the ability to generate at will certain inner states of consciousness and sustain them over time—and large-scale patterns of gamma frequency activity in the brain.

Given this relation, it should be possible to refine the correlation between conscious experience and gamma activity by working with highly experienced meditators. For example, if gamma phase synchrony is a neural correlate of consciousness, do moment-to-moment changes in gamma activity correspond to moment-to-moment changes in qualitative aspects of awareness?

Addressing this question would require work with individuals who can observe or witness their awareness from moment to moment and report on its changing qualities. Since focused attention and open awareness meditation train this kind of meta-awareness, expert practitioners of these meditation styles seem ideally suited for such an investigation.

Lutz and Davidson took precisely this approach in a follow-up study.[10] When the monks were participating in the 2004 *PNAS* study, they reported that it took them around five to fifteen seconds to go from the resting state into the meditative state. This had matched the time course for the gamma activity to begin increasing. In the follow-up investigation, Lutz and Davidson asked the monks to report whenever there was any noticeable change in the subjective "clarity" of their awareness during the meditative state. As it turned out, increases and decreases in clarity correlated with increases and decreases in the size or amplitude of the gamma brain waves. In other words, greater clarity in awareness was reflected by stronger gamma frequency activity in the EEG.

To appreciate this finding, it's important to know that the term "clarity" doesn't come from Western psychology; it belongs to the vocabulary Tibetan Buddhists use for talking about consciousness, especially in the context of meditation. Tibetan Buddhists describe states of consciousness in terms of the "clarity" and "stability" of awareness. "Clarity" refers to the subjective intensity or vividness of awareness. In focused attention meditation, "clarity" is the vividness of the chosen object, such as the breath or a mental image, whereas in "objectless" meditation, it means the vividness of the entire field of awareness and whatever arises in it. "Stability" refers to the calmness and steadiness of awareness. Clarity and stability contrast respectively with dullness and excitation. A dull state of awareness lacks clarity, and an excited state lacks stability. For example, in a meditation where your awareness is focused on the breath, your awareness may be unstable—you're repeatedly distracted and lose your focus— but nonetheless clear or vivid whenever you do follow the breath. Or your awareness may be stable but dull—you follow the breath, but in a dreamy or sleepy way. In inexperienced meditators, clarity and stability tend to work against each other: the greater the stability, the more likely the awareness is dull, the extreme being that you simply fall asleep; and the greater the clarity or vividness, the more likely you are to become excited and distracted. Focused attention and open awareness styles of meditation involve monitoring these qualities of awareness in various ways and learning how to balance them, so that neither dullness nor excitation impedes the mind.

When Lutz and Davidson asked the monks to report on a scale from 1 to 9 any change in the clarity of their awareness during meditation, they found a strong correlation, over a time course of several dozen seconds, between reports of increasing clarity and the emergence of the high-amplitude gamma oscillations. In other words, the way the gamma activity changed during the meditation period reflected changes in the felt quality of awareness from moment to moment.

We can look at this correlation in two ways. On the one hand, the EEG measure provides an objective confirmation of a subjective report about experience. On the other hand, the subjective report provides indispensable information that helps us to make sense of the EEG patterns. Without such first-person reports from highly experienced meditators, we wouldn't know how certain basic qualities of awareness can fluctuate from one second to another. And without this information about conscious experience, the changes in the amplitude of the gamma frequency signal from second to second would most likely be seen as meaningless noise.

Although the EEG signatures of changes in the quality of consciousness undoubtedly involve far more than just changes in the size of the gamma brain waves, this correlation nonetheless illustrates the idea that we should expect to find a stronger relation between subjective reports about consciousness and objective measures of brain activity for highly experienced meditators than for people without this kind of mental training. People without such meditative expertise probably wouldn't be able to report how the quality of their awareness changes from moment to moment, so they wouldn't be able to provide first-person information about conscious experience on a fine enough time scale for researchers to relate moment-to-moment fluctuations in the quality of consciousness to the millisecond changes in electrical brain activity that EEG records. It's precisely this kind of fine-grained phenomenological information, however, that we need in order to get a better picture of the relationship between consciousness and dynamic networks of the brainweb, especially on a moment-to-moment time scale.

THE VISION OF CONTEMPLATIVE NEUROSCIENCE

What inspires those of us who work in contemplative neuroscience, whether as scientists, philosophers, scholars of contemplative traditions, or contemplative practitioners, is the vision of a science of the mind with roots firmly planted in modern science and ancient contemplative wisdom.

"Contemplative wisdom" is hard to define. I use the term to describe the kind of knowledge and experience belonging to traditions that use practices of calming and focusing the mind and body in order to bring about human transformation according to a vision of a harmonious and benevolent life. Many contemplative movements from different cultures and historical periods may fit this general definition. This book focuses on the Indian yogic traditions and Tibetan Buddhism because they have been especially concerned with investigating the mind and understanding the nature of consciousness. In Buddhism, the term "insight" (*vipaśyanā* in Sanskrit; *vipassanā* in Pāli) signals this concern.

Contemplative neuroscience relies on meditative insight in order to investigate human mental development and to study aspects of consciousness not readily apparent to the mind untrained in meditation. In making a place for meditative insight within science, it strives to be more than just the brain science of meditation. Instead of treating meditation as merely another object of scientific study, contemplative neuroscience aims to create a new kind of mind science, where contemplative expertise plays an investigative role as central and indispensable as those of experimental observation and mathematical analysis.

And yet, despite this collaborative effort and its prospect of a new kind of self-knowledge, there remain deep and fundamental differences between the Indo-Tibetan and the Western scientific views of consciousness. For Western science, especially neuroscience and the rest of cognitive science, consciousness is a biological phenomenon wholly dependent on the brain. For Buddhism, especially Indo-Tibetan Buddhist philosophy, the fundamental nature of consciousness is not biological, and consciousness has an existence independent of the brain.

How should we think about these differences and evaluate them in relation to each other? What do they mean for the Buddhism-science collaboration and for the possibility of a contemplative science of consciousness? The time has come to confront these questions head on.

THE QUESTION OF CONSCIOUSNESS

It's April 2007, and I've arrived in Dharamsala for the fourteenth Mind and Life Dialogue. Our meeting is going to be different from all the other ones since the dialogues began in 1987. This time we've come to Dharamsala not to talk about some particular area of scientific research, such as emotion or brain plasticity, but instead to talk with the Dalai Lama about his recently published book, *The Universe in a Single Atom: The Convergence of Science and Spirituality*. Besides being his scientific autobiography—the story of how he came to interact increasingly with the world of science—the book presents the Dalai Lama's personal reflections on science from his side of the dialogue as a Buddhist leader and teacher, and especially as a Buddhist philosopher. So our dialogue will be philosophical as well as scientific, and we will probe not only where Buddhism and science converge but also especially where they diverge.

The Mind and Life Dialogues have always had a Western philosopher participate as one of the presenters in addition to the scientists. I'm the philosopher for this meeting, and my role includes opening the dialogue by considering the main "fracture points" that distinguish Western scientific accounts of the world from the Indo-Tibetan Buddhist worldview. We've decided that I should focus on the nature of consciousness and its relation to the brain.

If you've studied the history of Western philosophy, you know about the importance of great centers of learning, such as the University of Oxford or the College of the Sorbonne at the University of Paris. Around the time these European universities were being established in the Middle Ages, India's great Buddhist center of learning, Nālandā, had already existed for over 500 years but was about to be destroyed. Founded in the fifth century C.E., Nālandā was renowned for the study of every field of learning, from astronomy to medicine to logic. At the height of its influence, it had over a thousand residents and attracted scholars from as far away as China; its library was said to contain hundreds of thousands of volumes. Nālandā's curriculum focused on the systematic study of logic, epistemology, and metaphysics. Tibetan Buddhism, especially the Gelugpa school to which the Dalai Lama belongs, sees itself as the inheritor of this Indian Buddhist tradition. For this reason, the Dalai Lama often says that Tibetan Buddhists are the holders of the "Nālandā tradition."

As I sit down next to the Dalai Lama to begin my presentation, I remember when I first learned about Nālandā in my Indian philosophy class at Amherst College with Robert Thurman, who is here, sitting in the audience. It's a special privilege, I say to the Dalai Lama and the other *geshes* or scholar-monks sitting with us, to present Western philosophical ideas to representatives of the great Nālandā tradition.

To establish common ground, I begin by pointing out that Buddhist philosophy and Western science agree on the importance of observation and critical investigation, and that both view reality, including the mind, as being made up of interdependent events or processes that come to be and pass away according to various causes and conditions. Nevertheless, despite this convergence, Buddhist philosophy and Western science diverge over what the Dalai Lama himself in his book identifies as "the critical issue of the role of consciousness."[11]

On the one hand, Tibetan Buddhists, as the Dalai Lama explains in his book, believe that the fundamental nature of consciousness—the luminosity of pure awareness—isn't a physical state of the brain and body. In other words, according to Indo-Tibetan Buddhist philosophy, pure awareness isn't physical in nature.

On the other hand, the working assumption of neuroscience is that all mental phenomena, including conscious experiences, are physical phenomena. In other words, most neuroscientists believe that every mental state is identical to—one and the same as—some physical state of the brain or body of a human being or other animal. Some philosophers prefer to say that every mental state is instantiated in and directly determined by a physical state of the brain or body, instead of being strictly identical with that physical state. In either formulation, however, this poses a challenge to the Buddhist view.

At the same time, I point out, most neuroscientists recognize that there's an "explanatory gap" between consciousness and the brain. It isn't simply a matter of our not knowing exactly how each conscious state is correlated with some brain state. At a deeper level, the explanatory gap consists in our not understanding how physical processes could possibly give rise to conscious experiences, or at a certain level of complexity, could possibly be one and the same thing as conscious experiences. Put another way, the gap consists in our not understanding how something subjective or experiential could possibly arise from something that fundamentally lacks subjective or experiential properties (the standard assumption being that physical reality, at its most basic level, is devoid of anything subjective or experiential).

Western philosophers, I tell the Dalai Lama, disagree about the meaning of the explanatory gap. Some believe it reflects a gap in our knowledge of the brain; others believe it reflects a difference in the concepts we use to think about subjective experience versus the concepts we use to think about physical reality; and others believe it reflects a basic difference in the nature of the conscious mind versus the nature of physical reality.

I've organized my presentation so that I can end by asking the Dalai Lama about what he said at MIT—that even the subtlest states of pure awareness must have some physical base. Before we get to that point, however, a bit more background will be helpful.

You might be wondering exactly why Tibetan Buddhists believe that consciousness isn't physical in nature. The Dalai Lama states this in a short article he wrote for the *New Scientist* in 2003:

> There is no reason to believe that the innate mind, the very essential luminous nature of awareness, has neural correlates, because it is not physical, not contingent upon the brain. So while I agree with neuroscience that gross mental events correlate with brain activity, I also feel that on a more subtle level of consciousness, brain and mind are two separate entities.[12]

The belief that consciousness isn't physical goes back to early Indian Buddhist thought but was later forcefully stated by Dharmakīrti (ca. 600–660 C.E.), one of the most renowned logicians and philosophers of the Nālandā tradition, whose theories still form part of the core of the Tibetan Buddhist monastic curriculum.[13] Dharmakīrti reasoned as follows: matter and consciousness have totally different natures; an effect must be of the same nature as its cause; hence consciousness cannot arise from or be produced by matter (though material things can condition or influence consciousness).[14]

This argument depends on specific conceptions of the nature of matter and the nature of consciousness. Material things have spatial dimensions and obstruct other material objects. Consciousness—or more precisely, the luminous nature of pure awareness—is formless and unobstructed, like open space. Furthermore, consciousness has the capacity to cognize an object, like a light that illuminates space and what it contains; and consciousness is self-revealing, like a light that illuminates itself while illuminating the other things around it.

For these reasons, Dharmakīrti also argued that whereas physical capacities are limited, mental qualities, such as compassion, can be developed to a limitless degree. The physical body has certain natural limits on its abilities, but the qualities of the mind have the potential for limitless development.[15]

Dharmakīrti's argument also depends on a certain view of causation. Indo-Tibetan Buddhist philosophy distinguishes between the "substantial cause" and the "contributory conditions" of an event. For example, the substantial cause of the coming into existence of a clay pot is what constitutes or makes up the pot, namely, the clay, whereas the contributory conditions are all the other factors that enter into the pot's production—the potter, the potter's wheel, the kiln, and so on.

Dharmakīrti argued that since consciousness and matter have totally different natures, one can never be the substantial cause of the other; they can only condition or influence each other as contributory conditions. For example, light and the physical organ of the eye contribute to seeing, but a moment of visual consciousness must have a preceding moment of consciousness as its substantial cause, because light and the eye, being material, are insufficient to give rise to consciousness.

Dharmakīrti also used this reasoning to argue for "rebirth." In the Buddhist sense of the term, "rebirth" doesn't mean the reincarnation of one and the same self from one life to the next; Buddhists deny the existence of this kind of a self. Instead, rebirth means that the causal series of mental events continues after the death of the physical body and eventually comes to be associated with another new body. In other words, the death of the physical body doesn't interrupt the causally continuous series of moments of mental consciousness belonging to an individual stream of consciousness or mental continuum, though it does interrupt sensory consciousness, and this mental continuum eventually acts as a contributing condition in relation to another body in a future life. Part of the reasoning here is that since matter can't produce consciousness, the consciousness of the newborn infant must come about from a preceding moment of consciousness.

These concepts and arguments from Dharmakīrti are what lie behind the Dalai Lama's statement that the "innate nature of mind" or the "very essential luminous nature of awareness" isn't physical and isn't contingent upon the brain.[16] But what about his other statement that "even in such extremely subtle states of consciousness, the mental state must have some physical base, however subtle it may be"?

In *The Universe in a Single Atom*, the Dalai Lama presents both the philosophical arguments from Dharmakīrti and the view that subtle consciousness has a physical base. This view, however, comes not from Dharmakīrti's philosophy but from the system of thought and contemplative practice known as Vajrayana or Tantric Buddhism. The Dalai Lama explains the Vajrayana view as follows:

> At the most fundamental level, no absolute division can be made between mind and matter. Matter in its subtlest form is *prana*, a vital energy which is inseparable from consciousness. These two are different aspects of an indivisible reality. *Prana* is the aspect of mobility, dynamism, and cohesion, while consciousness is the aspect of cognition and the capacity for reflective thinking.[17]

I zero in on this at the end of my presentation. I ask the Dalai Lama, how do these two subtle aspects of matter and mind relate? Is this viewpoint open to scientific investigation?

SUBTLE CONSCIOUSNESS

The Dalai Lama's answers take up the rest of the morning session and set forth the framework for our dialogue over the next five days. Within Buddhism, he explains, we need to distinguish three things. First is the investigation of reality, or what we might call "Buddhist science." Second are Buddhist concepts, which are based on investigating reality. Third is Buddhist practice, including meditation. The Buddhism-science dialogue concerns the first part, whereas the other two, he says, laughing, are Buddhists' private business.

Buddhist science includes investigations of external physical phenomena and internal mental phenomena. Modern science, however, concentrates on the external. Your knowledge of these phenomena, the Dalai Lama tells us, is very advanced, so it's useful for Buddhism to learn from modern science. In Buddhism, however, the main reason for investigating reality isn't merely to get knowledge; it's to achieve peace of mind. Since the main obstacles to mental peace arise internally and belong to our own mind, the remedy must also come from within the mind. We need to understand which mental phenomena are disturbances and which ones counteract these disturbances. So in Buddhism, there are many detailed terms for different mental states, and there's more emphasis on investigating the mind than on investigating physical reality.

Yet different Buddhist schools, the Dalai Lama continues, give different explanations of reality. When these schools contradict each other, we have to rely on the highest perspective, the school that uses the most comprehensive means of investigation and so is the most reliable. We consider the Vajrayana system to be the highest standpoint, the most precise and reliable for investigating the mind. Especially for the difficult subject of how mind and matter relate to each other, the Vajrayana system is the most appropriate.

One issue here, he says, concerns the different levels of consciousness, from gross to subtle. Just as there are many different levels of consciousness, there are many different levels of energy. Broadly speaking, from the Vajrayana perspective, so long as an event or phenomenon is a conscious one, it is necessarily contingent upon a physical event or physical phenomenon. In general, we can make that broad statement, which will probably give greater comfort to the scientist. But it will have to be followed immediately with a "however" clause. The "however" or caveat is that the physical basis—the energy—for subtle consciousness is also of a very subtle kind. It carries all movement or excitation, even at the level of brain cells. Mental movement is also due to that energy. So the scientific concept of matter needs to be modified in order to appreciate this subtle energy.

The Dalai Lama now directs his words to the Tibetan scholars and monks in the room. We Buddhists, he points out, learn certain taxonomies of reality at a very early age before we fully know what they mean. We distinguish between permanent and impermanent phenomena, and within impermanent phenomena between material phenomena, mental phenomena, and abstract entities or ideas. Because we learn this taxonomy at an early age, we tend to assume that matter and mind are two independent and discrete realities. But we need to watch out for this tendency. From the Vajrayana perspective, any mental event will have a physical dimension. And, apart from Vajrayana practice, in day-to-day life we experience states of consciousness subtler than the waking state, such as the dream state and deep sleep, and these are contingent on the brain. So we Buddhists don't need to be alarmed when scientists talk about these kinds of consciousness as dependent on brain processes, because we recognize that these states of consciousness require the brain.

But these states of consciousness, the Dalai Lama continues, addressing the scientists again, belong to what Buddhism would call the gross level of mind. The Vajrayana perspective recognizes many subtler levels of consciousness that include different phases or stages. These phases can be discerned in the dying process. It is traditionally said that when a person dies, there are eighty types of thought states or "conceptions" that progressively dissolve once the five gross modes of sensory consciousness no longer function. These thought states are called "the eighty conceptions indicative of mental processes." As they dissolve, the Dalai Lama says, respiration ceases. When all eighty have completely dissolved, most probably brain function also ceases. So up to that point, all the mental processes, all the different types of sensory and mental conscious experiences, are contingent upon the brain. But once the eighty mental states have come to an end, the so-called phase of "appearance" begins, and this phase as well as those that follow transcend the brain.

I recognize that the Dalai Lama is drawing from the rich and sophisticated Tibetan Buddhist phenomenology of the dying process. According to this understanding, as the inner mind falls apart, the "appearance" phase, a subtler and deeper level of awareness, begins to manifest. The experience of this phase of consciousness is traditionally described as "the dawning of extreme clarity and vacuity as well as of light with a white aspect like a night sky pervaded by moonlight in the autumn when the sky is free of defilement."[18]

The Dalai Lama now turns to the question of whether this kind of consciousness can be scientifically investigated. One possible avenue, he says, is to experiment on meditators who are in what the Tibetan Buddhist tradition calls the "clear light state."

I'm reminded of the earliest Mind and Life dialogues organized by Varela, where the Dalai Lama first explained to the scientists the Tibetan Buddhist concept of the clear light mind.[19] According to his explanation, in the Vajrayana system, there are three levels of mind—the gross, the subtle, and the very subtle. The gross level is sensory experience. The subtle level comprises a host of conceptual and emotional mental states that dissolve in the dying process after the senses have stopped functioning. And the very subtle level is the clear light mind, which manifests at the end of this dissolution: "The experience of the clear light, which is said to be like a 'clear, cloudless autumn sky just before dawn,' represents the mind at its subtlest, and awareness of it is called the *natural clear light*. When the practitioner maintains awareness of it, she has realized the fundamental nature of the mind itself, in that the clear light is the subtle basis for all other mental content."[20]

I've often wondered whether the advanced Tibetan meditators who are said to be able to recognize the clear light and rest their awareness in it after their breathing has ceased are fully dead according to Western clinical criteria. The Dalai Lama, seeming to anticipate this thought, tells us that a few years back there was a machine in one of the Indian hospitals that was going to be used to monitor advanced meditators as they died, but while the machine was there nobody died, and then afterward, when someone remained in that clear light state for several weeks, the machine was no longer available.

At this point, Arthur Zajonc, a physicist from Amherst College and one of the co-organizers of our dialogue, enters the discussion. Is it only at death, he asks, that the clear light state emerges, or does it also emerge in deep meditation states?

I think again of the early Mind and Life dialogues, where Varela and the Dalai Lama talked about consciousness. In the first dialogue in 1987 the Dalai Lama had said, "According to Buddha, in tantric explanation, every living being can have the experience of clear light at the time of death, naturally, but this experience could be brought about through meditative techniques as well."[21] A few years later he explained in more detail how certain meditative techniques work.[22] In the first stage, one uses imagination to visualize going through the dying process, beginning with the waking conscious state, followed by the progressive dissolution of the subtler levels of mind, and culminating in the clear light state. Eventually, thanks to this practice, one will consciously experience an analogous process of mental dissolution as one falls asleep. This is the best method for learning to recognize the deep sleep state while it happens as deep and dreamless sleep. In the second stage, one actually brings about the mental dissolution process so that all cognitive activity subsides, instead of simply imagining it. At a certain level of this stage of practice, the clear light itself will manifest.

In answer to Arthur's question, the Dalai Lama tells us that for a long time he has thought that the Buddhist explanations of consciousness are like paper money in that whether they remain valuable depends on the gold reserve of these meditative techniques and the experiences they produce. For this reason, he has long encouraged some of the monks and yogis who live up in the mountains nearby to dedicate themselves to these practices. Some have had extraordinary experiences. This is our own lab where we reproduce experiences through meditation, he tells us. In order to do this, your single-pointed mental power, your single-pointed concentration, must be able to remain fully alert and stable on a chosen object without the slightest wavering for at least four hours. Then you can use that single-pointed mental power to meditate not on external things but on certain energy channels and areas within the body, and this will eventually make the energy move, with actual effects on the body. If we have that level of meditation practitioner with these kinds of experiences, then we can challenge you scientists, and that will give you new phenomena and a new understanding.

These comments draw Richard Davidson, the other co-organizer of our dialogue, into the discussion. One of the things we've done on this visit to Dharamsala, he says, is to bring with us a very sensitive thermal camera that can monitor the body temperature from a distance. It's our hope to do what you're encouraging now and have suggested to us before, which is to see whether we can begin to investigate the clear light state by monitoring great meditation practitioners as they die. We'll train people to use the camera and leave it here for them to use when we're gone. We think this is the best available state-of-the-art method to begin to examine this phenomenon in a rigorous scientific way.

Very good, the Dalai Lama responds. The best candidate, he says, would ideally be someone who is able to have the clear light experience while alive rather than having to wait until the time of death. What happens after death is a real mystery. So if there were a yogi who could deliberately bring about the mental dissolution process as a result of meditative training, this would be best—someone able to reach a point beyond the "eighty conceptions indicative of mental processes," reach the stage of "appearance." Such a yogi, the Dalai Lama suggests, should show the same physical expressions as someone who is dying—the respiration should cease and brain activity too should cease, exactly like experience at the time of death.

I wonder how the meditative state the Dalai Lama is describing could be exactly like death, since we know the brain decays quickly when deprived of oxygen. Maybe the state is more like hibernation, a dramatic slowing down of metabolic activity but, in this case, achievèd through deep meditation. Our morning session is almost over, however, and there's no time to raise this question.

The Dalai Lama ends by telling us with a chuckle that he believes the great practitioners in the past who've reported these experiences haven't all been liars. Some of them might have been lying, but not all of them. So there must be some basis to what they say. When he was thirty years old, he says, he was very keen to practice these techniques. But time didn't permit, and now it's too late. Now, he says laughing, he prefers death as an ordinary person.

IS PURE AWARENESS CONTINGENT ON THE BRAIN?

Since this discussion with the Dalai Lama, I've continued to grapple with two thoughts.

On the one hand, I'm convinced that many statements about subtler levels of mental consciousness as well as statements about the luminous nature of pure awareness are based on sincere reports by great meditation practitioners about their experiences of fundamental phenomenal qualities of consciousness. These statements provide what I like to think of as a kind of "quantum phenomenology," compared to "classical phenomenology." They're based on highly refined observation, and they describe subtle or microscopic phenomena that aren't apparent to less refined observation at coarser or macroscopic levels.

On the other hand, I also believe that these statements are shaped by the systems of thought and practice that produce them, and that it's important to try to distinguish within these systems between what derives from direct experience and what derives from the interpretation of experience. By "interpretation" I mean the ways experience is conceptualized, described, and inculcated through systems of thought and practice belonging to particular cultural traditions and communities. Moreover, it's an open question how and to what extent interpretation in this broad sense directly shapes what an individual experiences during meditation, and to what extent meditative experiences tap into universal aspects of consciousness that transcend these interpretive frameworks.[23]

I make these points because the view that the clear light state isn't contingent on the brain is a metaphysical interpretation of experience. I see no way that direct experience on its own could show or establish that pure awareness is independent of the brain. So the view must depend on inference—on drawing conclusions about how things are based on how they appear to be.

Moreover, the view that pure awareness isn't contingent on the brain, especially as stated by the Dalai Lama, is an interpretation shaped by the complex encounter of Buddhism with modernity.[24] Traditional Tibetan Buddhism did not know anything substantive about the brain, and our familiar concept of the brain as the organ of cognition had no place in their intellectual systems. So any statement about pure awareness and its standing in relation to the brain expresses a modern interpretation of traditional Buddhist concepts and the experiences they map. Of course, within Indo-Tibetan Buddhist philosophy, as we've seen, reasons can be found for believing that consciousness, at a fundamental level, isn't contingent on the brain—consciousness is held to be nonmaterial, but the brain is a material organ. Nevertheless, it's important not to lose sight of how present-day expressions of this view represent modern interpretations of traditional Buddhist beliefs.

Since the nineteenth century, one of the most striking features of the Buddhist encounter with modernity has been the way both Asian Buddhists and Western Buddhist converts have argued that Buddhism, unlike other religions, not only is compatible with modern science but also is itself "scientific" or a kind of "mind science."[25] But this strategy cuts both ways and requires that Buddhism be open to critical examination from the side of science. If we're going to assess the view that pure awareness is independent of the brain, then we need to evaluate the reasoning and evidence Buddhists use to support this view.

What I propose to do now is to look at the question about pure awareness from the perspective of neuroscience. Eventually, however, we'll see that we need to invert the whole framework of our discussion and go back to the perspective of direct experience.

From the neuroscience perspective, there's a basic conceptual question to raise about the reasoning used to support the view that pure awareness isn't contingent on the brain. The reasoning assumes that we can draw conclusions about the neuronal support for consciousness solely from our own inner experience. But this assumption is unwarranted, because subjective experience doesn't directly reveal anything about the brain and the role it plays in supporting consciousness. Consider that waking consciousness, dreaming consciousness, and deep and dreamless sleep all depend upon the brain—a point the Dalai Lama allows—yet our subjective experience of these states reveals nothing of this dependence; to discover it, we need neuroscience. Why doesn't the same hold for pure awareness? How could we tell simply from the experience itself how pure awareness is related to the brain? More to the point, how could we legitimately infer from the experience alone that pure awareness can occur independently of the brain?

Sometimes the Dalai Lama seems open to the possibility that pure awareness or the clear light state does have neural correlates, despite his statement to the contrary quoted earlier. Thus, in his book on the Tibetan meditative and philosophical system called *Dzogchen* ("Great Perfection"), he states:

> When that ultimate experience of clear light takes place, all the other types of consciousness, the coarse levels of mind—sensory faculties, sensory consciousness and the coarse levels of mental consciousness— are all dissolved, and the breathing process ceases. But one question which is not settled or certain yet is whether or not a very subtle functioning of the brain might still be present in that state. This is something we still have to discover, and I have discussed it with a number of brain scientists. Given the premise of neuroscience, that consciousness, awareness, or psychological states are states of the brain, we have to find out whether or not at that point of clear light the brain still retains some function.[26]

From the neuroscience perspective, if highly realized meditators with functioning brains can directly experience pure awareness, then the working assumption is that pure awareness is contingent on the brain. Of course, it must immediately be added that the specific neuronal systems and functions on which pure awareness is supposed to be contingent are unknown to us at present. Nevertheless, the neuronal support of this experience should, in principle, be discoverable.

Neuroscience can make the same argument in relation to the Vajrayana view of mind and body, according to which every moment of consciousness is contingent on a physical basis, where "physical" is understood to include both the material basis for gross or sensory consciousness and a subtle energetic basis for pure awareness. But given that this subtle energetic basis is identified by way of how it feels to the meditation practitioner, that is, by way of subjective experience, the same questions arise again: How could we tell simply from the experience itself how this felt energy is related to the brain and the rest of the living body? How could we legitimately infer from the experience alone that this subtle energy can occur independent of any biological support?

From the neuroscience perspective, if highly realized meditators directly experience a subtle energetic basis for pure awareness, then the working assumption is that this subtle energy is contingent on the living body. How to interpret this subtle energy in relation to our present scientific understanding of the living body remains an open question. Perhaps the energy corresponds to something we can already map using our present physiological models, or perhaps it requires revising or enriching those models. In either case, the physiological support for this felt energy should, in principle, be specifiable.

The Dalai Lama is often cited as an unusual example of a religious figure who wants to embrace science rather than contradict it. Consider these words from the first pages of *The Universe in a Single Atom*: "My confidence in venturing into science lies in my basic belief that as in science so in Buddhism, understanding the nature of reality is pursued by means of critical investigation: if scientific analysis were conclusively to demonstrate certain claims in Buddhism to be false, then we must accept the findings of science and abandon those claims."[27] At the same time, the Dalai Lama is always careful to add what Thupten Jinpa, his principal English translator and an important Buddhist scholar, calls "the caveat." In Jinpa's words:

> The Dalai Lama . . . argues that it is critical to understand the scope and application of the scientific method. By invoking an important methodological principle, first developed fully as a crucial principle by Tsong Khapa (1357–1419), the Dalai Lama underlines the need to distinguish between what is negated through scientific method and what has not been observed through such a method. In other words, he reminds us not to conflate the two processes of *not finding* something and *finding its nonexistence*. For example, through current scientific analysis so far we may have not found evidence for rebirth, but this does not imply by any means that science has somehow *negated* the existence of rebirth.[28]

The caveat enables Tibetan Buddhists to argue that although science may not have found evidence for forms of consciousness independent of the brain, it doesn't follow that science has negated the existence of such forms of consciousness. Another way to state this argument, using a distinction from the logic of experimental science, is that "absence of evidence" isn't equivalent to "evidence of absence," so absence of evidence for a form of consciousness not contingent on the brain isn't the same as evidence for the absence or nonexistence of such a consciousness.

From the neuroscience perspective, however, although this argument may be logically sound, it doesn't come to grips with the main challenge that neuroscience poses to the Tibetan Buddhist view: to give positive evidence that consciousness isn't contingent on the brain. First-person reports alone can't provide that evidence, and no reliable evidence has come from third-person observations.

On the one hand, first-person reports about pure awareness—specifically, that its luminous nature doesn't feel contingent on one's living body and hence (note the inference) doesn't seem to be contingent on the brain—don't constitute evidence that pure awareness is, in fact, not contingent on the brain, though they do provide important information about how the pure awareness experience feels. To use philosophical language, these reports give us information about the "phenomenal character" of the experience, but they can't tell us how the experience is physically embodied, because the phenomenal character of an experience isn't "transparent" with regard to its physical embodiment. One can't simply "look through" the experience from within and see completely into the nature of its physical substrate or support.

This line of thought also calls into question Dharmakīrti's view that matter and consciousness have totally different natures. If the phenomenal character of an experience isn't transparent with respect to how it's physically embodied, then we can't conclude that consciousness and matter have totally different natures just because consciousness as experienced from the inside through mental awareness seems different from matter as experienced from the outside through the senses. More importantly, we can't conclude that consciousness and physical reality have totally different natures. Not everything physical is material (fields and forces are physical but not material). So although consciousness from the inside doesn't seem to be material—because it seems open, unobstructed, and luminous—it may nonetheless be physical (these phenomenal qualities may nonetheless be physical ones).

Moreover, strictly speaking, according to the Dalai Lama, any first-person report about pure awareness is necessarily always an after-the-fact, retrospective report about a previous pure awareness experience, because pure awareness or the clear light state is nonconceptual and devoid of all thought and sense of self. The Dalai Lama even says that such first-person reports are in some ways like third-person reports, because they are made from a perspective outside the experience:

> When the very subtle energy-mind is manifest, it does not have the clear light as its object. It does not apprehend anything as an object. It itself is the clear light. Similarly when you abide in meditative equipoise, experiencing ultimate reality, you are not aware that you are in meditative equipoise. But if you are very well trained in such deep meditative insight, after such an experience you can look back on it

and think, "At that time I was experiencing the clear light." This is now, in a sense, a third-person perspective. It's an outsider's point of view looking back on your own experience of meditative equipoise at a former time. But you certainly do not think anything at all while you're in that state itself. You're not thinking in terms of existence, nonexistence, or any other conceptual category.[29]

It follows from this understanding that first-person reports about how pure awareness feels or seems to be not contingent on one's living body are retrospective judgments about past experience made from the standpoint of the conceptual mind by embodied individuals with functioning brains. Such judgments represent interpretations of experience and cannot by themselves show that pure awareness or the clear light state isn't contingent on the brain or the living body more generally.

On the other hand, third-person observations of individuals who are judged from the outside to be "abiding" in pure awareness while their bodily functions either cease (in death) or perhaps go into some kind of stasis (in certain types of meditative states) need to be supplemented with modern psychophysiological measurements for us to be able to determine the precise physiological correlates of pure awareness in these conditions.

When we consider the Dalai Lama's caveat as applied to the possibility of consciousness not being contingent on the brain, we also need to ask what "negating" something or "finding nonexistence" for it means.

On the one hand, if "negating" something means denying its existence on the basis of having evidence for its absence, then one could argue that since we're not able to detect any sign of consciousness in the absence of the brain, we do have evidence for the absence of such a consciousness.

The principle we're using is the one that says, given a reliable detection method, if something is regularly not detected, then it's legitimate to conclude that it's absent. For example, given a Geiger counter and readings not above those for the background radiation, we conclude that additional radiation isn't present.

Applied to consciousness, however, this argument presents a serious problem—it depends on the assumption that we have a reliable detection method for the presence of consciousness, such that we can say with a high degree of certainty whether consciousness is present or absent in any given case. In fact, we lack such a method. I'll come back to this important point in the next section.

On the other hand, if "negating" something means deductively proving its nonexistence, then one could argue that science rarely proceeds this way, and when it does, it's only in abstract, theoretical areas that are highly dependent on mathematics where deductive proof is possible. So, again, the issue isn't whether one can prove the nonexistence of a form of consciousness not contingent on the brain; the issue is whether there's compelling empirical evidence to believe in the existence of such a consciousness.

So far I've been voicing skepticism about the Tibetan Buddhist view that pure awareness isn't contingent on the brain and about using the Dalai Lama's caveat to argue for this view. In short, from the perspective of neuroscience, there is, at present, no compelling evidence—by which I mean no reliable empirical evidence—for consciousness not being contingent on the brain or living body.

(Some people think that good places to look for such evidence are "out-of-body experiences," "near-death experiences," and purported memories of past lives. I address these cases in chapters 7 and 9.)

But there's a deeper message in the Dalai Lama's caveat, concerning the scope and limits of the scientific method in relation to consciousness. Coming to grips with this deeper message requires inverting the whole framework of our discussion and recognizing what I will call the primacy of direct experience.

THE PRIMACY OF DIRECT EXPERIENCE

Let's go back to the way Thupten Jinpa states the caveat: "the Dalai Lama underlines the need to distinguish between what is negated *through scientific method* and what has not been observed *through such a method*" (my emphasis). Here's the crucial deeper message: *Consciousness itself has not been and cannot be observed through the scientific method, because the scientific method gives us no direct and independent access to consciousness itself. So the scientific method cannot have the final say on matters concerning consciousness.*

In order to unpack what this means, I need to explain first why the scientific method gives us no *direct* access to consciousness, and second why it gives us no *independent* access to consciousness.

We can begin by noting that the scientific method gives us no way to detect, let alone measure, consciousness itself. As philosopher David Chalmers remarks, "we have no consciousness meter."[30] The principal reason is a conceptual one—consciousness isn't the kind of phenomenon that can be measured in such a way. A meter records a publicly observable property, such as temperature, and assigns to it some magnitude or amount expressed in terms of standard units, such as degrees Fahrenheit, Celsius, or Kelvin. Conscious experience, however, isn't publicly observable and quantifiable; it's subjective and qualitative. Whenever we construct a meter to measure something, we must abstract away from any associated subjective and qualitative aspects, as we do when we abstract away from felt warmth in order to measure temperature. It simply makes no sense to suppose we could use a meter to measure what we're required to abstract away from in order to make any meter function.

Scientists sometimes speak about devising a "consciousness meter," but what they really mean is a way of measuring something they believe to be a publicly observable sign or indication of the presence of consciousness, such as some measurable aspect of brain activity. For example, the Bispectral Index (BIS) is an EEG measure used during general anesthesia to monitor the patient's level of consciousness and thus the depth of anesthesia. And new EEG and fMRI methods are being devised to assess cognitive functions in patients who have been diagnosed on the basis of behavioral criteria as being in the "vegetative state," in which they have periods of wakefulness but show little or no sign of being aware of themselves and their environment.[31] Some of these patients have been shown to be able to generate reliable EEG or fMRI responses to verbal commands, despite being otherwise unresponsive, and hence have been reclassified on the basis of these brain-response criteria as being in a so-called "minimally conscious state." Clearly, such research is of huge medical and scientific importance, but we shouldn't lose sight of the fact that its promise lies not in being able to measure consciousness itself—which, as we've seen, makes no sense—but in being able to provide refined measures of the neural correlates of consciousness.

We also need to remember that any neurophysiological criterion for the presence of consciousness ultimately depends on investigations where the principal criterion for assessing the level and contents of consciousness is either verbal reports or some other behavior (such as pressing a button according to instructions) indicating that individuals have cognitive access to their consciousness. Given that specific patterns of brain activity correlate with states deemed to be conscious because individuals have cognitive access to their contents, and thus can report those contents, we infer that these patterns of brain activity are reliable neural correlates of consciousness. So, strictly speaking, we're not correlating consciousness itself with brain activity; rather, we're correlating something that we already take to be a reliable indication or expression of consciousness—verbal reports or some other cognitive performance—with brain activity.

This approach is unavoidable, but we shouldn't lose sight of the difficult questions it raises for the experimental or clinical investigation of consciousness. When we're not able to obtain any verbal report or elicit any cognitive performance from someone, how can we be sure that the person isn't experiencing anything at all? Why couldn't there be cases where the brain systems needed to make a report about experience, or more generally the brain systems needed for cognitive access to experience, are either shut down or impaired, yet nonetheless the person is still in some way aware and feeling something—still living through some kind of subjective and affective state—and thus still conscious?

Moreover, trying to answer these questions gives rise to a serious puzzle, which philosopher Ned Block calls the "methodological puzzle" in the science of consciousness. In Block's words: "how can we find out whether there can be conscious experience without the cognitive accessibility required for reporting conscious experience, since any evidence would have to derive from reports that themselves derive from that cognitive access?"[32] Different ways of dealing with this are being debated by scientists and philosophers, but the point I want to emphasize now is that there's no way simply to bypass verbal reports or cognitive access and go straight to conscious experience itself or its neural correlates. Rather, the best we can do is to balance all the different criteria—the presence versus absence of verbal reports, of body movements, and of certain measures of brain activity—and try to infer the best explanation of what we observe.

In practice and in everyday life, of course, we don't proceed this way—we don't infer the inner presence of consciousness on the basis of outer criteria. Instead, prior to any kind of reflection or deliberation, we already implicitly recognize each other as conscious on the basis of empathy. Empathy, as philosophers in the phenomenological tradition have shown, is the direct perception of another being's actions and gestures as expresssive embodiments of consciousness.[33] We don't see facial expressions, for example, as outer signs of an inner consciousness, as we might see an EEG pattern; we see joy directly in the smiling face or sadness in the tearful eyes. Moreover, even in difficult or problematic cases where we're forced to consider outer criteria, their meaningfulness as indicators of consciousness ultimately depends on and presupposes our prior empathetic grasp of consciousness.

This brings me to the second main point about the scientific method, which is that it gives us no access to consciousness that's independent of consciousness. When we use the scientific method to investigate consciousness, we're always necessarily using and relying on consciousness itself. Perceptual observation, which is necessarily first-personal, and the intersubjective confirmation of perceptual experience, which necessarily presupposes empathy or the recognition of others as having the same kinds of experiences as oneself, are the bedrock of experimental science. In addition, the scientific method includes asking questions, formulating hypotheses, doing background research, analyzing data, and communicating results, none of which is possible or even intelligible as a human activity without consciousness.

The upshot is that there's no way to stand outside consciousness and look at it, in order to see how it fits into the rest of reality. Science always moves within the field of what consciousness reveals; it can enlarge this field and open up new vistas, but it can never get beyond the horizon set by consciousness. In this way, direct experience is primary and science secondary.

NOT ONE, NOT TWO

The primacy of direct experience requires that we reevaluate how we think about the project of relating consciousness and the brain. We can never step outside consciousness to see how it measures up to something else, and consciousness never appears or shows up apart from some context of embodiment. What we need to do is to hold together these two points of view without privileging one over the other.

To say that we can never step outside consciousness implies that consciousness has a kind of irreducible primacy.[34] This primacy is first and foremost existential. Consciousness is something we live, not something we have. We don't have consciousness in any sense that would allow us to lose it without our ceasing to be what we are. Consciousness is our way of being, and it cannot be objectified, that is, treated as just another kind of object out there in the world, because it is that by which any object shows up for us at all.

Consciousness also has an epistemological or methodological primacy in any scientific investigation. Consider how we go about getting objective knowledge through science. We extract from our experience those features on which we can reach some kind of consensus as observers and that we can model using tools such as logic and mathematics. For example, we set aside our individual, qualitative sensory experiences of hot and cold in order to extract temperature as a purely structural property of matter. Although these objective descriptions remove many contextually varying and idiosyncratic aspects of our experience, such as how the water feels warm or cold to me versus how it feels to you, they never leave consciousness in its entirety behind. They always map certain abstract and invariant structural features of how the world appears to us at various spatio-temporal scales of observation, and there can be no appearances or observations without consciousness. To use philosopher of science Michel Bitbol's words, these objective descriptions are constructed out of the "structural residues" of consciousness, that is, out of what we get when we abstract away from our concrete, qualitative experience of the world and use mathematics to represent what remains in our experience.[35]

Given that science must always proceed this way, it's nonsensical to think we could explain consciousness exhaustively without remainder in terms of its own structural residues. In other words, given that scientific models of the world are always distillations of our conscious experience as observers, it makes no sense to try to reduce consciousness to one or another of our scientific models, including our models of brain activity.

At this point, some neuroscientists and philosophers will want to argue that brain science shows or at least strongly suggests that consciousness is nothing other than a brain process. According to this "neurophysicalist" view, neuroscience provides overwhelming evidence that every conscious experience is identical to some pattern of brain activity.[36]

But neuroscience itself doesn't demonstrate this identity; rather, the identity is a metaphysical interpretation of what neuroscience does show, namely, the contingency or dependence of certain kinds of mental events on certain kinds of neuronal events. In every neuroscience experiment on consciousness, the evidence is always of the co-occurrence of mental events and neuronal events, and that isn't sufficient to establish their identity. Even the causal manipulations, strictly speaking, go both ways, from neuronal events to mental events and from mental events to neuronal ones. We can alter a person's mental states by acting on her brain (through direct electrical brain stimulation, drugs, surgery, and so on), and we can alter a person's brain activity by acting on her mental states (by asking her to imagine something, by having her direct her attention in a certain way, and so on).

Moreover, the brain is always embodied, and its functioning as a support for consciousness can't be understood apart from its place in a relational system involving the rest of the body and the environment. The physical substrate of mind is this embodied, embedded, and relational network, not the brain as an isolated system.[37]

At the same time, we can't infer from the existential or epistemological primacy of consciousness that consciousness has ontological primacy in the sense of being the primary reality out of which everything is composed or the ground from which everything is generated. One reason we can't jump to this conclusion is that it doesn't logically follow. That the world as we know it is always a world for consciousness doesn't logically entail that the world is made out of consciousness. Another reason is that thinking that consciousness has ontological primacy goes against the testimony of direct experience, which speaks to the contingency of our consciousness on the world, specifically on our living body and environment.

The view that consciousness is the primary reality out of which everything is composed is found in the *Upanishads*. The early Buddhist view, in contrast, as we saw in chapter 2, is that consciousness is contingent on "name-and-form"—on the entire body-mind-environment complex or the makeup of a person—while "name-and-form" in turn is contingent on consciousness. In the Buddha's words, "Well, then, friend, I will make up a simile for you. . . . Just as two sheaves of reeds might stand leaning against each other, so too, with name-and-form as condition, consciousness [comes to be]; with consciousness as condition, name-and-form [comes to be]."[38] To my mind, this view makes better sense of our experience than the view that jumps from the existential primacy of consciousness to its ontological primacy. We can never step outside consciousness, but consciousness always shows up contingent on our embodiment, and we can never step outside our embodiment, but our embodiment always shows up contingent on consciousness. Like the two sheaves of reeds propping each other up, consciousness and embodiment are not one, not two.

Philosophical readers may be feeling unsatisfied at this point. Is the view I'm presenting a version of "emergentism," according to which consciousness is a higher-level property of living beings that emerges from lower-level biological and physical processes?[39] And if so, am I supposing that consciousness is scientifically explainable or presenting a version of "mysterianism," the doctrine that says that the human mind is incapable of understanding how consciousness fits into the natural world?[40]

My view can be described as an emergentist one, in the following sense. I hold that consciousness is a natural phenomenon and that the cognitive complexity of consciousness increases as a function of the increasing complexity of living beings. Consciousness depends on physical or biological processes, but it also influences the physical or biological processes on which it depends. I also think the human mind is capable of understanding how consciousness arises as a natural phenomenon, so I'm not a mysterian.

Nevertheless, my view differs from emergentism in its standard form, according to which physical nature, in itself, is fundamentally nonmental, yet when it's organized in the right way, consciousness emerges. This view works with a concept of physical being that inherently excludes mental or experiential being, and then tries to show how consciousness could arise as a higher-level property of physical being so conceived. In my view, however, no concept of nature or physical being that by design excludes mental or experiential being will work to account for consciousness and its place in nature.

I take this conclusion to follow from the primacy of consciousness discussed above. Since consciousness by nature is experiential, and experience is primary and ineliminable, consciousness cannot be reductively explained in terms of what is fundamentally or essentially nonexperiential. Yet classical science, as well as much (though not all) of modern science, has conceived of physical phenomena as being fundamentally, in themselves, essentially devoid of any relation to anything experiential. Bridging from nature so conceived to consciousness is an impossible task, for the two concepts mutually exclude each other. Hence, understanding how consciousness is a natural phenomenon is going to require a radical revision of our scientific concepts of nature or physical being.

With this thought, we connect back to one of the things the Dalai Lama said at our Mind and Life meeting in Dharamsala—that the scientific concept of matter may need to be modified in order to account for the ultimately nondual relationship between consciousness and matter, that is, for the way that "subtle consciousness" and "subtle energy" are contingent upon each other. In my view, this proposal is best understood not as pointing back to something like nineteenth-century style "vitalism," according to which living things possess a special nonphysical element or substance, but instead as pointing toward the need to rethink what we mean by "physical" so that physical being is understood as naturally including, at its most fundamental level, the potential for consciousness or experiential being.

This line of thought also enables us to go back to Dharmakīrti and retrieve a crucial insight from his argument against materialism. Recall that Dharmakīrti reasoned in the following way: matter and consciousness have totally different natures; an effect must be of the same nature as its cause; therefore, consciousness cannot arise from or be produced by matter. This argument rejects emergentism in its standard form, for it denies that physical nature, understood as being fundamentally or essentially nonmental, is sufficient to produce or give rise to consciousness. This aspect of Dharmakīrti's argument is hardly antiquated, for a similar and very forceful argument against standard emergentism has recently been given by the philosopher Galen Strawson.[41] In both cases, the crucial insight is that the emergence of experiential being from physical being is unintelligible, given a concept of physical phenomena as fundamentally or essentially excluding anything mental or experiential.

Whereas Dharmakīrti used his argument to support a version of mind-body dualism, Strawson uses his argument to support "panpsychism." Dualism says that matter and consciousness have totally different natures; panpsychism says that every physical phenomenon possesses some measure of experience as part of its intrinsic nature. Neither position is attractive to me. Dharmakīrti's dualism, which allows for modes of consciousness not dependent on the brain, doesn't sit well with the scientific evidence, as we've seen. Neither does Strawson's panpsychism. This position attributes "micro-experiences" to microphysical phenomena, despite there being no evidence for protons or electrons having any experiences of their own. Not only does ascribing "micro-experiences" to physical particles seem ad hoc, but also it gives rise to the so-called "combination problem" of how it's possible for "micro-experiences" to coexist or combine coherently in a human or other kind of animal subject.

I take a different message from the primacy of consciousness and the failure of standard emergentism: that we need to work our way to a new understanding of what it means for something to be physical, in which "physical" no longer means essentially nonmental or nonexperiential. This would be different from postulating in a speculative and ad hoc way that everything physical—as we presently conceive of the physical—has an extra experiential ingredient as part of its inner nature. Instead, such an understanding would replace our present dualistic concepts of consciousness and physical being, which exclude each other from the start, with a nondualistic framework in which physical being and experiential being imply each other or derive from something that is neutral between them.[42]

Where does this leave us with regard to the project of relating consciousness to the brain, and more comprehensively, to our bodily being? Since we can't step outside of consciousness or our embodiment, we need to work within them carefully and precisely. Instead of focusing mainly on the outward behavioral and physiological expressions of consciousness, we need to give equal attention to cultivating our experience from within. Concretely, this means we need to work with contemplative forms of mental training and embed them within a larger framework that also includes the experimental sciences of mind and life. Such a framework, in which both contemplative practice and scientific observation and measurement are seen as grounded in direct experience, is precisely what Francisco

Varela had in mind when he introduced his research program of "neurophenomenology."[43]

Neurophenomenology provides the framework for the rest of this book. In the coming chapters, I'll use it to examine a wide spectrum of human modes of consciousness and the sense of self. We begin with what happens to consciousness and the sense of self as we fall asleep and dream.

4

DREAMING

Who Am I?

t's three o'clock in the morning in Dharamsala, India. The village dogs are refusing to let the monkeys outside my room have the last screech. I'm jet lagged, and the commotion woke me up. I listen for a while, enjoying their raucous contest, and then put in earplugs and settle back to bed. The earplugs muffle the racket but amplify the sound of my own breathing, so I focus on this inner rhythm in order to fall asleep. Since I'm in a place known for meditation—my room overlooks Namgyal Monastery, the Dalai Lama's home, and behind the hotel stand the Himalayas, where yogis still meditate in caves—I decide to practice *yoga nidrā* or sleep meditation and watch my mind move from waking into sleep. I've always wanted, since I was a kid, to catch the exact moment when sleep arrives and notice when I begin to dream. With the rising and falling of my breath, colored shapes start to float on the inside of my eyelids. They hover just beyond my gaze, turning into cows and shacks and mules, like the ones I saw this morning on the bus ride up the mountain. As I watch these images, trying not to tamper with them so they don't fall apart, I find myself thinking of how Jean-Paul Sartre explains the presleep state in his book *The Imaginary*, which we read in my philosophy class a week before I left for India.[1] When we're conscious of drifting off to sleep, Sartre says, we delay the process and create a peculiar state of conscious fascination. The dancing lights captivate our drowsy consciousness, and from them we fashion images—but these shift with each eye movement and refuse to settle into dreams.

The next thing I know, I'm flying over a large, tree-filled valley. I must be dreaming, I tell myself. From the memory of trying to watch myself fall asleep—still fresh in the dream—and the lack of memory for what came after, I realize I must have lost awareness during my drowsy reverie and reawakened in the dream. I'm having a lucid dream—the kind of dream where you know you're dreaming. Indian and Tibetan traditions say that meditating in the lucid dream state can make it easier to see the consciousness beneath waking and dreaming, so I try to sit down, cross-legged, and meditate. But my intention to sit this way won't translate into action and I wind up kneeling instead. Then I lose the intention entirely and I'm flying again, still aware that I'm dreaming. . . .

Later, when I woke up again, I wrote down the dream in a notebook my sister Hilary had given me for this trip, my first to India. A lucid dream seemed like an auspicious way to begin my stay in Dharamsala. I was there for a Mind and Life Institute conference with the Dalai Lama and Western neuroscientists on how experience can change the brain.[2] I'd been working on my talk for months, and we'd all arrived a day early to go over each other's presentations before the meeting started.

We were there to talk about "neuroplasticity," or the brain's ability to reshape itself throughout life.[3] Changing the brain affects how and what we experience, but changing how and what we experience also affects the brain. My lucid dream seemed an appropriate illustration of one kind of plasticity. Both sorts of influence—brain to mind and mind to brain—seem to happen in lucid dreaming. Certain events in my sleeping brain were necessary to trigger the realization that I was dreaming, yet my awareness of the dream state and my partial ability to guide the dream content must have influenced what my brain was doing as the dream progressed. This idea of affecting the brain through intention and volition—what some scientists and philosophers call "downward causation"—was precisely what I was going to talk about at this conference.[4]

But lucid dreaming also raises deeper questions. When I talk about "my" awareness, who exactly is this self? When I say, "I realized I was dreaming," who is "I"? When I describe "my" partial dream control, who is the agent or controller? And how is all of this—the awareness, the sense of self, and the feeling of agency—related to my sleeping brain and body? To put these questions in general terms: Who or what is the experiential subject of lucid dreaming, and how is this subject related to the brain and body?

My answers will take this chapter and the next two chapters to develop. But here, in outline, is what I will propose.

We need to distinguish between the dreaming self and the dream ego—between the self-as-dreamer and the self-within-the-dream. In a nonlucid dream, we identify with our dream ego and think, "I'm flying." In a lucid dream, we think, "I'm dreaming," and we recognize that the dreaming self isn't the same as the dream ego, or how we appear within the dream. The dream ego is like an avatar in a virtual world; the dreaming self is its user. To generate virtual worlds with avatars we need computers, but to create dream egos in dream worlds we need only our imagination—our ability to fashion mental images and to envision other worlds. When we fully appreciate that dreaming has its source in the imagination, we can see that the view of dreaming dominant in neuroscience today—that dreams are delusional hallucinations thrown up randomly by the brain—isn't right. When we dream, we don't hallucinate at random; we spontaneously imagine things, but without realizing that this is what we're doing. Imagination already propels us in waking life—when we daydream, but also when we mentally rehearse what to say or do, when we project ourselves into the remembrance of things past or the anticipation of things future, and when we shape what we perceive with personal meaning. Often these kinds of spontaneous imaginings completely absorb us, especially as we fall asleep and dream. In a nonlucid dream, we lose the awareness that we're imagining things and identify with the dream ego as the I. We're like gamers who identify so completely with their avatars they forget they're gaming.[5] In a lucid dream, we regain awareness of our imagining consciousness. Nonlucid dreams frame experience from the imagin*ed* perspective of the dream ego; lucid dreams reframe experience from the perspective of the imagin*ing* and dreaming self. Lucidity can enable the dreaming self to act consciously and deliberately in the dream state through the persona of the dream ego, who becomes like an avatar in a role-playing game, and these intentional dream actions have measurable effects on the sleeping brain and body. For these reasons, lucid dreaming offers a model of "downward causation"; it illustrates the causal influence of the mind on the brain and body in the dream state. So when neuroscientists say the dreaming mind reflects the sleeping brain, they are not wrong, but they get only half the story, for the opposite is equally true—the sleeping brain reflects the dreaming mind. A new dream science that includes dream yoga or meditative practices of lucid dreaming can deepen our understanding of this complementary relationship.

To work toward these ideas, we need to examine the special features of dream consciousness compared to waking consciousness. Dreaming isn't uniform, however, for there are different kinds of dream experience involving different ways of experiencing the self. A good place to start is the hypnagogic state, the state leading into sleep.

ON THE POINT OF DREAMING

In everyday life we tend to think of waking and dreaming as two distinct and discrete states. If we're dreaming, then we're not awake; and if we're awake, then we're not dreaming. Yet the ancient Indian image from the *Upanishads* suggests otherwise: like a great fish swimming back and forth between the banks of a wide river, we journey between waking and dreaming. This image hints of deeper currents beneath the surface while allowing for intermediate areas and eddies where waking and dreaming flow into each other. One place where this confluence happens is the hypnagogic state.

"Hypnagogic" means "leading into sleep" (from the Greek words *hypnos*, meaning sleep, and *agogos*, meaning leading). The nineteenth-century French scholar Alfred Maury (1817–92), a pioneer in the systematic study of dreams, coined the term to refer to the sensory images we experience as we fall asleep.[6] Some years later, in a book published posthumously in 1903, the English writer Frederick Myers—a founder of the Society for Psychical Research in England and coiner of the term "telepathy"—introduced the word "hypnopompic" (from the Greek word *pompe*, meaning leading away) for the experiences on the other, waking side of sleep, in particular the persistence of dream images in the early waking moments.[7] To complicate matters, hypnagogic thoughts and imagery are not necessarily sleep-inducing, for they can be experienced with open eyes and can occur in certain contexts—notably meditation—without leading into sleep.

Here's a brief example of hypnagogic imagery from my dream notebook: "In bed, going to sleep. A sudden image—an old woman's face, purple eyes, in profile, looking to my left. Her eyes attract mine, but as I try to make eye contact she transforms into a jumble of shapes and I wake up."

Often these images carry over from things we've seen or felt during the day—like the strange animals I saw when I went back to sleep that night in Dharamsala, or the way my inner field of view sways back and forth after being out on the water sailing or canoeing. Robert Frost, in his 1914 poem "After Apple-Picking," describes this experience of falling asleep and reliving the day's activity in the form of hypnagogic imagery:

And I could tell
What form my dreaming was about to take.
Magnified apples appear and disappear,
Stem end and blossom end,
And every fleck of russet showing clear.
My instep arch not only keeps the ache,
It keeps the pressure of a ladder-round.
I feel the ladder sway as the boughs bend.

And I keep hearing from the cellar bin
The rumbling sound
Of load on load of apples coming in.[8]

Synesthesia—experiencing sounds giving rise to colors, or associating letters and numbers with colors and personalities—is another common feature of the hypnagogic state. Shapes can speak, geometric patterns have personalities, and ideas are colored. In his poem "The Garden," Andrew Marvell (1621–78) describes a colored thought while lying on the grass in a dreamy state:

Meanwhile the mind, from pleasure less,
Withdraws into its happiness;
The mind, that ocean where each kind
Does straight its own resemblance find;
Yet it creates, transcending these,
Far other worlds and other seas,
Annihilating all that's made
To a green thought in a green shade.[9]

When we have these drowsy images and sleepy thoughts, we're hovering on the point of dreaming. We've entered a liminal zone where we're neither fully awake nor fully asleep. We experience the elements of dreams, but without their coalescing into full-blown dream narratives. Sometimes we're still aware of our immediate surroundings and our mental state; other times we're so completely absorbed in imagistic thought that the boundary between ourselves and what we're imagining seems to disappear.[10]

Marcel Proust begins the first book of his seven-volume novel, *In Search of Lost Time* (*À la recherche du temps perdu*, also known in English as *Remembrance of Things Past*), with an account of this liminal experience, told by the unnamed first-person narrator, who recalls falling asleep while reading:

For a long time, I went to bed early. Sometimes, my candle scarcely out, my eyes would close so quickly that I did not have time to say to myself, "I'm going to sleep." And, half an hour later, the thought that it was time to try to sleep would wake me; I wanted to put down the book I thought I still had in my hands and blow out my light; I had not ceased while sleeping to form reflections on what I had just read, but these reflections had taken a rather peculiar turn; it seemed to me that I myself was what the book was talking about: a church, a quartet, the rivalry between François I and Charles V.[11]

Proust depicts falling asleep as dissolving the boundaries between the self ("I myself") and what is not the self ("what the book was talking about"); later we'll see that this "loosening of ego boundaries" is a key feature of the hypnagogic state. Awakening restores those boundaries:

This belief lived on for a few seconds after my waking; it did not shock my reason but lay heavy like scales on my eyes and kept them from realizing that the candlestick was no longer lit. Then it began to grow unintelligible to me, as after metempsychosis do the thoughts of an earlier existence; the subject of the book detached itself from me, I was free to apply myself to it or not; immediately I recovered my sight and I was amazed to find a darkness around me soft and restful for my eyes, but perhaps even more so for my mind, to which it appeared a thing without cause, incomprehensible, a thing truly dark.[12]

Waking detaches the enthralled and drowsy thinking from what it thinks about; memory loses hold of the felt meaning of the dreamy thoughts; and outer perception again takes over consciousness.

Proust's narrative focuses on the changing qualities of thought, memory, and felt meaning at the borderland between sleep and waking; most scientific studies of the hypnagogic state, however, beginning with Maury, focus on the peculiar sensory imagery that arises in the early sleeping moments. Twenty years before Maury, the German physiologist Johannes Müller (1801–58), one of the founding fathers of modern neurophysiology, carefully observed these sensory qualities of the hypnagogic state:

Before falling asleep, with my eyes closed, I usually see in the darkness of the visual field various bright images; it happens rarely that I
do not see them. I remember these phenomena since my early youth. I
was always able to distinguish them from true dream imagery because
I could reflect on them, often for a long time, before I fell asleep. . . .

Not only in the night, at any time I am capable of [perceiving] such
phenomena. Indeed, I spent many quiet hours with my eyes closed,
being far from sleeping, with such observations. Often I need only
to sit back, close my eyes, take abstraction from all things, and then
these images, known intimately to me since my youth, appear spontaneously.[13]

Although Müller describes the hypnagogic experience as effort-
less, anyone who has played with it knows that being able to observe
the images come and go is a kind of balancing act. To stay in the hypnagogic state you need to perch between drowsiness and alertness.
Too drowsy and you fall asleep; too alert and you block the images.
Balance must be struck between absorption in the images and awareness of them as only images. Too much absorption diminishes the
observant or witnessing aspect of awareness and leads directly into
sleep; too much sharp-eyed awareness compromises the vividness
and liveliness of the images. You also need to poise between attention and passive reception. Getting the images to arise requires being
open and receptive, while seeing them takes a certain kind of diffuse
attentiveness. Scrutinizing them too directly makes them fall apart.
Finally, the images must be recorded right away, because they tend
to be forgotten quickly.

Sleep scientists have come up with various ways to deal with these
difficulties. The standard experimental method is to wake people at
various stages of falling asleep and ask them to report what was in
their minds just before awakening.

Using this method, scientists have tried to determine where hypnagogic imagery comes from. In a landmark study, Robert Stickgold, a professor of psychiatry at the Harvard Medical School,
asked 27 people to play the video game Tetris for 7 hours over 3
days.[14] Ten of the participants were expert players, 12 had never

played, and 5 were amnesic patients who could not form short-term memories because they had suffered damage to a part of the brain (the medial temporal lobe) crucial for "declarative memory"—the memory of consciously reportable facts, including autobiographical events. Stickgold woke the participants at sleep onset and asked them what they were seeing; 17 of the 27 (63 percent) said they saw falling Tetris pieces. The images were most common on the second night of playing, not the first. Two expert players reported imagery that incorporated memories of Tetris experiences that had taken place one to five years earlier. These findings show that memories of both recent events and older related memories can form hypnagogic imagery.[15] The surprising finding, however, was that three of the five amnesiacs—who could not remember the Tetris game or the experimenter from one session to the next—reported Tetris-like hypnagogic imagery. This finding shows that hypnagogic imagery can arise from memories that are not consciously accessible, that is, from what psychologists call "implicit memory." Stickgold believes that most hypnagogic imagery comes from "procedural memory," the kind of implicit memory that we use to learn and remember skills (like apple-picking) and can't easily access at the level of reflective consciousness.

Another way to study the hypnagogic state is to train people to observe their own experience as they fall asleep. In Western science, this method goes back to Maury in the nineteenth century; he seems to have been the first scientist to use systematic self-observation to study hypnagogic imagery and dreams. The methods used for this today resemble focused attention and open awareness forms of meditation, for they rely on the same mental functions that these forms of meditation train: intention, attention, meta-awareness, and memory. By training people in self-observation methods, sleep scientists can get richer and more reliable reports about the hypnagogic state. The reports can then be related to measures of brain activity in order to cast light on what goes on in your brain as you slide from waking into sleep. This marriage between first-person methods for observing experience and third-person methods for measuring brain activity is central to neurophenomenology.

Sleep scientist Tore Nielsen, who directs the Dream and Nightmare Laboratory at Montreal's Sacre-Coeur Hospital, has taken this neurophenomenological approach in his experiments on the hypnagogic state.[16] He has developed a systematic self-observation method and put it to work in experiments measuring the electrical brain activity associated with hypnagogic imagery. As we'll see, his findings suggest there may be a regular sequence to the formation of hypnagogic imagery as we fall asleep.

To follow Nielsen's self-observation method, sit with your head upright and unsupported, and your eyes closed. Establish your "observational intent," which may be general—"I'll observe whatever comes to mind"—or specific—"I'll observe only color features of fleeting images." Direct your attention inward and observe whatever arises in the field of your experience as you allow yourself to enter a drowsy state. When your attention wanders, notice what type of change provoked the wandering. If the change is relevant to your observational intent, immediately review its details; if it's not relevant, then return your attention to the field of your experience. Eventually, as you become sleepy, your head will fall forward and you'll suddenly wake up. Try to remember the immediately preceding mental activity and examine it. With practice you'll be able to remember more detail and retain it well enough to describe it. If you're interested in memory associations, then you can reconsider your observations while carrying out a memory scan of the previous seconds, minutes, hours, or days.

The psychologist Charles Tart, well known for his pioneering scientific work on altered states of consciousness, has a slightly different method.[17] Lie flat on your back but keep your arm raised in a vertical position, resting on the elbow. You can drift far into the hypnagogic state before your muscle tone decreases and your arm falls down and wakes you up. Then practice holding the hypnagogic material in memory so your recall strengthens.

If you try these methods, or if you practice "lucid sleep onset"[18]—relaxing your mind and body and watching what happens to them as you fall asleep—you'll notice all sorts of things easily missed or absent if you just crash when you go to bed.

Sleepiness feelings—fullness inside the eyes, thickening inside the head—usually arrive first, followed by "sleep starts" or sudden muscle contractions accompanied by falling sensations.

Next come kinesthetic sensations or feelings of movement, along with fleeting visual images, sounds, fully formed images of faces and scenes, and trains of thought of the sort Proust's narrator describes.

Sometimes these trains of thought take the form of thought-image amalgams that seem to represent symbolically what you're thinking about or feeling. Here are two examples from Herbert Silberer (1882–1923), a Viennese psychoanalyst associated with Freud, who coined the term "autosymbolic phenomenon" to describe this kind of experience: "My thought is: I am to improve a halting passage in an essay. *Symbol*: I see myself planing a piece of wood." "I take a deep breath, my chest expands. *Symbol*: With the help of someone, I lift a table high."[19]

Finally, you may experience mini-dream episodes or dream fragments, which some researchers call "dreamlets,"[20] and sometimes you'll even experience full-fledged dreams.

THE HYPNAGOGIC BRAIN

What's going on in the brain during this hypnagogic journey from waking to sleeping?

Since the 1950s, sleep scientists have known that the brain cycles repeatedly through five main stages during sleep.[21] Each stage shows up as a distinct brain-wave pattern when we use EEG (the electroencephalograph) to record electrical activity at the scalp. Stages 1–4 are collectively known as non-REM (non-rapid-eye-movement) or NREM sleep; stage 5 is REM sleep and comes at the end of an NREM-REM sleep cycle, many of which occur throughout the night. As the brain proceeds through stages 1 to 4 of NREM sleep, the EEG waves become progressively slower (lower frequency) and larger (higher voltage). Neuroscientists divide brain rhythms into various frequencies, measured according to hertz (Hz), or the number of times an event repeats per second (1 hertz equals once per second, 2 hertz equals twice per second, and so on). In the waking state, fast rhythms in the beta (12–30 Hz) and gamma (30–80 Hz) frequency ranges predominate.

The transition from waking to sleeping happens at stage 1. As you relax and close your eyes, slow alpha waves (8–12 Hz) predominate; then as you nod off, the alpha waves subside, your eyes make slow rolling movements, and slower theta waves (4–8 Hz) come on the scene. This mixture of alpha and theta waves is the mark of stage 1 sleep. Here you're likely to experience a host of hypnagogic phenomena, and you're easily awakened—for example, by hearing your name called out as you doze off in a warm seminar room after a full lunch.

At stage 2, sleep really gets under way. You rest without much movement, and waking up is more difficult. A dramatic electrical display of "sleep spindles" and "K-complexes" announces the arrival of this stage in the EEG. Sleep spindles are rapid 12- to 14-hertz bursts lasting around 1.5 seconds; they are often preceded by K-complexes— brief high-voltage waves that resemble the letter K because of their high and sharp peaks.

Stages 1 and 2 make up light sleep; deep sleep happens in stages 3 and 4. Stage 3 contains a mixture of spindles and high-amplitude, slow-frequency delta waves (0.5–4 Hz). In stage 4, the delta waves increase to more than 50 percent. Because of the predominance of these slow delta frequencies, stages 3 and 4 are also known as "slow-wave sleep."

During the first hour of the night, we descend rapidly through the four stages of NREM sleep, arriving at stage 4 after about half an hour. Then we ascend through the stages in reverse order, but instead of returning to stage 1 after stage 2, we enter the entirely new stage of REM sleep. Here the EEG resembles the waking-state EEG, with fast-frequency/low-amplitude waves replacing the slow-frequency/ high-amplitude waves of slow-wave sleep. Yet the eyes remain closed and dart about rapidly under the eyelids, and our limb muscles are paralyzed. When woken up during REM sleep, we're likely to report an imagistic dream, more likely than during slow-wave sleep.

Nevertheless, reports of conscious experience as well as imagistic dreams can be elicited at all sleep stages, so the exact relationship between dreaming and REM sleep remains unresolved and a topic of controversy within sleep science.[22]

Although these stages have formed the bedrock for sleep science, recent research is moving away from fixed stages.[23] Traditional sleep staging was based on ways of measuring brain activity that treated it as uniform in space and time in ways we now know it is not. Scientists placed only a few electrodes far apart on the scalp, and they analyzed the electrical activity in thirty-second intervals. But we now know that different brain regions are in different states at the same time, and that brain activity varies subtly in numerous ways over milliseconds. Thus new methods that use many electrodes (so-called high-density EEG) and refined spectral analysis (measurements of the amplitude and phase of electrical activity across many frequencies and time scales), along with other neuroimaging methods, may soon offer a new conceptual framework for understanding the subtle dynamics of the sleeping brain.

The hypnagogic state is a case in point. The traditional sleep stages, in particular the categories "waking state" and "stage 1 sleep," are coarse compared to the fine-grained progression of hypnagogic events that can be observed in the waking-sleeping transition. The most detailed study of these experiences is still the 1987 book *Hypnagogia: The Unique State of Consciousness Between Wakefulness and Sleep*, written by psychologist Andreas Mavromatis.[24] He distinguished four main stages of hypnagogic experience: flashes of light and color; floating and drifting faces and nature scenes; autosymbolic phenomena (thought-image amalgams); and hypnagogic dreams. These stages are phenomenological and don't map onto any known EEG stages. Nevertheless, sleep scientists have tried to link the experiential progression of the hypnagogic state to the way brain events unfold during the waking-sleeping transition.

In the 1990s, the Japanese scientist Tadao Hori and his colleagues mapped the progression of electrical brain events within the hypnagogic period into nine distinct EEG stages.[25] These begin prior to stage 1 sleep and continue for a few minutes after stage 2 sleep. Hori's team also woke people and asked them to report what they were experiencing during these stages. Based on these reports, the scientists proposed that the frequency of kinesthetic images peaks in the first sleep-onset stage and then decreases steadily in later stages, whereas the opposite happens for visual images. Dreamlike experiences seem to occur most frequently when the alpha waves of sleep-onset stages 1–3 subside at stage 4 and the slower theta-wave ripples marking stage 5 arrive.

Tore Nielsen's neurophenomenological approach takes off from these findings. In a study with Anne Germain (now a professor of psychiatry at the University of Pittsburgh School of Medicine), Nielsen trained subjects in his self-observation method and asked them to signal the occurrence of spontaneous hypnagogic imagery, and then to report and describe it.[26] Nielsen and Germain found that kinesthetic images are linked to increased slow-frequency delta waves at the front of the head, whereas visual images are linked to increased delta-wave activity at the back of the head on the left side.

This suggests that hypnagogic imagery unfolds in a regular rather than haphazard way. For example, have you ever experienced a sudden jerk or contraction of your arm or leg as you start to fall asleep? These "sleep starts" often involve strong kinesthetic images of falling or stepping off into space. One EEG marker of these images seems to be a sharp increase in delta-wave activity in certain frontal areas of the brain (the prefrontal cortex) that are known to be involved in the ability to imagine performing motor actions. Visual images ensue when the delta-wave activity spreads to the posterior brain areas crucial for vision (visual association areas). This image-generation sequence seems to be one way that hypnagogic experience reflects the shifting patterns of brain activity as we fall asleep.

Nielsen's research illustrates the value of combining careful self-observation methods with neuroscience in order to study consciousness. Without reports about experience based on trained self-observation, the distinct brain patterns associated with kinesthetic versus visual hypnagogic imagery would remain buried in the EEG recordings. By helping to excavate these patterns, phenomenology, or the careful study of experience from within, casts light on the workings of the brain.

MEDITATION

The next step, which scientists have yet to take, will be to use meditators to investigate the hypnagogic state.

In meditation, especially during long periods, hypnagogic phenomena often arise as one slips in and out of a drowsy state.[27] Usually on the first day of a silent meditation retreat, I experience mostly hypnagogic images and thoughts, but without ever seeming to fall asleep. Although far less frequent by the second day, they will periodically arise at certain points throughout the week while my awareness of them seems to become progressively more acute.

At a Vipassanā meditation retreat for scientists held at the Insight Meditation Society in Barre, Massachusetts, I once slipped into a state where for some time I watched a parade of unfamiliar faces while I remained aware of the subtle feeling of the air entering and exiting my nostrils, and of my whole body expanding and contracting with each in-breath and out-breath. The faces seemed imaginary—creations of my mind—but also perceptual—they appeared in the outside space in front of my half-open and lightly focused eyes.

I attribute this kind of experience to a combination of factors occurring in insight meditation: an overall state of relaxation; the high arousal needed to maintain an upright sitting posture; a complex interplay among attention (focused and diffuse), absorption, and meta-awareness; and the changing relations between what Tibetan meditation texts call the mental qualities of "excitation versus laxity" (restlessness versus sluggishness) and "clarity versus dullness"—the ideal being a state that's both stable (neither restless nor sluggish) and vivid or clear (not dull).[28]

Given these features of meditative experience, neuroscientists should take advantage of meditation as a way to probe the hypnagogic state. As neurologist and longtime Zen practitioner James Austin writes in his groundbreaking book, *Zen and the Brain*: "No study both accurately describes in graphic detail the full content of hypnagogic hallucinations while also correlating them with contemporary polygraphic measurements [such as EEG]. When meditators are doubly monitored in this manner in the future, it will then be possible to specify which hallucination occurred during that interval when they were 'descending' from waking (hypnagogic), and which arose during that other interval when they were 'ascending' from sleep toward waking consciousness."[29]

This kind of research uses meditation to identify more precisely the brain events associated with certain states of consciousness. Yet we should demand more from neurophenomenology than simply better correlations between states of consciousness and the brain. We also want to understand the meaning of these states of consciousness for human experience. What does the hypnagogic state tell us about the self and consciousness? Neuroscience alone can't answer this question; we need phenomenology.

SPELLBOUND

Two key features mark the hypnagogic state—a slackening of the sense of self and a spellbound identification of consciousness with what it spontaneously imagines. Together they show that the sense of self is not fixed and that consciousness outstrips the waking ego.

Vladimir Nabokov's report of his hypnagogic experiences, told in his autobiography, *Speak, Memory*, shows these two features—the alteration to the felt unity of the waking self and the absorbed interest in the strange images consciousness concocts:

Just before falling asleep, I often become aware of a kind of one-sided conversation going on in an adjacent section of my mind, quite independently from the actual trend of my thoughts. It is a neutral, detached, anonymous voice, which I catch saying words of no importance to me whatever—an English or a Russian sentence, not even addressed to me, and so trivial that I hardly dare give samples, lest the flatness I wish to convey be marred by a molehill of sense. This silly phenomenon seems to be the auditory counterpart of certain praedormitary visions, which I also know well. . . . They come and go, without the drowsy observer's participation, but are essentially different from dream pictures for he is still master of his senses. They are often grotesque. I am pestered by roguish profiles, by some coarse-featured and florid dwarf with a swelling nostril or ear. At times, however, my photisms take on a rather soothing *flou* quality, and then I see—projected, as it were, upon the inside of the eyelid—gray figures walking between beehives, or small black parrots gradually vanishing among mountain snows, or a mauve remoteness melting beyond moving masts.[30]

Nabokov contrasts these "silly" phenomena with "the bright mental image (as, for instance, the face of a beloved parent long dead), conjured up by a wing-stroke of the will"; "*that*," he says, "is one of the bravest movements a human spirit can make." Yet we shouldn't be so quick to dismiss the spontaneous images of early sleep, for they show that imagination exceeds the waking ego and its will, and so can draw from deeper wellsprings of consciousness.

Ordinary waking consciousness is conditioned by what Austin, in *Zen and the Brain*, calls the "I-Me-Mine."[31] "I" is the self as thinker, feeler, and actor; "Me" is the self as affected and acted upon; "Mine" is the self as possessor, the appropriator of thoughts, feelings, body characteristics, personality traits, and material possessions. This tightly knit and mutually reinforcing triad constitutes our sense of self as ego, our deep-seated impression of being a distinct and bounded self standing over against the world.

Yet this impression is distorted. The boundaries we take to define our self aren't fixed but fluid and subject to constant change. Although the "I-Me-Mine" may seem to be an independent thing, it's really a shifting and impermanent construct (as will be discussed further in chapter 10).

One way to see this shiftiness up close is to watch what happens to the ego in the transitional zone between waking and sleeping. Here the waking ego starts to lose its grip on consciousness—thoughts turn colors, you overhear conversations happening inside you, and strange images make their way before your eyes. This breakaway of consciousness is one of the key features of the hypnagogic state.

Normal waking consciousness is ego-structured, conditioned by the appearance of a sharp difference between a bounded self and an outside world. Dreams typically re-create this structure, for often one has a dream body in which one participates in the dream world. Even when one instead experiences oneself as an observational point of view, one still experiences oneself as a subject situated in relation to the dreamscape.[32]

The hypnagogic state, however, doesn't fit this structure. There's no hypnagogic world; the images occur on their own, one following another, with no ties to anything beyond them. A cow dissolves into the scattering of a flock of birds that transforms into a woman's face, but none of these images participates in any larger universe. At the same time, there's no hypnagogic ego in the sense of an I who acts as a participant within what's being imagined. I can affect the images while watching them, but I don't experience myself as immersed within the scene—if I do, then I've entered into full-blown dreaming.

The hypnagogic state blurs the boundaries between inside and outside, self and world. The flashing lights and colors seem to occupy a space around me, but this space appears within my eyes and reconfigures with each eye movement. The strange faces and distorted scenes look other to me, yet my drifting gaze creates them. I behold these images, but they absorb me. In these ways, the distance between me and them diminishes or even seems to disappear. As Proust's narrator says, "It seemed to me that I myself was what the book was talking about: a church, a quartet, the rivalry between François I and Charles V."

Sometimes, in this borderland state, a peculiar kind of double consciousness ensues; we retain awareness of the outside world while watching the inner mental scene usurp its place. Swedish poet Tomas Tranströmer captures this moment in his poem "Dream Seminar," when he describes dozing off at the theater so that the stage before him is "outmaneuvered by a dream" while another theater and "overworked director" take over from within.[33]

Scientists too have found a source of insight in this intermediate state. The classic case is how German chemist Friedrich August Kekulé (1829–96) discovered the ring form of the benzene molecule during a hypnagogic reverie:

I was sitting, writing at my text-book; but the work did not progress; my thoughts were elsewhere. I turned my chair to the fire and dozed. Again the atoms were gamboling before my eyes. This time the smaller groups kept modestly in the background. My mental eyes, rendered more acute by repeated visions of the kind, could now distinguish larger structures, of manifold conformation: long rows, sometimes more closely fitted together; all twining and twisting in snakelike motion. But look! What was that? One of the snakes had seized hold of its own tail, and the form whirled mockingly before my eyes. As if by a flash of lightning I awoke; and this time also I spent the rest of the night working out the consequences of the hypothesis.[34]

The hypnagogic state offers a unique concoction of relaxation, absorption, diffuse and receptive attention, ego dissolution, reactivation of recent memories as well as older ones, synesthesia, and hyperassociative and symbolic thinking. By outstripping the waking "I-Me-Mine," this mode of consciousness can tap deeper sources of creative thinking and intuitive problem solving.

Some psychologists, viewing this alteration to consciousness through Freudian lenses, see it as a throwback or "regression" to earlier childhood modes of experience.[35] As we grow from child to adult, according to Freud, we build up an ego capable of "reality testing" (differentiating among hallucination, imagination, and perception) and of obeying the "reality principle" (postponing immediate gratification in the face of the world's demands). These capabilities diminish or disappear in the hypnagogic state. Carried along by inner events, we eventually lose awareness of the outer world and the volitional control of thought; we also lose the reflective awareness that what we see is imaginary. Because the images are often bizarre, distorted, and unrealistic, Freudians describe them as "regressive" in their content. Early childhood memories can also surface. Thus, from a Freudian perspective, the hypnagogic state appears to be a state of altered ego functioning marked by regressive tendencies.

Psychologist Andreas Mavromatis, in his *Hypnagogia*, follows this Freudian line but gives it a transpersonal psychology twist.[36] He writes, "The core psychological phenomenon out of which springs the whole gamut of hypnagogic experiences is the loosening of the ego boundaries of the subject."[37] Yet this ego loosening, he argues, need not be regressive; it can also be a progressive step toward a new and more integrated mode of consciousness—a "double consciousness" in which you can watch your dreaming mind while staying awake and aware of your surroundings.

Meditation often gives rise to this double consciousness, especially during long sitting periods when a certain degree of relaxed concentration has been achieved. Earlier I described one of my own meditative experiences of this. Yet focused attention and open awareness forms of meditation don't aim to produce this kind of consciousness and don't value it for its own sake. It's seen as a natural outgrowth of increasing relaxation and concentration, as well as a potential source of creativity, but also as posing challenges to the cultivation of mindfulness. Thus Theravada Vipassanā meditation counsels against getting stuck in "sinking mind"[38]—a pleasant dreamlike state that lacks the clarity of mindfulness—and Zen teaches that makyō[39]—spontaneously arising images—should be treated as illusions and ignored. The hypnagogic procession of faces I saw while practicing mindfulness of breathing is an example of makyō arising from sinking mind.

Mindfulness or insight forms of meditation also make clear that this transitional zone between waking and dreaming, despite its relaxed ego boundaries and creativity, is hardly free from the identifying and appropriating functions of the "I-Me-Mine." On the contrary, watching images come and go during meditation can bring into relief the other key feature of the hypnagogic state—the spellbound identification of consciousness with its images and thoughts.

My favorite examination of this spellbound consciousness comes not from a meditation text but from Jean-Paul Sartre's book *The Imaginary*, published in 1940. Sartre writes, "Hypnagogic phenomena are not 'contemplated by consciousness': they *are consciousness*."[40] By this he means that we should not think of mental images as pictures inside the head, but instead as consciousness itself in its activity of imagining. When you imagine the Eiffel Tower, you don't look at a picture of it in your head; you mentally simulate seeing the Eiffel Tower from a certain vantage point. So what we really mean when we say, "I can't get that image out of my mind," is "I can't stop visualizing that thing."[41] In the hypnagogic state, consciousness spontaneously visualizes while holding attention spellbound. What underlies and sustains the state is thus spontaneous imagination combined with an alteration to the functioning of attention. Rather than being distracted, attention is fascinated; it loses itself in the image—in the imagining—and remains caught by it. The hypnagogic state is a state of "captive consciousness." "But of course," Sartre tells us, "this consciousness is not captive to its objects, it is captive to itself."[42]

We can now see how the two key features of the hypnagogic state—the relaxation of ego boundaries and the spellbound identification of consciousness with what it imagines—are really one. They are the unified result of the way attention functions in relation to imagination in the liminal zone between waking and sleeping.

IMMERSION IN THE DREAM

What's the difference between the hypnagogic state and dreaming? If ego boundaries weaken in the hypnagogic state, what happens to the self in dreaming?

The core feature of full-blown dreaming is the experience of immersion in the dream world:

Dreams

Here we are all, by day; by night, we're hurled
By dreams, each one into a several world.
—Robert Herrick (1591–1674)[43]

In the hypnagogic state, we look at visual patterns and they absorb us. When we dream, we experience being in the dream; more precisely, we experience being in the dream world. The experience of being a self in the world, which marks the waking state but diminishes in the hypnagogic state, reappears in dreams.[44]

We know from developmental psychology that our sense of self emerges from infancy to early childhood to adolescence to adulthood. Similarly, studies of children's dreams indicate that dreaming of yourself as an active participant in the dream takes time to develop. According to psychologist and dream researcher David Foulkes, preschoolers don't usually report self-participation dreams, though they report dream imagery, whereas children around seven or eight do report this kind of dream.[45]

The earliest of my own dreams that I know anything about, I don't remember myself. My parents tell me it happened when we lived in Ireland and I was two years old. I saw my father reading an article about Marc Chagall's paintings and stained-glass windows. I looked at the pictures and said, "That's Evan sleeping." I suspect the colored images looked like what I saw when I fell asleep.

In the earliest dream I do remember, I was an active participant. I had the dream when we lived in Watertown, Massachusetts, so I must have been four or five years old. I dreamt I was in our backyard. The sun had set and some kind of party was going on. My mother was there, wearing her nightgown, skipping and dancing in a circle. A bright full moon was up in the sky. The moon came down until it was floating next to me. I reached out and touched it with my finger. The moon turned into some kind of mechanical monster, and I woke up in terror. In this dream turned nightmare, I was fully immersed in the dream world, experienced myself as my dream body, and saw the dream world through my dream body's eyes.

But there was another way I experienced myself in dreams when I was a kid. When I was still young enough to take an afternoon nap, but old enough to be bored and restless, I'd watch the hypnagogic images turn into a story, and sometimes one of the characters would be me. When I was a bit older and my father would read me chapters from *The Hobbit* or *The Chronicles of Narnia* as a bedtime story, I'd have dreams where I saw myself from the outside as a dream character in a dream adventure.

Around this time I realized there was a difference between dreams where I experienced things from the inside and dreams where I saw myself from the outside. Seeing yourself from the outside in a dream is like watching an actor play a role on screen—a role that happens to be you. Maybe the dream person looks like you—like your mirror image—or maybe there's no resemblance. The person could be someone you've never seen before, yet you as onlooker know or have the feeling of knowing that this person is you.

Sartre describes this experience in his chapter on "The Dream" in *The Imaginary:* "It also frequently happens that dreams—mine, for example—are given at first as a story that I am reading or being told. And then, all of a sudden, I am identified with one of the people of the story, which becomes *my* story."[46]

If you play online role-playing games (like World of Warcraft) or participate in virtual worlds (like Second Life), you'll be familiar with these two perspectives. Seeing yourself from the outside is like viewing your avatar from a third-person perspective; seeing things from the inside is like being immersed in the virtual world and viewing it through your avatar's eyes.

One of the oldest references to dreams in Western literature can help to illustrate these different dream perspectives. In Book Twenty-Two of Homer's *Iliad*, the Trojan hero Hector tries to face the Greek warrior Achilles but cannot help running away in fear:

> As in a dream a man chasing another
> cannot catch him, nor can he in flight
> escape from his pursuer, so Akhilleus
> could not by his swiftness overtake him,
> nor could Hektor pull away.[47]

Fleeing while being chased is a common dream theme. When you feel your pursuer at your back and see him coming as you look over your shoulder, you're experiencing what's happening from the first-person perspective. When you're watching the chase from above and you see yourself running away, you're experiencing the chase from the third-person perspective. These perspectives can also shift back and forth as you dream.

Third-person perspective dreams feel different from looking at hypnagogic imagery. You're still immersed in the dream, for you as onlooker occupy a space or have a spatiotemporal point of view inside the dream. Your point of view, however, is separated or dissociated from the dream body you see and identify as yours. First-person perspective dreams lack this kind of dissociation, because you occupy your dream body and see the dream world through its eyes.

MEMORY AND DREAMS

These two perspectives—first-person and third-person—also appear in memory. Take a moment to remember your last birthday or some other memorable event. Do you see yourself from an outside point of view (the way someone else would see you), or do you see things from the perspective of your own eyes (as you do now reading these words)?

Psychologists use the terms "observer memory" and "field memory" to describe these different ways of remembering.[48] In the observer mode you're an outside spectator on yourself; in the field mode you recall what happened from within. The involuntary transformation of first-person-perspective experiences into third-person-perspective memories shows that we don't retrieve a photocopy of the past when we remember; we actively reshape and consolidate our relation to the past from the standpoint of the present. As my father wrote in his poem "Chicago 1943–45":

Why does our memory trick us
with imagery from later on?
Why do we turn old memories
we keep playing over again
into movies we've filmed outside
ourselves and not inside our eyes,
adding images we've never seen,
changing point of view and mind?[49]

One answer is that this kind of perspective switching enables us to imagine how we look to others, and the mental ability to get an outside view of ourselves is required for empathy and social cognition. If I want to understand how my behavior affects you, then I need to be able to imagine how I appear to you; I need to be able to envision your outside perspective on me.

Phenomenologists call this kind of experience "self-othering."[50] In memory, but also in dreams, imagination, fantasy, and anticipation or prospection, you can simulate being an other to yourself by visualizing yourself from the outside.[51] When you remember yourself from the outside as a little kid taking a trip with your parents, or when you prepare for an upcoming speech by imagining how you'll look in front of the audience, you're simulating being an other to yourself within your own stream of consciousness. We need this kind of "self-othering" in order to project ourselves mentally into our personal past and future—to have what psychologists call an autobiographical sense of self—and to experience ourselves in relation to others—to have a social sense of self.

Another kind of answer to my father's question is that what we experience in memory—how our past appears to us—depends on how we going about recalling it. Studies show that asking people to focus on their feelings (How did you feel taking that trip with your parents?) produces more field memories, while asking them to focus on objective circumstances (Where did you sit on the ferry?) produces more observer memories.[52] Field memories include more information about our emotional reactions, physical sensations, and psychological states; observer memories include more information about how we looked, what we did, and where things were.

These differences are also reflected in the brain.[53] Taking the outside perspective in observer memory (compared to the inside perspective in field memory for the same events) significantly decreases the activity in regions of the cortex associated with internal body awareness (the insula and primary and secondary somatosensory cortices); moreover, the amygdala—a structure known to play a role in memory for emotional events—shows greater activity in field memory compared to observer memory. Thus, at a neurophysiological level, field memories are more emotional and bodily, while observer memories are more distanced and perceptual.

These considerations about memory take on added significance when we realize that we can't inspect our (nonlucid) dreams directly; we can inspect only our waking memories of dreams.[54] As Jorge Luis Borges observes in his essay "Nightmares:"

> The study of dreams is particularly difficult, for we cannot examine dreams directly, we can speak only of the memory of dreams. And it is possible that the memory of dreams does not correspond exactly to the dreams themselves. A great writer of the eighteenth century, Sir Thomas Browne, believed that our memory of dreams is more impoverished than the splendor of reality. Others, in turn, believe that we improve our dreams. If we think of the dream as a work of fiction—and I think it is—it may be that we continue to spin tales when we wake and later when we recount them. . . . Suppose I dream of a man, simply the image of a man—I'm using a very poor dream—and then, immediately after, I dream the image of a tree. Waking, I can give this dream a complexity it does not have: I can think I have dreamed that a man has been changed into a tree, that he was a tree. Modifying the facts, I spin a tale.[55]

Yet even before we spin a tale and give a narrative structure to our dream—making something sequential out of something multiple—the elements of the dream already appear to us in memory in certain ways that will condition how we spin the tale in its telling or recording.

Think of a recent dream and ask yourself: In which mode of memory—field or observer—does your dream memory appear to you? If you remember feeling your pursuer behind you and seeing him chase you as you looked back, then your memory is in the field mode, and you remember your dream as one where you experienced things through your dream body's eyes. If you remember seeing yourself flee, then your memory is in the observer mode, and you remember your dream as one where you were watching yourself from the outside.

Over time, however, dream memories, like all memories, change. A dream remembered in the field mode when you wake up might later be remembered in the observer mode. When I recall childhood dreams, they often appear to me first in the observer mode (I see myself as someone in the dream). I need to make a special effort to evoke them in the field mode, to relive them from within. Yet I have reason to believe that in the original dreams I was fully immersed in the dream world, and that I remembered the dreams first in the field mode when I woke up, for many were nightmares that wouldn't have been so terrifying if they weren't experienced and remembered from within.

SENSES OF SELF

The features of dreaming and memory we've been considering illustrate something crucial about the self. We experience our self both as an object of awareness and as the subject of awareness. William James called these two forms the "Me" and the "I," the self as known and the self as knower. Phenomenologists call them the "self-as-object" and the "self-as-subject."[56] These terms refer not to two different entities but to two ways of experiencing yourself—two modes of self-experience.

When you look at yourself in a mirror, you-as-subject see yourself-as-object. You experience yourself both as the subject of this perception—it's your seeing and no one else's—and as its object, as what you see—you recognize the mirror image as your own. We don't need a mirror, however, to illustrate this distinction. If you look down at your hands, they are the objects of your visual attention and you immediately recognize them as yours. In this experience, the self-as-object or Me includes the hands you see as yours; the self-as-subject or I includes your implicit awareness of this visual perception as your experience.

Let's now turn to memory. When I remember myself from the outside taking the ferry from Mull to the island of Iona in Scotland when I was eleven, the remembered self I see from a third-person perspective is an example of the self-as-object, while the self-as-subject consists in my being aware of myself now remembering this event. In this way, my observer memory contains both the Me remembered from the outside and the I who is doing the remembering.

When I remember the same trip from the inside, in the field mode, I project myself back into the past situation and relive it from within. Instead of seeing myself on the ferry, I see the waves and the shoreline from where I sat, and I relive the fear I felt during that stormy crossing. In this way, I project myself back into a previous self-as-subject or first-person perspective. Because I remember the past experience from within, it shows up for me in memory automatically as my former experience, as an experience had by me, despite however much I've changed or however reconstructive or revisionist my memory might be.

The self-as-subject and the self-as-object—the I and the Me—carry over into dreaming, but the way they relate to each other takes different forms depending on the kind of dream.

When you see yourself from the outside running from your pursuer in a dream, the character you identify with constitutes your dream Me, or self-as-object. Your spatiotemporal point of view as onlooker in the dream constitutes your dream I, or self-as-subject. Your dream Me and I are both aspects of your dream ego or your self-as-dreamed (by contrast with your self-as-dreamer, who appears when you have a lucid dream). Thus the Me and the I dissociate in this kind of third-person perspective dream, for you see yourself (Me) as located in a different place from your observational point of view (I) within the dream. (A special form of this dissociation happens in out-of-body experiences, which I examine in chapter 7.)

When you experience your dream body from within, however, your awareness seems located at the place of your dream body, and you see your pursuer through your dream body's eyes. The I (self-as-subject) and the Me (self-as-object) coincide in their felt location.

These perspectives can shift or alternate as you dream. You might first see yourself from the outside and then all of a sudden experience yourself from within. You might see another person or being—an animal, perhaps—and then abruptly feel yourself to be that being acting in the dream world.

One of the salient features of the dream state (salient in retrospect) is that we seldom take notice of these striking shifts in viewpoint, and if we do, we rapidly lose track of them. This brings us to the last main question of this chapter. What are the distinctive psychological features of the I or self-as-subject in the dream state, and how are these features related to the brain?

THE DREAM SUBJECT

If the hypnagogic state is a state of captivated consciousness with little or no sense of immersion in a world, the dream state is a state of captivated consciousness where we feel immersed in the dream world, for even when we observe ourselves from the outside, we do so from some vantage point within the dream. What we need to examine now is how the fascinated attention that holds consciousness captive to its images functions in relation to the other cognitive and emotional processes characteristic of the dream state. Adding the view from neuroscience will bring us back to this chapter's opening questions about lucid dreaming.

Here's a.short exercise to bring out some of the features of waking attention that vanish for the most part when we have a nonlucid dream. Look straight ahead, fix your gaze at a particular spot, and then without moving your eyes, direct your attention to something in the middle-left periphery of your visual field and hold it there for about fifteen seconds.

A lot that we take for granted goes into doing this little exercise. We follow the instructions, form an intentional plan, volitionally direct our visual attention straight ahead, and then simultaneously direct and try to sustain our mental attention off to the side. We remember what we're supposed to be doing for the allotted time, we ignore distractions, and we redirect our attention if we notice it waning or wandering. By relying on metacognition—our ability to think about and keep track of our mental activity—we monitor and regulate our attention. Furthermore, we can observe certain felt qualities of this whole attentional endeavor—for example, how our attention seems to fluctuate or compete between the two demands or how effortful the task seems.

Let's compare these mental abilities to the way attention functions in a nonlucid dream. Here's a dream I had the other night:

> I'm on the subway in Toronto. The train is above street level, and I see Paris streets below me through the window. I'm with a former girlfriend from many years ago. I'm anxious waiting for the next stop, where I know I have to get off. Then the stop is past and she's out in the street. I'm more anxious and look for my suitcases. One is missing. Maybe she took it, but the train's moved on and she's gone. I wake up feeling anxious and thinking I need to find my suitcase.

This dream illustrates how attention and emotion usually work in nonlucid dreams. First, I fail to notice or pay attention to a number of striking discrepancies and discontinuities: I think I'm on the subway, but I can see I'm above street level; I think I'm in Toronto, but I can see the streets belong to Paris; I'm traveling with a friend, but I haven't seen or talked to her in years; I'm waiting for the next stop, but then it's already past; my friend is on the street, but I didn't see her get off the train; I remember having suitcases, but I had no thought of them before; I look for the missing suitcase, but don't notice the train moving on. Second, certain feelings and action tendencies or behaviors pervade and dominate each episode of the dream—in this case, the emotion of anxiety and the behavior of seeking or searching for something.

Because of these attentional and emotional features, the nonlucid dream state lacks the kind of cognitive control that's present in the waking state. Working memory is weak, so keeping track of what's going on is difficult. Distraction happens constantly and attention can't be volitionally sustained. Metacognition is unsteady, so we have difficulty monitoring thoughts and feelings. At the same time, emotions intensify—sometimes fear, anxiety, or anger; sometimes joy and elation—while basic behaviors such as seeking and fleeing often dominate what we do.

For these reasons—as philosophers Jennifer Windt and Thomas Metzinger discuss in a groundbreaking article on the dream state[57]—the dreaming subject isn't an effective metacognitive subject of experience. In a dream, it's difficult to conceptualize and experience yourself as a self in the act of deciding (a volitional subject), a self in the act of attending (an attentional subject), a self in the act of thinking (a cognitive subject), or a self in the grip of emotion (an affective subject). Lacking insight into the nature of your ongoing conscious state, you can't experience yourself as a dreaming subject.

Recent cognitive neuroscience has given us a new look at this psychological profile of the dreaming subject in the form of images of brain activity during REM sleep—the sleep stage most closely (but not exclusively) associated with imagistic dreaming.[58] Certain formal features of dreams that remain present through wide variations in specific dream contents correlate with certain areas of the brain being either selectively activated or deactivated in REM sleep, compared to the waking state.

For example, the instability in dreams between experiencing yourself from the inside (seeing through your own eyes someone chasing you) and seeing yourself from the outside (seeing from above someone chasing a dream character you identify as you) seems related to the reduced activity in REM sleep in an area called the temporoparietal junction (where the temporal and parietal lobes meet). This area has been associated with the ability to shift between taking a first-person perspective and a third-person perspective on a situation. The temporoparietal junction has also been implicated in out-of-body experiences, in which you observe yourself from a vantage point outside your body.

Other links between the psychology and neuroscience of dreaming are starting to be established. On the one hand, dreams contain strong visual imagery, intense emotions such as fear or elation, and fragmented and transformed memories of events experienced in the past (episodic memories). The activations in visual brain areas (visual association cortex), emotion-related areas (amygdala, orbitofrontal cortex, and anterior cingulate cortex), and memory systems (hippocampus, entorhinal cortex, and other parahippocampal regions) reflect these formal features of dreams. On the other hand, an area at the front of the brain known as the dorsolateral prefrontal cortex shows greatly reduced activity during REM sleep.[59] This area is crucial for working memory, reflective self-awareness, volition, and directed thought. Hence the diminished activity in this area may reflect the weakness in these mental capacities when we dream.

In some dreams, however, like the one described at the beginning of this chapter, these mental capacities, especially meta-awareness and the ability to direct attention intentionally, remain present or revive. As a result, we're able to know we're dreaming while we're in the dream state. We think about ourselves as dreamers and experience what we see and hear and feel as dream images. In short, we're able to experience ourselves as dreaming subjects while we're dreaming. This kind of self-experience defines the state of lucid dreaming.

But now we have new questions: If we're spellbound when we dream, does lucid dreaming break the spell? If dreaming is a kind of captive consciousness, does lucid dreaming offer a kind of freedom? If we identify with our dream ego when we dream, then who are we when we lucid dream? And what's going on in the brain when we lucid dream? If frontal areas of the brain crucial for working memory and reflective self-awareness decrease their activity during dreaming, do they increase in lucid dreaming?

We've come back to where we started. I began this chapter with questions about lucid dreaming that we're now ready to confront head on: Who is the subject of lucid dreaming? And what is this subject's relation to the brain and body?

5

WITNESSING

Is This a Dream?

I'm walking down the hallway of the Sheraton Vancouver Wall Centre Hotel and suddenly remember that I'm lying in bed in my hotel room. I've gone to sleep; I'm dreaming. Elated, I decide to test my state. A red exit sign hangs on the wall; I look away, then back again, and now the letters are all jumbled. Yes! I'm dreaming! I leap up to where the wall meets the ceiling and crouch there like Spiderman. I fly down the hall at ceiling level and go straight through the wall at the other end. As I pass through the wall, the room disappears and the whole visual scene goes gray, with color flashes, like the lights that appear on my inner eyelids when I'm falling asleep. I can't get back a stable visual world, and I start to feel claustrophobic. A sense of paralysis takes hold—not in my dream body, but in my real body, which seems to be displacing my dream body from my consciousness. For a moment it almost feels like I have a double body, one lying in bed and the other floating in dream space, but then, unable to stop fighting the paralysis, I manage to move my real body, and that effort wakes me up.

AWAKENING THE DREAMER

Dreams like this one—lucid dreams—reveal an important distinction we need to make in order to deal with the questions raised at the end of the last chapter: between the dreaming self and the dream ego, between the "I as dreamer" and the "I as dreamed." These are not two entities or things; they're two kinds of self-awareness, two modes of self-experience. In a strong lucid dream, your sense of self shifts as you become aware of yourself both as dreamer—"I'm dreaming"—and as dreamed—"I'm flying." In contrast, in a nonlucid dream your sense of self takes the perspective of the dream ego, who often appears in the form of the dream body. Lucid dreams reframe experience from the perspective of the dreamer who can recognize the dream state, influence or guide the dream ego, and remember waking life.[1]

Let me illustrate these features with another lucid dream. Writer Dorion Sagan told me about this a few days after we both gave talks at the Upaya Zen Center in Santa Fe, New Mexico. We were there for the 2010 annual "Fellows Conference" of the Lindisfarne Association, a group of scientists, artists, ecologists, and contemplatives, founded by my father, William Irwin Thompson. In my talk, I presented some of my ideas about lucid dreaming and read the section from this book where I describe the lucid dream I had on my first night in Dharamsala. Two days later, Dorion had a lucid dream, which he wrote down the next morning while we were sitting by the Japanese hot baths at the Ten Thousand Waves Spa in the mountains above Santa Fe. Here's the dream, transcribed from the back of the walking map where he wrote it down:

> I took porcine melatonin (derived from a pig's pineal gland) before going to sleep. I woke up at 3:40 a.m. and took another small amount. I brought myself back to sleep by trying to remember my talk, especially my opening comments, word for word. . . . In my dream, I was to go on a trip and was at a dilapidated bus stop. It was dusty and there was a bus I was to get on, but I realized I had a small amount of marijuana I needed to hide before I got on, due to security issues. Hurriedly, I hid it in my blue jeans. I rolled the weed between my fingers, letting a dry stem or two fall to the unpaved ground, and stuffed the rest into the corner of my change pocket and patted it down. I was aware even of the little metal nib, the stud near that pocket (a lot of

haptic sensations). Now, I also had to get money for the trip. I pushed my ATM card into the machine but encountered difficulties. Finally, I noticed that a credit card had jammed the machine, making it impossible to remove cash. At some point an indigent person—a kind of youngish bag lady—scrambled by, shifting her wares in a hurry. Figuring that I could pay on the bus, I ran to it (I believe my mom and others I knew were on it), but it was gone. At this point—where loss manifested itself—I realized I was in a dream. I remembered Evan's distinction between knowing you're dreaming and being able to take control of actions in the dream. I spread my arms and began to fly at about a forty-degree angle to the ground. I was conscious of my pineal area or "third eye," and looked through it, upward, to the sky. (I had done a kundalini yoga exercise that included pineal visualization in a yoga class two weeks before.) The lucid control part of the dream was the decision to fly, but the most spectacular part was now what I saw. I saw the stars, manifold and white against the black depths of space. The vision was exquisite. Still conscious that this was a dream, I marveled at the realism of the night sky. Now I saw—at the same time—beautiful white clouds roll across a blue sky, again with striking realism. Certainly, this is what made this dream unique—the combination of a night sky of stars and a blue sky of day perceived at the same time. It was as if I were experiencing some sort of quantum superposition of day and night. Then I woke up.

This dream illustrates the striking shift in perspective that happens when you realize you're dreaming. Dorion's dream ego or self-in-the-dream tries to get on the bus, but his feeling of loss at missing it somehow triggers the realization that what's happening is a dream. The dream reframes itself so that now the dreamer knows he's dreaming.

In my talk, I had explained that what defines the lucid dream state isn't being able to control or actively guide the dream, which you can do sometimes when you don't fully realize you're dreaming, but rather being able to direct your attention to the dream with the knowledge that it's a dream. Dorion's remembering my distinction between recognizing the dream state and guiding the dream content shows how memories from waking life can become accessible in a lucid dream.

Although Dorion was able to guide his dream ego by deciding to fly, "the most spectacular part" of his dream was precisely what he didn't deliberately control or create but still recognized as something he was dreaming—the vivid night-day sky. Its "superposition" of opposite features seems akin to other kinds of strange dream combinations—a face that's both your ex-girlfriend's and an unknown woman's, a city that's both New York and Paris, a room that's both a basement and an attic. But the "striking realism" and "exquisite" vision of the dream, as we'll see later, are typical features of strong lucid dreams.

According to anecdotal reports, various factors seem to make lucid dreams more likely to happen, though more research needs to be done on which ones are reliable.[2] Some of those I've noticed or that people have told me about are changes to the sleep cycle (jet lag), taking melatonin, doing presleep memory exercises (like Dorion's rehearsing his talk), carefully observing hypnagogic imagery as you fall asleep, and waking up around 3 a.m. and reading about dreams for half an hour, then going back to sleep.

Lucid dreams can also be induced through various techniques. One basic method is to form a clear intention to recognize the dream state when it happens and tell yourself every night as you fall asleep to be on the lookout for anything unusual that will give away the fact that you're dreaming. This method depends on what psychologists call "prospective memory" (remembering to remember) and "auto-suggestion." Telling yourself each night that you're going to remember to recognize when you're dreaming is a form of prospective memory with auto-suggestion that helps to bring about lucid dreaming.

Although Indian and Tibetan traditions have practiced meditative lucid dreaming for millennia, Western science has only just begun to investigate this unique state. So, from a scientific standpoint, it amounts to largely uncharted waters in the stream of consciousness. To navigate them, we need to distinguish within consciousness between the witnessing aspect of awareness, the changeable contents of awareness, and ways of experiencing particular contents of awareness as the self.

When we dream, we see the dreamscape from the perspective of our dream ego. Although the entire world of the dream exists only as the content of our awareness, we identify our self with only a portion of that content—the dream ego that centers our experience of the dream world and presents itself as the locus of our awareness.

In a lucid dream, however, we experience another kind of awareness with a different locus. This awareness witnesses the dream state, but without being immersed in the dream world the way the dream ego is. No matter what contents come and go, including the forms taken by the dream ego, we can tell they're not the same as the awareness witnessing the dream state. From the vantage point this provides, we can observe dream images precisely as dream images, that is, as manifestations of the mind—not the mind of the dream ego but the mind of the dreamer who imagines the dream ego. In this way, we no longer identify only with the ego within the dream; our sense of self now encompasses the witness awareness of the whole dream state.

HOW FREUD MISSED LUCID DREAMING

One reason lucid dreaming hasn't gotten the attention it deserves in Western psychology is the towering influence of Sigmund Freud and his masterwork, *The Interpretation of Dreams*, first published in 1899.[3] Freud says nothing there about lucid dreaming, except for two paragraphs, one added in 1909 and the other in 1914.[4] Even there, he doesn't mention lucid dreaming by name, despite having met and corresponded with Frederik van Eeden (1860–1932), the Dutch psychiatrist who coined the term "lucid dream" in his seminal 1913 article, "A Study of Dreams."[5] Freud presents many analyses of his own dreams, but he doesn't seem to have been a lucid dreamer himself (he doesn't report any lucid dreams). Moreover, the theoretical edifice of his psychoanalytic dream psychology prevented him from recognizing the significance of lucid dreaming.

What did catch Freud's interest—not in the first edition of *The Interpretation of Dreams*, but in material he added later—was the judgment, "It's only a dream," made while dreaming, as well as the "enigma of the 'dream within a dream.'"[6] For Freud, the judgment is basically defensive: you think, "It's only a dream," when you feel threatened by something in the dream and want to reduce its importance. Trying to escape from someone chasing you, you cry out, "It's only a dream," and the chase becomes less fearful. Similarly, a dream within a dream serves "to detract from the importance of what is 'dreamt' in the dream, to rob it of its reality," while what you dream after waking from a dream within a dream is "what the dream-wish seeks to put in place of an obliterated reality." Thus, in the case of Dorion's lucid dream, Freud would say that the thought that he was dreaming served to detract from the feeling of loss at missing the bus (and his mother, who was on it) in favor of some other wish represented in the urge to fly and gaze at the beauty of the sky. In these ways, thinking "It's only a dream" serves to keep you asleep by denying the importance of something that would otherwise disturb you and wake you up.

Freud takes this view because he maintains that "a dream is the fulfillment of a wish." Suppose you go to sleep after eating salty food and you dream of drinking a glass of cold water; the thirst gives rise to a wish and the dream shows you that wish fulfilled. Freud calls dreams of this kind "dreams of convenience." Often the wish appears in a disguised way, as in this dream from Proust's novel *The Way by Swann's*: "When we are asleep and a raging toothache is as yet perceived by us only in the form of a girl whom we attempt two hundred times in a row to pull out of the water or a line by Molière that we repeat to ourselves incessantly, it is a great relief to wake up so that our intelligence can divest the idea of raging toothache of its disguise of heroism or cadence."[7]

For unconscious and forbidden desires especially, the dream can't show the wish without disguising it, because otherwise the wish would upset you and wake you up. So the dream you recall and describe consists of disguised wish fulfillments; they constitute the dream's "manifest content," while the "latent content" or hidden dream thoughts constitute the underlying psychological meaning of the dream.

Freud calls the process that transforms the latent content into the manifest content the "dream work." Since the latent dream thoughts would wake you if not disguised, the dream work functions to keep you sleeping by creating the dream. Thus dreaming is the mind's way of fulfilling the wish to sleep. For this reason, Freud writes, "All dreams are in some sense dreams of convenience: they serve the purpose of prolonging sleep instead of waking up. *Dreams are the GUARDIANS of sleep and not its disturbers.*"[8]

Yet, if our primary purpose in dreaming is to keep reality at a distance and prolong sleep, then we must in some sense know while dreaming that what we're experiencing is only a dream, not reality. Freud accepts this implication: "I am driven to conclude that *throughout our whole sleeping state we know just as certainly that we are dreaming as we know that we are sleeping.*"[9] Of course, this is an implicit, preconscious knowing, not an explicit, conscious one (otherwise every dream would be a lucid dream).

On this point, Friedrich Nietzsche (1844–1900) had already anticipated Freud. In *The Birth of Tragedy*, published twenty-seven years before Freud's *The Interpretation of Dreams*, he wrote that when we dream we have the sense that what we see is "semblance" (*Schein*):

> Every human being is fully an artist when creating the worlds of dream, and the lovely semblance of dream is the precondition of all the arts of image-making. . . . We take pleasure in dreaming, understanding its figures without mediation; all forms speak to us; nothing is indifferent or unnecessary. *Yet even while this dream-reality is most alive, we nevertheless retain a pervasive sense that it is semblance* [my italics]; at least this is my experience, and I could adduce a good deal of evidence and the statements of poets to attest to the frequency, indeed normality, of my experience.[10]

When we dream, according to Nietzsche, we have the feeling that what we're seeing is an appearance or a likeness—an image—and not a real thing. Nietzsche also thought that sometimes we know explicitly we're dreaming—in other words, that sometimes we have lucid dreams:

> A person with artistic sensibility relates to the reality of dream in the same way as a philosopher relates to the reality of existence: he attends to it closely and with pleasure, using these images to interpret life, and practising for life with the help of these events. Not that it is only the pleasant and friendly images which give him this feeling of complete intelligibility; he also sees passing before him things which are grave, gloomy, sad, dark, sudden blocks, teasings of chance, anxious expectations, in short the entire "Divine Comedy" of life, including the Inferno, but not like some mere shadow-play— for he, too, lives in these scenes and shares in the suffering—*and yet never without that fleeting sense of its character as semblance. Perhaps others will recall, as I do, shouting out, sometimes successfully, words of encouragement in the midst of the perils and terrors of a dream: "It is a dream! I will dream on!"* [my italics] I have even heard of people who were capable of continuing the causality of one and the same dream through three and more successive nights. All of these facts are clear evidence that our innermost being, the deep ground common to all our lives, experiences the state of dreaming with profound pleasure and joyous necessity.[11]

When we dream, we're caught up in the images our minds create, yet we also sense that the dream is semblance, that what it offers are appearances. Sometimes we know explicitly that we're dreaming. With this knowledge we can let the dream proceed, releasing ourselves to it while watching it unfold. These facts tell us that dreaming belongs to our nature and is a necessary and deeply pleasurable part of human life.

For Freud, however, the thought, "It's a dream! I will dream on!" can't be taken at face value. It belongs to the manifest content of the dream, and so must be seen as an upshot of the dream work, in particular a result of how the dream work transforms and disguises the latent dream thoughts. When something troubling in the latent dream thoughts manages to break through into consciousness—into the dream's manifest content, thereby escaping "censorship"—the dream work inserts the thought, "It's only a dream," so the sleeper can tolerate what follows. In this way, Freud interprets the thought as a "contemptuous critical judgement" or "criticism of the dream."[12]

But this way of looking at things is one-sided. The thoughts, "It's only a dream," "I'm dreaming," and "This is a dream," need not express criticism of the dream (for presenting something unreal); nor need they be contemptuous. On the contrary, as Nietzsche saw, they can express appreciation or admiration of the dream state for its own sake as a deeply pleasurable and meaningful experience: "If we imagine the dreamer calling out to himself in the midst of the illusory dream world, 'It is a dream, I will dream on' . . . this compels us to conclude that he is deriving intense inward pleasure from looking at the dream."[13] Moreover, as psychological studies of lucid dreaming show, the thought, "This is a dream," often expresses a joyous affirmation of the dream arising precisely from the awareness that one is dreaming.

Freud acknowledges this kind of dream awareness, but only as an afterthought, and he says nothing interesting about it. He notes, "There are some people who are quite clearly aware during the night that they are asleep and dreaming and who thus seem to possess the faculty of consciously directing their dreams."[14] He mentions the French dream research pioneer and scholar of Chinese, the Marquis d'Hervey de Saint-Denys (1823–92), whose book, *Dreams and How to Guide Them,* records many lucid dreams and offers instructions on how to become conscious of the dream state.[15] But all Freud goes on to say is that, "It seems as though in his case the wish to sleep had given place to another preconscious wish, namely to observe his dreams and enjoy them. Sleep is just as compatible with a wish of this sort as it is with a mental reservation to wake up if some particular condition is fulfilled."[16] In short, Freud sees lucid dreaming as just another way the dream is the fulfillment of a wish.

But the problem for Freud's view isn't that lucid dreaming contradicts his thesis that dreaming is wish fulfillment. On the contrary, the wish either to observe or to consciously guide the dream makes lucid dreaming a more obvious case of wish fulfillment than many ordinary dreams are. Rather, the problem is that the wish isn't hidden and disguised; it's consciously available to the dreamer and readily accessible at the level of the manifest dream. For this reason, there's no basis for viewing a lucid dream as composed through mechanisms of disguise and censorship, or more generally for imposing onto a lucid dream Freud's distinction between manifest content and latent content.

At the same time, Freud discounts the manifest content of a dream, and this creates another problem. What's important for Freud is not how you experience the manifest dream—in particular, whether you experience it as a dream or not—but what the manifest content symbolizes or means in relation to the latent dream thoughts. In this way, Freud devalues the manifest dream experience (a point that contributed to his break with Jung, who refused to discount the manifest dream, and who rejected Freud's view that it presents a disguised and distorted version of the unconscious dream thoughts).

This devaluing of dream experience leads Freud to ignore, rather than examine carefully, any of the distinctive features that make lucid dreaming a qualitatively different state from nonlucid dreaming. As we've seen, these include greater clarity or vividness of the dream experience, emotional exhilaration and a sense of freedom, and the defining feature of a lucid dream—being able to think about and direct your attention to the dreamlike quality of your state. By neglecting these features—already detailed by de Saint-Denys and van Eeden—Freud missed the opportunity to investigate lucid dreaming as a unique state, both on its own terms and from a psychoanalytical perspective. This neglect seems to have carried over into psychoanalysis in general, which to date has paid little attention to lucid dreaming.

KNOWING YOU'RE DREAMING OR DREAMING YOU'RE DREAMING?

Why isn't a lucid dream just a dream within a dream? After all, when I'm flying and think, "I'm dreaming," I'm in a dream state, so why isn't this just my dreaming that I'm dreaming? When Dorion dreamt the bus was gone and realized he was dreaming, why wasn't he having a dream that he was having a dream? If there's a difference between knowing you're dreaming and dreaming you're dreaming, then what exactly is it?

Thirty years ago the philosopher Daniel Dennett posed the following challenge:

> If someone gives a waking report of the lucid dream variety, there will be two hypotheses consistent with the report; the subject had, as she believes she had, a lucid dream; or the subject had an ordinary dream in which she was aware she was dreaming, decided to fly in her dream, etc. She wasn't really aware she was dreaming, of course; she just dreamt she was aware she was dreaming. So subjective testimony . . . cannot establish that lucid dreams are anything other than a variety of ordinary dreams, viz., dreams of having lucid dreams.[17]

Today two things can be said to meet this challenge. First, knowing you're dreaming and dreaming you're dreaming don't feel the same and seem different to memory when you wake up. The reason, as we'll see, is that knowing you're dreaming involves a certain kind of attention that's missing when you dream you're dreaming. Second, we now have more evidence for lucid dreaming than just what people tell us when they wake up; we also have physiological and brain-imaging verifications of the lucid dream state.

To see how lucid dreaming feels different from dreaming you're dreaming, let's go back to the hypnagogic state. The great lucid dream pioneer, the Marquis d'Hervey de Saint-Denys, claimed he could observe, without any break in consciousness, how hypnagogic imagery turns into full-blown dreams. Here's one of his dream reports:

> I close my eyes to go to sleep, thinking of some objects I noticed this evening in a shop in the rue de Rivoli. I remember the arcades of the street in question, and I catch the glimpse of something like luminous arcades forming repeatedly in the distance. Soon a serpent covered with phosphorescent scales appears before my mind's eye, surrounded

by innumerable ill-defined images. I am still in the period when things are confused. The images fade and re-form very quickly. The long fiery serpent has turned into a long dusty road burning under the summer sun. I immediately see myself travelling along it, and my memories of Spain are revived. I talk to a muleteer carrying a *manta* (blanket) on his shoulder; I hear the bells of the mules; I listen as he tells me a story. The countryside matches the central figure; at this moment the transition from waking to sleep is complete. I am completely taken up in the illusion of a clearly defined dream. I was offering the muleteer a knife, which he seemed to like, in exchange for a fine antique medallion he showed me when I was suddenly brought out of my sleep by an external cause. I had been asleep for some ten minutes, as far as the person who woke me could tell.[18]

Here we see the transition from looking at hypnagogic imagery to being immersed in the dream world, with the third-person or outside view of the dream ego as an intermediate state between the two.

The hypnagogic-imagery technique for bringing about a lucid dream works with this kind of transition by taking advantage of the fact that waking awareness continues for a while in the hypnagogic state. The technique aims to lead you from the hypnagogic state into the dream state without losing the witness awareness of your own consciousness. Adopting the perspective of a detached observer, you watch the images arise, transform, and subside, without trying to control or hold onto them. In the words of lucid dream researcher Stephen LaBerge:

> When the imagery becomes a moving, vivid scenario, you should allow yourself to be passively drawn into the dream world. Do not try to actively enter the dream scene, but instead continue to take a detached interest in the imagery. Let your involvement with what is happening draw you into the dream. But be careful of too much involvement and too little attention. Don't forget that you are dreaming now![19]

The experience of falling asleep is a human birthright, so it's not surprising that many individuals from different times and cultures have hit upon this method of carrying awareness into the dream state. Besides de Saint-Denys and LaBerge, Russian philosopher P. D. Ouspensky (1848–1947), American psychiatrist Nathan Rapport, and German lucid dream researcher Paul Tholey used this technique for lucid dreaming.[20] But an older and more practiced form belongs to Tibetan Buddhist dream yoga, as described in a seventeenth-century Tibetan text:

> Before you fall into a deep sleep, there are so-called thoughts between falling asleep and dreaming. Before you actually fall asleep and you are still in the process of falling asleep, thoughts arise and sounds are faintly heard. You have a sense of the body's becoming very torpid and a sense of becoming pressed into darkness. You also have a sense of the experience of deep breathing as you begin to relax. Right after that, there is a sensation of numbness at the point midway between the eyes. At that time, you will begin to feel vague impressions of people, animals, environments, or whatever your recent mental impressions are. These vague mental impressions are the cause for the dream. The dream you will have actually arises as the result of those impressions. If you recognize this, it is your chance to recognize the dream, like threading a needle right through the eye, and you will immediately enter the dream and apprehend it.[21]

What makes this technique difficult is the spellbound quality of attention in the hypnagogic state. Too much absorption and you're no longer able to direct your attention to the dreamlike quality of the images; your consciousness becomes captive, so it's easy to fall into a state where you wind up dreaming you're dreaming.

Here's the crucial point: *This state of dreaming you're dreaming lacks the mental clarity of being able to observe and direct your attention to your dreaming.* As Andreas Mavromatis notes in his book *Hypnagogia*: "The experience of being conscious in one's dream has a very different status from that of dreaming that one is conscious in one's dream."[22] The difference has to do precisely with attention and meta-awareness: *to be conscious of your dream requires that the dreamlike quality of the state be directly available to your attention.* Only if you can observe or witness the dream state this way can you know clearly that you're dreaming.

Consider this classic lucid dream report from Oliver Fox, the pen name of Hugh Calloway, an early twentieth-century British researcher of dreams and out-of-body experiences:

> I was standing on the pavement outside my London home . . . I was about to enter the house when, on glancing casually at these [pavement] stones, my attention became riveted by a passing strange phenomenon, so extraordinary that I could not believe my eyes—they had seemingly all changed their position in the night, and the long sides were parallel to the curb!
>
> Then the solution flashed upon me: though this glorious summer morning seemed as real as real could be, I was dreaming! With the realization of this fact, the quality of the dream changed in a manner very difficult to convey to one who has not had this experience. Instantly, the vividness of life increased a hundred-fold. Never had sea and sky and trees shone with such glamourous beauty; even the commonplace houses seemed alive and mystically beautiful. Never had I felt so absolutely well, so clear-brained and inexpressibly "free"! The sensation was exquisite beyond words; but it lasted only a few minutes and I awoke.[23]

This report illustrates many of the special features of strong lucid dreams (also seen in Dorion's dream). The dreamer remembers something from waking life—that the paving stones lie perpendicular, not parallel, to the curb. Noticing the discrepancy, he correctly realizes he's dreaming. The realization is immediate, not abstract; the dreamer doesn't merely think, "I'm dreaming," but directly experiences his state as a dream. The onset of lucidity increases the state's sensory vividness without diminishing the clarity of the state awareness (the clear comprehension of being in the dream state). Finally, the dreamer doesn't suffer from the cognitive deficiencies typical of nonlucid dreams (disorientation, poor working memory, inability to direct and sustain attention) but has insight into the nature of his ongoing conscious state. In these ways, the dreamer now experiences himself as a dreaming subject.

As mentioned earlier, and as in Dorion's lucid dream, experiencing yourself as a dreaming subject doesn't necessarily mean you're able to control the dream in the sense of actively and deliberately guiding the dream content. Such control comes in degrees and may be more or less present in any given lucid dream. Of course, being able to direct your attention to the dreamlike quality of your state is a kind of cognitive control, and it does affect the dream content because what you dream depends on how you direct your attention in the dream. But one of the things that makes lucid dreaming so compelling is precisely the inability to control many dream features while recognizing that you're dreaming them. In this way, you're able to see—far more vividly than in waking life, unless you're a vivid daydreamer—how your imagination lives beyond your ego's willful control.

Jorge Luis Borges describes this quality of lucid dreaming in his parable "Dreamtigers."[24] When he was a child, Borges writes, he had a fascination with tigers. Although the fascination faded as he grew up and he eventually became blind, he would still see tigers in his dreams. Sometimes, when he knew he was dreaming, he would try "to dream up a tiger" as an exercise of the unlimited powers of his will. A tiger would appear, yet never the one he wanted; instead it would be an "enfeebled tiger," or a "stuffed tiger," or a tiger "looking something like a dog or a bird."

"Dreamtigers" depicts how being able to direct your attention to the dream as a dream often goes along with being unable to control the exact dream content. The former ability is the crucial feature of a lucid dream. As philosophers Jennifer Windt and Thomas Metzinger explain, the fact that you're dreaming is not only cognitively available—you're able to think about the fact that you're currently dreaming—it's also attentionally available—you can attend directly to the dreamlike character of your experience.[25] In the case of a strong lucid dream, you also know you're lying in bed and can recall events from waking life, sometimes even previous dreams. These features make lucid dreaming feel very different from dreaming you're dreaming.

Critical readers will rightly point out that these observations still aren't enough to deflect Dennett's challenge. For one thing, the observations are based on subjective testimony alone. For another, they're based on waking recall, and so strictly speaking count as reports of the waking memories of lucid dreams.

But what if the lucid dreamer could signal to the outside world that she's having a lucid dream? What if she could signal the presence of certain features or qualities of her lucid dream while she's dreaming? And what if these signals could be correlated with behavioral and physiological measurements in the sleep lab?

SIGNALS FROM THE DREAM WORLD

One of the striking scientific findings about REM sleep—the finding that made possible the physiological verification of lucid dreaming—is that eye movements of the dream ego often correspond to physical eye movements of the dreamer.[26] The direction of your eye gaze within the dream often matches the direction of your eye movements during REM sleep: when your dream ego looks left, your physical eyes move left; when your dream ego looks right, your eyes move right. This indicates there's a close relation between your dream ego and your living brain and body.

In the late 1970s and early 1980s, two dream researchers—Keith Hearne in England and Stephen LaBerge in the United States—independently took advantage of this correlation in order to verify lucid dreaming in the sleep lab.[27] They reasoned that if lucid dreamers can act voluntarily in the dream state, then they should be able to indicate when they become lucid by making prearranged eye movements within the dream that would be externally observable on a polygraph. Lucid dreamers were instructed to look right and left a specific number of times, and their eye movements were recorded. When they woke up, the subjects reported how many deliberate eye-movement signals they had made, and the number matched the number of recorded eye movements on the polygraph.

These ingenious experiments were momentous. They showed that the subjects were genuinely asleep when they signaled their lucid dreams; they enabled subjects to report on their dreams while the dreams were happening, instead of just retrospectively; they effected a kind of transworld communication from the private dream world to the public waking world; and they opened up a new way to investigate consciousness whose prospects we've barely begun to tap.

Before these experiments, many scientists believed that lucid dreams were most likely experiences that happened during brief moments of awakening but were then later misremembered as having occurred during sleep. When the subjects signaled the onset of a lucid dream, however, the EEG showed brain waves indicating REM sleep. So, according to standard sleep-science criteria, the subjects were unequivocally asleep. This kind of lucid dream in the sleep lab is now known as a "signal-verified lucid dream."

Further analysis by LaBerge and his colleagues showed that lucid dreaming typically happens during the most active period of REM sleep, called "phasic" REM sleep.[28] During both wakefulness and REM sleep, neurons fire at a high rate in many areas of the brain. In REM sleep this high level of activity can be broken down into two components—"tonic" or continuous background activity and "phasic" or transient activity. Phasic activity consists of short, periodic bursts of firing, which coincide with rapid eye movements, muscle twitches, and surges of activity in the middle ear muscles (middle ear muscle activity, or MIMA, being the ear equivalent of REM). In short, lucidity seems to occur when there are brief and rapid increases to the already high level of cortical activity in REM sleep.

Having verified lucid dreaming from the sleep-science perspective, these researchers could take the further step of asking people to report on features of their lucid dreams while they were dreaming. In this way, sleep science could go beyond relying only on the dream reports subjects make after they wake up.

Taking exactly this approach, LaBerge investigated time in lucid dreams.[29] Do dream actions take the same amount of time as waking actions, or do they take more or less time (as the movie *Inception* suggests)? Lucid dreamers estimated ten-second intervals by counting "one-thousand one," "one-thousand two," and so on, and marked the beginning and end of the intervals with eye-movement signals.

In all cases, the time it took to count from one to ten during the lucid dreams corresponded closely to the time it took during wakefulness. Recently, Daniel Erlacher and Michael Schredl in Germany replicated these results.[30] They also found, however, that performing squats (deep knee bends) took roughly 40 percent more time in lucid dreams than in the waking state. Erlacher and Schredl speculate that longer tasks or motor tasks (such as squatting compared to counting) could require disproportionately more time in lucid dreams, but it's not clear what accounts for this result.

In another recent study by Martin Dresler and his colleagues at the Max Planck Institute for Psychiatry in Munich, two experienced lucid dreamers were able to carry out the predefined motor task of clenching their dream hands after making eye movements to signal that they were in a lucid dream state.[31] These scientists compared the brain activity during lucid dream performance of this task with the brain activity when the same individuals clenched their real hands while awake. In both the lucid dream state and the waking state, their brains showed activations in the same sensorimotor regions, though the activations were weaker in the lucid dream state. This finding supports the idea that brain areas active when we carry out a motor action are also active when we imagine carrying out that action.

These studies rely on what I like to think of as a kind of transworld communication. A link is set up between the private dream world and the public waking world. The sender writes messages in the phenomenal medium of dream ego eye gazes; the receiver reads the messages in the physical medium of the sleeper's eye movements. An even more amazing method uses Morse code. When the lucid dreamer clenches his left and right dream ego fists, the polygraph picks up a corresponding sequence of left and right forearm twitches. LaBerge (himself a lucid dreamer) sent a Morse code signal spelling the initials of his name, with left dream fist clenches corresponding to dots and right dream fist clenches corresponding to dashes.[32]

Can the communication go the other way, from the dream scientist to the dreamer? Can there be two-way communication between the waking world and the dream world?

The answer seems to be yes. In another study, lucid dreamers were stimulated with repeated low tones interrupted by the occasional high tone.[33] The subjects were asked to listen for the high tone and to signal hearing it with a single left-right eye movement. Three lucid dreamers were able to carry out this task, and the EEG showed a waveform known to indicate detection of this kind of target stimulus. In the words of a recent article, "This is but one example of a simple way of communication between the 'real world setting' (research in the sleep lab) and the lucid dreamer in the dream world. Future exploration of two-way communication between the dreamed and waking world should provide very promising research opportunities."[34]

THE LUCID BRAIN

What happens in the brain during a lucid dream? Since lucid dreamers can signal when they become lucid and scientists can monitor what's going on in the brain at the same time, this question is now testable.

In a recent EEG study, German sleep scientist Ursula Voss and Harvard University psychiatrist and dream scientist Allan Hobson, compared brain activity during three states—waking, signal-verified lucid dreaming in REM sleep, and nonlucid REM sleep.[35] They found that lucid dreaming resembles waking in showing increased activity in the gamma frequency range (36–45 Hz) in frontal and frontolateral areas, compared to the reduced activity in nonlucid REM sleep. They also found that the EEG "coherence"—a measure of the degree to which the brain waves at various frequencies are synchronized across areas—is much higher during lucid dreaming than during nonlucid REM sleep. As we saw in chapters 2 and 3, this kind of "neural synchrony" is thought to be a way that the brain coordinates or integrates processes across multiple areas. In short, this study shows that during lucid dreaming the brain adopts a novel, coherent pattern of activity marked by features characteristic of alert waking consciousness, but without leaving REM sleep.

What's the significance of the frontal and frontolateral regions where these electrical patterns predominate? As mentioned at the end of the last chapter, an area at the front of the brain called the dorsolateral prefrontal cortex, which is crucial for working memory, volition, and reflective self-awareness, shows greatly reduced activity during REM sleep. We might hypothesize, therefore, that activity in the dorsolateral prefrontal cortex must increase in order for dream lucidity to occur. Hobson has long advocated this hypothesis.[36] But although the increased gamma activity and coherence in frontal regions during lucid dreaming are consistent with this hypothesis, using EEG to measure electrical activity at the scalp can't tell us whether the dorsolateral prefrontal cortex is the underlying source of these brain wave patterns.

More direct evidence comes from another recent study that used fMRI (functional magnetic resonance imaging), a different kind of neuroimaging method that can localize activity inside the brain. In this study, conducted by Martin Dresler and Michael Czisch at the Max Planck Institute for Psychiatry in Munich, four lucid dreamers slept in an MRI scanner and were told to signal continuously their state of lucidity using quick left-right-left-right eye movements.[37] When the scientists compared lucid dreaming to nonlucid REM sleep in one subject, they found increased activity in a range of regions belonging to the neocortex (the exclusively mammalian part of the cortex), including the dorsolateral prefrontal cortex, during periods of lucidity. These activated regions are larger and more developed in human beings than in other primates. They include frontal and parietal areas that are thought to form an intentional control system for cognition and decision making. This "frontoparietal control system" links brain networks that support the ability to pay attention to the outside world with other networks supporting long-term memory and the ability to think about oneself—to remember one's personal past and project oneself into an anticipated future.[38] Thus lucid dreaming reactivates not just the dorsolateral prefrontal cortex but also a wide and distributed network of brain areas crucial for attention and our human sense of self and personal identity through time.

This finding is significant in two ways. First, it supports my proposal that lucid dreaming reframes experience from the perspective of the dreaming self rather than the dream ego, that is, from the perspective of the dreamer who can remember waking life and think about and direct attention to the dreamlike quality of experience. Second, it indicates that there's no "lucid dream spot" in the brain, no one brain area where lucid dreaming happens. Rather, lucid dreaming—like all modes of consciousness—reflects large-scale and distributed patterns of brain activity.

DISSOCIATION OR INTEGRATION?

Is lucid dreaming a "dissociative" state? In other words, do mental processes that are normally integrated split apart and function separately or in some unusual combination? Or is lucid dreaming an "integrative" state? Does it effect some kind of integration or reintegration of mental functioning?

Ever since Frederik van Eeden introduced the term "lucid dream" in 1913, these questions have provoked debate. One of van Eeden's contemporaries, Havelock Ellis (1859–1939), an English physician and psychologist who pioneered the scientific study of sex, proposed in his 1911 book, *The World of Dreams*, that dreaming rests on dissociation because the usually coherent elements of waking mental life—memories, thoughts, perceptions, and emotions—are split up and reconstituted in the dream. Van Eeden agreed with Ellis that "during sleep the psychical functions enter into a condition of *dissociation*," but proposed that "it is not dissociation, but, on the contrary, *reintegration*, after the dissociation of sleep, that is the essential feature of dreams." In strong lucid dreams, he wrote, "the reintegration of the psychic functions is so complete that the sleeper remembers day-life and his own condition, reaches a state of perfect awareness, and is able to direct his attention, and to attempt different acts of free volition. Yet the sleep, as I am able to confidently state, is undisturbed, deep and refreshing."[39]

Allan Hobson and Ursula Voss have rekindled this debate by proposing (without taking into consideration van Eeden's view) that lucid dreaming is a "dissociated state."[40] Hobson thinks lucid dreaming is "paradoxical, even at a subjective level, in containing elements of both waking and dreaming consciousness." His way of dealing with the apparent paradox is to view lucid dreaming as "an example of dissociation" where certain elements of waking consciousness detach from the waking state and combine with dreaming. He believes the EEG results show that during lucid dreaming the brain enters a "hybrid state" that combines the posterior brain activity of dreaming with the frontal activity of waking. He and Voss also suggest that lucid dreaming is not a REM sleep phenomenon; instead, "lucidity occurs in a state with features of both REM sleep and waking. In order to move from non-lucid REM sleep to lucid REM sleep dreaming, there must be a shift in brain activity in the direction of waking."

I think this view of lucid dreaming goes astray in several ways—it misunderstands lucid dream experience, overinterprets the EEG findings, and works with a limited conception of dissociation.

Lucid dreaming seems paradoxical only if we assume both that dreaming and awareness of dreaming are generally incompatible, and that the only state we're normally aware of being in is the waking state. But both assumptions seem wrong.

On the one hand, Aristotle already recognized that being asleep often involves awareness of being asleep and dreaming: "Often, when one is asleep, there is something in consciousness which declares that what then presents itself is but a dream."[41] Moreover, Nietzsche and Freud, as we've seen, thought that there's a sense in which we always know we're dreaming when we dream. Finally, both spontaneous lucid dreaming and the fact that you can train yourself to have lucid dreams show clearly that dreaming and awareness of dreaming are perfectly compatible.

On the other hand, how often are we explicitly aware of being awake when we're in the waking state? How often do we direct attention to our waking consciousness in everyday life? I suspect not much, only when something forces us to reflect on being awake or when we practice mindfulness meditation and carefully observe moment-to-moment experience.

In short, dreaming and awareness of dreaming aren't generally incompatible, and normally we seem to pay little attention to the waking state itself.

Instead of trying to analyze lucid dreaming in terms of waking consciousness versus dreaming consciousness, we need to go back to the threefold framework of the witnessing aspect of awareness, the changeable contents of awareness, and ways of experiencing particular contents of awareness as the self. What marks a strong lucid dream is the felt presence of witnessing awareness, which can observe or witness the dream precisely as a dream. Its presence can inhibit the automatic identification with the dream ego that characterizes dreaming. The same witnessing awareness can be felt in the waking state in moments of heightened mindfulness; its presence can inhibit the automatic identification with the "I-Me-Mine" that characterizes the waking state. When we look at waking and dreaming consciousness from this vantage point, there's nothing paradoxical about lucid dreaming, even at an experiential level.

Hobson and Voss propose that lucid dreaming is a hybrid state, not an REM-sleep state, but as Stephen LaBerge points out, signal-verified lucid dreaming always occurs during REM sleep (as operationally defined by rapid eye movements and fast-frequency/low-amplitude brain waves).[42] So the EEG brain waves in Voss's study shouldn't be categorized as "REM sleep" versus "lucid dreaming," but instead as "nonlucid REM sleep" versus "lucid REM sleep." It follows that the more parsimonious interpretation of the EEG results isn't that lucid dreaming is a hybrid or dissociative combination of waking and dreaming, but rather that REM sleep, under the right conditions, can support the witnessing awareness of dreaming.

Finally, the concept of dissociation in this discussion is too limited. "Dissociation" sounds negative because the word usually means the pathological disruption of normal mental functioning; in this sense, "dissociative" is opposed to "integrated" and "coherent." But some kinds of dissociation are positive and normal, not negative and pathological, and dissociation and integration dynamically interrelate rather than being statically opposed.

Take memory, for example. When I remember myself from the outside as a kid on the ferry in Scotland, I see myself as an other from a third-person perspective and thereby distance myself from my past experience (I was scared while being tossed about in the little boat). At the same time, I represent that past self as me and appropriate its experience to my life story, my own sense of personal history. I can also try to relive the past experience from within, though always as a new and reconstructive memory in the present. In these ways, any memory of personally experienced past events involves both dissociation and (re)integration. As New York University psychologist and psychoanalyst Philip M. Bromberg, writes:

Self-states [states of self-experience] are what the mind comprises. Dissociation is what the mind does. The relationship between self-states and dissociation is what the mind is. It is the stability of that relationship that enables a person to experience continuity as "I." A flexible relationship among self-states through the use of ordinary dissociation is what allows a human being to engage the ever-shifting requirements of life's complexities with creativity and spontaneity.[43]

As Bromberg goes on to note, with reference to lucid dreaming, "our 'dreaming' reality is simply a different state of consciousness and . . . its taking place during sleep does not make it more unbridgeable to waking reality than *any* two dissociated self-states are to each other. That is to say that it is possible to be awake and dreaming at the same time."[44] Being awake and dreaming at the same time is what happens during lucid dreaming, and what bridges waking and dreaming is witnessing awareness, whose presence can be felt in both states.

From this vantage point, lucid dreaming is both dissociative and integrative—it's a state in which dissociation and integration relate dynamically as two aspects of self-experience. We thus come back to something like van Eeden's original view that in a lucid dream the mind reintegrates during the ordinary dissociation of sleep.

DREAM QUESTIONS

From a scientific perspective, lucid dreaming offers a kind of distilled consciousness, a way to examine consciousness in a preparation unmixed with current sensory input. This purified or concentrated form of consciousness has special features that raise new questions.

One unresolved question is what makes strong lucid dreams so vivid. During phasic REM sleep, when lucid dreaming seems most likely to occur, attention turns away from the senses and dream imagery derives from memory and imagination. Yet lucid dreamers often report that their dreams contain a high degree of vividness and clarity, sometimes even surpassing that of waking life. Recall Dorion Sagan's lucid dream: "I saw the stars, manifold and white against the black depths of space. The vision was exquisite. Still conscious that this was a dream, I marveled at the realism of the night sky. Now I saw—at the same time—beautiful white clouds roll across a blue sky—again with striking realism." Or Oliver Fox's: "Instantly, the vividness of life increased a hundred-fold. Never had sea and sky and trees shone with such glamourous beauty; even the commonplace houses seemed alive and mystically beautiful." Here's one more example, from a dream report Stephen LaBerge collected:

> I was standing in a field in an open area when my wife pointed in the direction of the sunset. I looked at it and thought, "How odd; I've never see colors like that before." Then it dawned on me: "I must be dreaming!" Never had I experienced such clarity and perception—the colors were so beautiful and the sense of freedom so exhilarating that I started racing through this beautiful golden wheat field waving my hands in the air and yelling at the top of my voice, "I'm dreaming! I'm

dreaming!" Suddenly, I started to lose the dream; it must have been the excitement. I instantly woke up.[45]

How much of this vividness is sensory—a sharpness of the image—and how much of it is cognitive—a clear and collected comprehension of the imagery as dream imagery?

One way to address this question would be to have lucid dreamers use the eye-signaling method to communicate their subjective ratings of dream vividness. With this kind of information, we could also investigate scientifically the following related issues: Are most lucid dreams especially vivid? Do spontaneous lucid dreamers have dreams with greater vividness and clarity than those of individuals who learn to lucid dream? Do these qualities increase over time, depending on the training method for inducing lucid dreams? Can we find dynamic brain patterns corresponding to the variations in vividness and clarity throughout the dream state? What are the psychological and neurophysiological sources of the apparent vividness and clarity of lucid dreams?

We could also investigate something impossible to study with nonlucid dreams—the relation between waking memories of dreams and dream experiences as they happen. For example, some lucid dreamers report that while they were dreaming they took the dream state to be as realistic as waking perception, yet once they awoke they changed their minds—they decided that the dream state only seemed to be so realistic but was not really.[46] Both assessments, however, are made retrospectively from the waking state. Subjective ratings made during the lucid dream state would provide real-time information about dream experiences and cast light on their relationship to waking memories.

For investigating all these questions and probing the depths of lucid dream consciousness, individuals with training in meditative lucid dreaming will be invaluable. Meditative lucid dreaming or "dream yoga" brings contemplative insight into the dream state. For centuries Tibetan Buddhists have cultivated dream yoga as a practice of mental transformation.[47] Combining their ancient practices with modern methods of lucid dream communication, we can envision a new kind of dream science that integrates dream psychology, neuro-science, and dream yoga.[48]

But dream yoga doesn't just offer new tools for dream science. It strikes deeper by challenging the assumption that reality is indepen-dent of the mind. Dream yoga asks us to view waking experience as a dream while also teaching us how to wake up within the dream state. In this way, as we'll see in the next chapter, dream yoga tries to show us how the waking world isn't outside and separate from our minds; it's brought forth and enacted through our imaginative perception of it.

6

IMAGINING

Are We Real?

When I was four years old, my parents took me to Disneyland. Underwater in Captain Nemo's submarine, I looked out through the portals. Monsters of the deep were everywhere. A shark lunged at the window; I was terrified. My mother tried to reassure me, saying, "It's all right, Evan, it isn't real." Then a giant squid attacked us; my father said it also wasn't real. At this point, my parents tell me, I looked up and asked, "Are we real?" My father says he smiled and answered, "Well, there are some mystics who would say, 'Not really.'"

I don't remember what I thought about this answer. But I do remember when I was a kid being fascinated by the idea that life might be a dream.

What if waking experience were a kind of dream? If you woke up, who would you be? What exactly is a dream? Is it a kind of hallucination? Or is it a form of imagination? If life were a dream, would we be hallucinating the waking world or imaginatively perceiving it?

DREAM YOGA

Apart from the power of mental imprints, (phenomena) do
not exist.
All avenues of appearances, negative and affirmative,

Are dreamlike, though they are apprehended as external phenomena.
Without distraction, earnestly and continually sustain your
    mindfulness and attention (to this truth).
—Lochen Dharma Shri (1654–1718)[1]

These lines come from the beginning of a well-known Tibetan Buddhist dream yoga manual. They describe a practice called "sustaining mindfulness without distraction during the daytime experience," which is the first stage of the training for the lucid apprehension of dreams. The heart of the practice is to remind yourself again and again that waking experience is a dream. In this way, you create a habit that will carry over into the dream state.

This is the opposite of the Western lucid dream induction technique known as "reality testing."[2] Here you create a habit of testing things to see whether they're real or you're dreaming. One well-known test, depicted in Richard Linklater's 2001 movie *Waking Life*, is to turn a light switch off and on; in a dream, flipping a switch doesn't seem to change the light.[3] Another test is to find some text and see whether the writing stays the same when you look away and then look back again. If you're dreaming, the writing most likely will have changed.

The premise of the Western approach is that waking reality is stable and mind-independent, but the dream state is unstable and suggestible. By getting in the habit of doing "state checks" when you're awake, you'll eventually remember to do them when you're dreaming. In this way, you can confirm that you're dreaming and what you're experiencing isn't real.

Dream yoga rests on the opposite premise—that the waking state is ultimately no more real or unreal than the dream state.[4] According to the Buddhist philosophical schools underlying dream yoga—the Madhyamaka or middle-way school and the Yogācāra or yoga-practice school—although distinguishing between waking and dreaming is valid on an everyday or conventional level, from the standpoint of an ultimate truth-seeking analysis, waking phenomena lack substantiality and independent existence, and in this way are dreamlike. Thus, instead of doing reality testing and state checks, dream yoga encourages you to view all waking events as dreamlike.

When I was in Dharamsala for the Mind and Life Institute meeting on neuroplasticity, I had the opportunity to talk about dream yoga with Alan Wallace, a Buddhist scholar and meditation teacher, who translated Lochen Dharma Shri's dream yoga manual. Alan teaches dream yoga and has led lucid dream retreats with Stephen LaBerge. At our Mind and Life meeting, he was acting as a Buddhist discussant in dialogue with the scientists, and also as a translator for the Dalai Lama.

On the last day of our meeting, I told Alan about the lucid dream I'd had on our arrival night in Dharamsala and asked him about how to meditate in the dream state. As we sat outside on the terrace of Chonor House overlooking Namgyal Monastery, with the Himalayas to our backs and monkeys in the trees above us, Alan gave me a crash course in his approach to combining Tibetan Buddhist dream yoga with modern lucid dreaming techniques.[5]

The first step, he said, is to *recognize the dream state*—in other words, to learn to have lucid dreams. Here the techniques developed by lucid dream researchers, such as LaBerge, are probably more helpful than traditional Tibetan methods (which involve complex visualizations and sleeping postures), at least for modern Westerners.

Once you've ascertained you're dreaming, you need to sustain that lucidity without waking up. Modern research indicates that the best way to do this is to get really involved in the dreamscape. Keep the dream going by filling your awareness with its content. Fly or whirl around. Don't try to meditate, as I did, by being mindful of your breathing. That will reduce the dream content and dissipate the dream. And don't sit still, because stillness will dissolve the dream. Keep moving.

After you can sustain lucidity for a while, the second step is to *transform the dream state*. Use your imagination to manipulate the dream. Be playful. Change things and transform them. Exchange people for animals and animals for people; change day into night and night into day; fly, go through walls, and so on. Explore the plasticity of the dream. In this way, the mind's supple nature will manifest, and you'll gain a deeper understanding of the dreamscape as a mental construct, a product of imagination.

If nightmares happen, take them as further opportunities to transform the dream. Change something threatening and fearful—someone chasing you—into something peaceful and soothing—a trusted friend. Even better, calmly allow the nightmare to unfold and accept it by entering into it. Remember that you aren't your dream body, so the dream can't harm you. In this way, you'll learn to see the dream as illusory in nature, and you'll start to see through the dream state to what underlies it.

If your lucidity is stable and you've gotten good at transforming the dream, then the third step is to *see through the dream state*. Transforming the dream prepares the way for seeing through it by breaking down the objectification of the dream—the impulse to treat it as something independent of your mind—that dominates the ordinary dreaming consciousness.

In transforming the dream, you can sometimes gain a glimpse of the awareness that lies behind any dream content, a kind of ground state of consciousness, which the Indian Yogācāra and Tibetan Dzogchen schools of Buddhism call the "substrate consciousness" or "store consciousness."[6] But to really see through the dream, instead of transforming the dreamscape, try to dissolve it completely. Release the dream and simply be aware of being aware. To be aware this way, with no dream imagery, is to experience lucidly the state of deep and dreamless sleep. And here the radiance of consciousness in its ground state can more easily manifest.

JOURNAL ENTRY

The night after my conversation with Alan, our last night in Dharamsala, I had another lucid dream.

I'm flying indoors, down a long hallway, at the midpoint between the floor and the ceiling. I become aware that I'm dreaming, and I remember Alan's advice to keep moving. So I start to fly by swimming through the air, using a breaststroke and scissor kick. He said to play with the dream—for example, by going through walls. Many people apparently find it difficult to go through face forward, so you can try to back yourself through instead. Along each wall is a row of mirrors, in which I can see my own reflection as I move. I turn toward the wall on my left and see my face smiling back. I decide I'm going to try to fly through the wall face first. I experience no difficulty and go straight through my reflection in the mirror. But once through, I find myself in an empty space of complete blackness with no visual dream body. I become uneasy and can't stop myself from waking up.

MIND THE DREAM

What significance do dream yoga and these sorts of lucid dream experiences have for contemporary neuroscience conceptions of dreaming?

Dream yoga shows that it's one-sided to think about dreaming as nothing but a way the brain activates itself in sleep. This conception gives primacy to the brain side of dreaming and ignores or underestimates the mind side. When I remembered Alan's advice to keep moving and to play with the dream, and when I decided to go through the wall face first instead of backing through it, these mental events in my stream of consciousness affected the course of my dream. I have no doubt that if neuroscientists knew enough about the dreaming brain and had more precise neuroimaging tools, they'd be able to track the brain events corresponding to this kind of intentional guiding of the dream. But when they say, "The dreaming mind reflects the sleeping brain," they leave out half the story, for the opposite is equally true—the sleeping brain reflects the dreaming mind. Lucid dreaming already serves to make this point, and dream yoga drives it home: how the mind dreams affects what the brain does.

This idea that the mind influences the brain is sometimes called "downward causation."[7] "Downward" describes causal influence that goes from "higher" to "lower" levels—the psychological to the biological, the biological to the chemical, and so on. Philosophers often use the term as shorthand for "downward mental causation"—the idea that when you form an intention or make a decision, your mind affects your brain and body. In my view, however, the term "downward causation" is misleading because it forces us into thinking of the mind and body as hierarchically related, with the mind being, in some unclear sense, "above" or "higher" than the body. Nevertheless, if we set aside the difficult mind-body problem, then we can think of the idea of downward causation as an attempt to make sense of the following undeniable fact: phenomena we understand in psychological terms—willing, intending, remembering, imagining, and so on—influence and change phenomena we understand in physiological terms—neuron firings, brain activities, gene expressions, immune responses, and so forth.

Here's a classic example from neuroscience: paying attention to a stimulus increases the firing rates of visual neurons sensitive to that stimulus.[8] ("Attention" is a psychological concept, "firing rate" a neurophysiological one.) Another example comes from human brain imaging: individuals who suffer chronic pain can learn to control voluntarily the activity of a certain brain region involved in pain perception (the anterior cingulate cortex) when they're shown that region's activity in real time while they're in the MRI scanner.[9] By intentionally changing their brain activity, they can control their pain perception, including lessening the magnitude of their felt pain.

Lucid dreaming provides another example. In a strong lucid dream, you the dreamer, acting intentionally through your dream ego, can influence what happens in your sleeping brain and body. Looking left and right with your dream ego eyes corresponds to making your physical eyes move left and right. A signal-verified lucid dream—where the dreamer uses eye movements to signal to the outside world she's having a lucid dream—constitutes a case of the mind affecting the body in the sleep lab. One of the techniques to wake yourself up from a lucid dream is to stare at something in the dream so you stop your dream ego from making eye movements.[10] Holding your dream gaze fixed suppresses your physical eye movements in REM sleep and causes you to wake up. If you don't want to wake up, then spinning your dream body so you feel the sensation of movement will prolong the dream state.[11]

These techniques indicate that your intention and decision to move your dream body, as well as your execution of dream actions, must be closely tied to the systems in your brain that enable you to plan and act in waking life. To activate these systems, you don't need to be awake and moving; you can prompt their activation in the dream state.

Cautious readers will insist that all we can really say is that intentional actions you perform through your dream ego correlate with measurable changes in your brain and body, but correlation is not causation. Nevertheless, even without knowing the exact nature of the relation between the psychological level of lucid dream intention and the physiological level of brain activity, we can still hold to the following point: by acting through your dream body, you can influence your physical body. For example, growing evidence indicates that practicing motor skills with your dream body in a lucid dream improves your performance in waking life.[12] Training your dream body changes your physical body.

Lucid dreaming is itself a learnable skill, a way you can change your brain by training your mind. As LaBerge observes:

> Lucid dreaming represents in this view what ought to be a normal ability in adults. If this is correct, why are lucid dreams so rare, especially in cases such as nightmares, where lucidity should be extremely helpful and rewarding? I think a partial answer can be seen by comparing lucid dreaming with another learnable cognitive skill, namely, language. All normal adults speak and understand at least one language. But how many would do so if they were never taught? Unfortunately, in this culture, with few exceptions, we are not taught to dream.[13]

Dream yoga belongs to traditions that do teach how to dream. The instructions, however, aim not merely to teach dreaming but also to transform the waking mind, and this is meant to happen at a deeper level than anything currently envisioned in the Western psychological literature on lucid dreaming.

Dream yoga targets what Tibetan Buddhism sees as certain basic delusions or mistaken instinctual views about what's real, as well as the negative mind states that spring from these delusions. In an ordinary dream, we identify with our dream ego and take what we experience to be real (even if, as Nietzsche supposes, we may sense at some deeper level that it's appearance, not reality). Whatever we see or feel seems to exist apart from us with its own being or intrinsic nature. This confused state of mind serves as a model for our waking ignorance of the nature of reality. We think our waking ego exists with its own separate and essential nature, but this belief is delusional, for our waking ego is no less an imaginative construction than our dream ego, formed by imaginatively projecting ourselves into the past in memory and into the future in anticipation. The "I-Me-Mine" standing over against the world as a separate thing or entity can function as a distorted mental construct in the waking state, not just in dreams. The dream world and the waking world both seem real and solid, yet in neither case do we realize that whatever we take to be real and solid is always a mode of appearance—something that appears real in one way or another—and that modes of appearance by their very nature can't be separated from the mind.

By contrast, full and complete lucidity—where we wake up in the dream state and directly experience it as luminous appearance, empty of substantiality—offers a traditional metaphor for liberation and enlightenment (the title "Buddha" means the fully awakened one). This metaphor isn't meant to deny the conventional reality of the waking world. It aims rather to effect a fundamental shift in our understanding of what it means for something to be real. "Real" is the name we give to certain stable ways that things appear and continue to appear when we test them, not the name for some essence hidden behind or within appearances.

Although this idea might seem abstract or theoretical, contemplative traditions report that it's easier to transform negative emotions such as fear or anger into positive mental states when we can see those emotions as modes of appearance rather than substantial realities. The dream state offers a special setting for cultivating this skill because it contains powerful and intense emotions, while also affording a view of the deeper reaches of consciousness beyond such mental turbulence.

Although scientific studies to investigate the transformative effects of dream yoga on the mind and body have yet to be done, we can already use the dream yoga perspective as a way to take a critical look at certain neuroscience models of the dreaming mind and brain.

THE DREAMING BRAIN

One of the most influential neuroscience models of dreaming comes from Harvard psychiatrist Allan Hobson.[14] It treats dreaming as the state of consciousness the "brain-mind" enters when the evolutionarily older brainstem activates the newer forebrain in REM sleep.

According to Hobson's "activation-synthesis" model, first presented with Robert McCarley in 1977, certain nuclei within a part of the brainstem called the pons cause REM-sleep dreaming by activating the forebrain through bursts of firing and the release of acetylcholine (a type of molecule that changes the way neurons fire and alters the brain's global state). The essentially random electrochemical stimulation arriving from the brainstem below activates the visual, motor, emotional, and memory areas of the forebrain above, except for the prefrontal cortex, which, Hobson proposes, is directly inhibited by acetylcholine in REM sleep. The forebrain must synthesize the haphazardly provoked images, memories, emotions, and thoughts, but it's deprived of its usual sensory and motor links to the outside world, as well as its prefrontal cognitive capacities, and it lacks sufficient amounts of certain molecules (norepinephrine and serotonin) essential for maintaining an awake and alert state. What results is a semicoherent mental representation or scenario—a dream. In Hobson and McCarley's phrase, the forebrain makes "the best of a bad job" in producing a dream from the chaotic barrage of brainstem stimulation.

According to this model, lucid dreaming happens when the floodwaters of acetylcholine begin to recede and the dorsolateral prefrontal cortex reactivates, but not so strongly as to suppress the brainstem signals it receives. Then the spellbound quality of dreaming weakens and the dream images can be seen for what they are.

Hobson has long used the activation-synthesis model to debunk Freud.[15] In Hobson's view, wishes don't instigate dreams; dreams result from the brainstem mechanisms that turn on REM sleep. What makes dream content strange isn't the disguising of unconscious wishes, it's the "hyperassociative synthesis" that the forebrain—minus a full-capacity prefrontal cortex—winds up carrying out in response to brainstem stimulation.

One of the core ideas of the activation-synthesis model in its original formulation—an idea Hobson still holds onto—is that dreams are epiphenomena of REM sleep: they are froth on the waves but do nothing to affect the currents.[16] The brainstem mechanisms that cause REM sleep produce dreams as a byproduct, and dreaming itself exerts no significant influence on brain activity during sleep. In Hobson's words: "We can see that, when the brain self-activates in sleep, it changes its chemical self-instructions. The mind has no choice but to go along with the programme."[17]

Hobson's views, though influential, are controversial; a number of dream scientists take issue with his approach for a variety of reasons.[18] My approach also implies a different way of thinking about the dreaming mind's relation to the sleeping brain.

Lucid dreaming provides a clear counterexample to the view of dreams as epiphenomenal to REM sleep. Not all REM-sleep eye movements arise randomly from the brainstem; lucid dreamers can make them voluntarily, an ability that depends on parts of the frontal cortex (specifically, a region called the frontal eye field). Lucid dreamers can volitionally affect their dreams. First, simply directing your attention to the dream state is a kind of volitional cognitive control, and it directly affects the dream content because what you dream depends on how you direct your attention in the dream. Second, you can explicitly guide the dream content to varying degrees. You can choose whether to intervene or to let the dream unfold spontaneously. These cognitive abilities in the dream state depend on fronto-parietal networks that interconnect many areas of the cortex; hence the cortex can influence how REM sleep progresses. Guiding the dream and observing it affect the dream state and thereby modify the course of REM sleep. The mind doesn't just go along with the program but has a hand in leading it.

Dream yoga puts this view of dreaming in the service of radical transformative ends whose effects on the sleeping brain and body we currently don't know. It seems likely, however, that these effects include not simply greater skill at recognizing and transforming the dream state but also changes to the kinds of nonlucid dreams we have. Many contemplative traditions distinguish between ordinary dreams that recycle recent memories and special kinds of dreams arising from deeper wellsprings of the mind. Tibetan Buddhist dream yoga, for example, distinguishes between ordinary nonlucid dreams and nonlucid "dreams of clarity." The latter are especially vivid; they arise from the quality of mental clarity cultivated in meditation and manifest the basic luminosity of awareness. In the words of Tibetan teacher Chögyal Namkhai Norbu: "Although the sun shines constantly, sometimes we cannot see it due to cloud covering, while at other times, we can see between the clouds for a few moments. Similarly, sometimes the individual's clarity spontaneously appears. One result of this is the appearance of dreams of clarity."[19]

In sum, from the lucid dreaming and dream yoga perspectives, dreaming is flexible and trainable in a variety of ways, and this fact implies that the mind can lead the way by altering how the brain and body sleep.

Hobson seems willing to acknowledge some of these ideas. In a recent interview with philosopher Thomas Metzinger, he said the following about lucid dreaming:

> The occasional awareness that one is in fact dreaming is an extremely informative detail of modern dream science. The fact that such insight can be cultivated thickens the plot considerably. Taken together, the data suggest that the conscious state accompanying brain activation in sleep is both plastic and causal. It is plastic because self-reflective awareness occasionally does arise spontaneously, and because with practice its incidence—and its power—can be increased. It is causal because lucidity can be amplified to command scene changes in

dreams and even to command awakening, the better to remember, and enjoy, occasional dream-plot control. My position about lucidity is that it *is* real, it *is* powerful, and it *is* informative.[20]

Although Hobson doesn't make the inference, these remarks imply that the sleeping brain reflects the dreaming mind no less than the dreaming mind reflects the sleeping brain. If the conscious state accompanying brain activation in sleep is both plastic and causal, then dreams cannot be froth on the waves of REM sleep; they must be dynamic and self-organizing patterns of the flow. Dreaming channels brain activity. With lucidity, it can even guide the brain in sleep, as perception does in waking.

SUCH STUFF AS DREAMS ARE MADE ON

We're now ready for the hard questions: What exactly is a dream? Is it some kind of false perception, or a form of imagination? Is waking life a kind of dream? How do you know you're not dreaming right now?

Among dream scientists, we couldn't find two more opposed theorists than Allan Hobson, the biological psychiatrist discussed above who's highly critical of Freud, and Mark Solms, a neuropsychologist and Freudian psychoanalyst at the University of Cape Town in South Africa. Whereas Hobson argues that the brainstem generates REM-sleep dreams, Solms counters that patients who've suffered damage to the REM-sleep generating regions of the brainstem still report dreaming, while other patients who've suffered damage to particular forebrain regions undergo REM sleep but report they have no dreams.[21] Dreaming and REM sleep can occur independently, so the cause of REM sleep can't be the cause of dreaming. Instead, according to Solms, dreams originate from forebrain mechanisms related to arousal, emotion, and mental imagery; furthermore, they aren't motivationally neutral, as Hobson claims, because the neurochemical systems of the forebrain that drive dreaming also underlie appetitive interests and seeking behaviors—an idea close to Freud's notion of wish fulfillment.

Yet, despite these disagreements, Solms agrees with Hobson that when we dream we hallucinate without knowing it, and we believe falsely that what we experience is real. For both scientists, dreaming is delusional hallucination. In Solms's words: "For our purposes, dreaming may be defined as *the subjective experience of a complex hallucinatory episode during sleep*."[22] Hobson and his Harvard colleagues, Edward F. Pace-Schott and Robert Stickgold, describe dreams the same way: "Dreams contain formed hallucinatory perceptions. . . . Dreams are delusional; we are consistently duped into believing we are awake unless we cultivate lucidity."[23]

I think this way of thinking about dreams is misguided. To dream is to imagine, not to have false perceptions. When we dream, we don't have pseudo-perceptions that form the basis for false beliefs; we imagine a dream world and we identify with our dream ego. In a nonlucid dream we lack the meta-awareness that we're imagining things; in a lucid dream we regain this awareness. Dream yoga is a way of training the imagination across the waking-dreaming continuum. If we are such stuff as dreams are made on, then this is the stuff of imagination, not pseudo-perception.[24]

To be clear about how these two conceptions of dreaming differ, we need to be clear about the differences between hallucination and imagination.

When you hallucinate, you seem to perceive something that's not there. Shakespeare's Macbeth seems to see before him a dagger with its handle pointed toward his hand, but there's no dagger there; he's hallucinating.

When you imagine, you evoke something absent (something not directly stimulating you at that moment) and make it mentally present to your attention. You can do this in a sensory or perceptionlike way or a cognitive or suppositionlike way. You can imagine something by forming a mental image, as when you visualize a tiger, or by supposing it to be a certain way, as when you imagine that tigers can talk. To form a mental image of a tiger is to imagine perceiving a tiger. The tiger appears from a certain angle or perspective, as it would if you were seeing it. To imagine that tigers can talk is to envision the world as being a certain way, namely, as containing talking tigers.

Some philosophers and psychologists describe sensory imagination as the mental simulation of perception: to visualize a tiger is to simulate seeing one. Notice that when you mentally simulate seeing a tiger, you're not hallucinating a tiger; you're imagining one. This holds true even if the tiger you simulate seeing appears especially vivid and clear to your mind's eye.

You might think neuroscience can tell us whether dreaming is hallucinatory perception or imagination. After all, if two experts in the neuroscience of dreaming who disagree on so much else agree that dreams are hallucinations, then this must be because that's what the evidence says.

But the evidence says no such thing. The hallucinatory perception conception of dreaming isn't a scientific finding; it's a conceptual model used to make sense of the scientific findings. The issue is conceptual and phenomenological, so neuroscience on its own can't make this call. When Hobson writes that brain-imaging studies "show an *increase* in activation of just those multimodal regions of the brain that one would expect to be activated in hallucinatory perception"[25] (cortical regions of the parietal lobe), he's unquestioningly relying on this model to interpret the brain-imaging data. But the data themselves don't show that dreams are hallucinatory perceptions and not imaginings, for one would expect increased activation in just those regions for sensory imagination too.

So what we need to know first is the answer to this phenomenological question: Is the experience of dreaming subjectively more like hallucinatory perceiving or more like imagining?

If dreaming is more like hallucinatory perceiving, then when you dream you're flying, you have the same kind of sensory experience you'd have if you were awake and flying. You have a sensory experience that subjectively seems to you like a perception. Since you're not flying but lying in bed asleep, your sensory experience misleads you, making it seem as if certain things are present that aren't really there. So your sensory experience is a sham perception, a hallucination.

If dreaming is more like imagining, then when you dream you're flying, you form a mental image of yourself flying and imagine that you're flying. In other words, you simulate the sensory experience of flying by forming visual and kinesthetic images, and you suppose that you can fly. Your experience isn't hallucinatory perception; it's imagination.

Many people, scientists included, think that vividness differentiates perception from imagination, that when you perceive something—like the full moon rising in the east—it appears more vividly than when you imagine it. So if what you dream seems as vivid as what you see, then dreaming must be more like perception than imagination.

But this line of thought is based on a misconception. Some people have highly vivid imagery in both memory and waking imagination, yet without hallucinating, and some people report having pale and sketchy dreams. Lack of vividness compared to perception isn't essential to imagination, for an imagined object—like a visualized full moon—can appear as vivid as a perceived one. Rather, what marks imagination is that it evokes or calls forth its object, regardless of how pale or vivid it seems. To form a mental image of the moon, you need to bring the moon to mind, and depending on your powers of imagination, it can seem pale or vivid to your mind's eye.

As we'll see, this feature of imagination makes its relation to attention and volition different from that of perception. Dreaming seems to bear the same kind of relation to attention and volition as imagination does, and this commonality alone seems enough to show that dreaming resembles imagining more than hallucinatory perceiving.[26]

Consider that dream imagery seems to depend directly on attention in a way that perception doesn't and sensory imagination does. We can consciously perceive something without devoting all of our selective attention to it—the background hum of a fan on a hot summer day or the crickets chirping at night. Yet when we dream or imagine, what we dream or imagine seems to be directly determined by the focus of our attention.[27]

Take sensory imagination first. If I visually imagine a harvest moon, I need to attend to the moon as my imaged object. The moment my attention wanders, I'm no longer imaging the moon; the mental image falls apart. If I don't notice the horizon in my imaging of the moon, the reason isn't that I image the horizon but fail to notice it in my mental image. The reason is that I don't visualize the horizon at all; it's not part of my mental image. Not noticing something in a mental image is no different from its not belonging to the image; it's the same as not imaging it.

Of course, I can suppose that my visualized harvest moon sits just above the horizon in the east at sunset. Using cognitive imagination, I can make this supposition without having to image the horizon. In this way, cognitive imagination frames and permeates what I visualize with meaning.

Now take dreaming. If I don't notice trees in my flying dream, is that because I fail to notice their presence or because I'm not generating any sensory imagery of trees? It strikes me as more phenomenologically accurate to say that the reason I don't notice them is that I'm not visually imagining them at that moment. If I look down and notice the treetops, then now I dream of trees because now I visualize their presence. In other words, not noticing something in a dream seems no different from its not being part of the dream imagery; it seems the same as not dreaming it, not imaging it in the dream state.

Of course, even if there's no dream imagery of trees until I see them, I can still dream that the trees were there all along but I didn't notice them at first. In other words, in my dream I can suppose that the dream world contains those trees (and many other things) even when I don't image them. Using cognitive imagination, I can make this supposition and thereby constitute the dream world as being a certain way and as framing and permeating my dream imagery with meaning.

Dreaming and imagining can also be voluntarily guided in a way perception can't. You can choose where to look, but you can't choose what's there for you to see. You can choose how to interpret what you see—this kind of seeing is imaginative perception—but you can't create the object that's there to be seen. You can choose what to imagine, however, and you can sometimes choose what to dream, as lucid dreaming and hypnagogic imagery show.

Although most dreams are involuntary, so too are most acts of everyday imagination—think of daydreaming, which usually happens spontaneously, though you can intentionally pursue it. Furthermore, the fact that most dreams are involuntary doesn't count against their being subject to voluntary guidance. If some dreams can be voluntarily guided, then dreaming must be the kind of experience whose content is directly subject to volitional influence. So dreaming and imagining, but not perceiving, are subject to volition in similar ways.

The imagination conception of dreaming I've been sketching also fits the evidence from neuroscience and psychology.[28] First, the neuroscience evidence. Loss of vision doesn't necessarily lead to loss of visual dream imagery, but mental imagery losses and dream losses go together. Although people who are completely blind from birth have no visual imagery in waking sensory imagination or dreams, patients who later in life suffer damage to the primary visual cortex, which connects more or less directly to the retina, lose vision but have no problem forming visual images in their dreams. Patients with damage to a much higher region of the visual system—the zone where the occipital, parietal, and temporal lobes meet (the occipito-temporoparietal junction)—report a complete loss of dreaming while still being able to see. Damage to this area also impairs mental imagery, including the ability to visualize first-person versus third-person perspectives (the same perspectives in field versus observer memory, and inner versus outer perspectives on the dream ego). So dreaming and imagining seem to depend directly on the same brain areas.

Now the psychology evidence. In children, the development of dreaming goes hand in hand with the development of visual imagination.[29] For example, dream recall correlates better with mental imagery abilities than with language proficiency. In adults too, visual-spatial imagination is the cognitive skill that maximally correlates with dream recall.

The imagination conception also fits evidence from studies of lucid dreaming.[30] It's well known that performing a body action (such as jumping) and imagining performing the same action engage common brain systems. Dreamed body actions carried out in lucid dreams also engage these same brain systems.[31] According to the simulation theory of motor action, to imagine performing an action is mentally to simulate performing that action with an imagined body. This theory suggests that performing a dream action consists in simulating the performance of that action with an imagined dream body.

So what exactly is a dream? A dream isn't a random false perception; it's a spontaneous mental simulation, a way of imagining ourselves a world. With this conception of dreaming now in hand, I want to revisit briefly a few ideas from the previous two chapters.

IMAGINING THINGS

Let's begin by going back to the difference between the hypnagogic state and the dream state. In the hypnagogic state, we feel little or no ego immersion in a world (we're looking at the images). In the dream state, we imagine the presence of a world and identify with our dream ego immersed in that world (we're in the dream world).

Sensory and cognitive imagination both contribute to how we identify with our dream ego. Sometimes the dream ego appears in the form of an imaged dream body (sensory imagination), which we can experience from within or see from without. Sometimes we don't image our dream ego at all; instead we see the imaged dream world from a bodiless, observational point of view. In either case, we identify with our dream ego by imaginatively supposing that this is who we are (cognitive imagination).

Imagination also contributes to our waking sense of self through the role it plays in memory. To remember things we've experienced requires imagining them, either from the first-person perspective (field memory) or the outside-spectator perspective (observer memory). To anticipate things also requires imagining them in these ways. Both forms of imagination—retrospective and prospective—depend on overlapping areas and networks of the brain.[32] We imaginatively construct an ego by appropriating and identifying with the remembrance of things past and the anticipation of things future. Imagination lies at the core of the "I-Me-Mine."

When we remember and anticipate things, we have a sense of agency, the feeling of either purposely pursuing some train of thought or being caught up in one that we can stop, or at least try to stop. When we dream, we often lose this kind of agency; we lose touch with our role as the ones imagining. Completely absorbed in what we imagine, we fail to recognize that we're imagining things, and so can't voluntarily guide or influence them.

All this changes in a strong lucid dream. Here we wake up to ourselves as the ones imagining and to the dream state as an imaginative creation of our minds.

Using the imagination conception of dreaming, we can also enlarge the answer to one of our earlier questions about lucid dreaming: How can you tell the difference between dreaming you're dreaming and knowing you're dreaming?

When you dream you're dreaming, you use cognitive imagination to suppose that you're in a dream. To be more precise, you imaginatively suppose that your dream ego is dreaming, that its experiences are a dream, while you nonlucidly identify with this ego.

When you know you're dreaming, you become aware of your imaging during sleep and can attend to it precisely as dream imaging. You may also be able to guide volitionally your sensory imaging as well as your cognitive imagining. These mental abilities mark the difference between being fully lucid in a dream and dreaming you're dreaming.

Dream yoga works directly with creative imagination by training these abilities. *Recognizing the dream state* means being aware of the dream state as an imaginative creation and sustaining that awareness with stability and vividness; *transforming the dream state* means exploring how dream appearances—including the dream ego—can be altered through sensory and cognitive imagination; and *seeing through the dream state* means suspending or releasing the imagination and resting in imageless awareness without waking up from sleep.

IS WAKEFULNESS A DREAMLIKE STATE?

The imagination conception of dreaming can also help us think about the idea, proposed by neuroscientist Rodolfo Llinás of the New York University Medical Center, that the waking state is like a dream, or in his formulation, that wakefulness is nothing other than a dreamlike state constrained by external sensory inputs.[33]

In REM sleep, the EEG patterns are similar to those of the waking state, yet it is often harder to rouse someone from REM sleep than from slow-wave (NREM) sleep. For this reason, the French sleep science pioneer Michel Jouvet introduced the term "paradoxical sleep" to describe REM sleep. Here's the puzzle: If the brain is so highly active in REM sleep, how can it manage to ignore strong outside stimuli and not wake up?

According to Llinás, the answer lies in how attention functions during REM sleep. Consciousness is very much still present—hence the similar EEG patterns in REM sleep and the waking state—but attention has turned away from outer sensory events and inward toward memories—hence outside stimuli go unnoticed. In REM sleep, the brain is hyperattentive to its own internal state. Sensory processing continues at the periphery, but it can't penetrate the core systems actively sustaining consciousness.

Llínas suggests that the brain sustains the same core state of consciousness during REM sleep and wakefulness, but the sensory and motor systems we use to perceive and act can't affect this consciousness in regular ways when we're REM-sleep dreaming. Consciousness itself doesn't arise from sensory inputs; it's generated within the brain by an ongoing dialogue between the cortex and the thalamus (a central structure that relays sensory and motor signals to the cerebral cortex and regulates levels of consciousness and sleep). The difference between wakefulness and REM sleep lies in the degree to which sensory and motor information can influence this thalamocortical conversation. During REM sleep, sensory inputs are kept from entering the dialogue, while motor systems are shut down (you're paralyzed except for eye movements) and attention fastens onto memories. Simply put, when sensory inputs participate in the thalamocortical dialogue generating consciousness, they constrain what we experience and we have waking perception. When sensory inputs can't participate in this dialogue in sleep, we dream. To put the idea another way, from the brain's perspective—or rather from the perspective of the thalamocortical system sustaining consciousness—wakefulness is a case of dreaming with sensorimotor constraints, and dreaming is a case of perceiving without sensorimotor constraints.[34]

The idea that wakefulness is a dreamlike state influenced by the senses, and dreaming a conscious state sheltered from the senses and fed by memory, recalls the ancient Indian view of waking and dreaming from the *Upanishads*. According to this view, when we dream we see an inner world created from materials preserved in memory. We take sense impressions from the waking state, break them down, and remake them into mental images. But neither outer sense impressions nor inner images produce consciousness; rather, one and the same consciousness illuminates the sense world in waking and the dream world in sleep.

These ancient Indian and modern neuroscience views share a deep insight: sense impressions don't produce consciousness; they modulate it in the waking state. Sense impressions change the tone, pitch, or volume of consciousness, as do memories and emotions in the waking and dream states, but they don't create the vibration of consciousness itself. We perceive the world thanks to our senses and our ability to move around, but sensing and moving don't fashion consciousness anew each time we wake up; they affect an already existing consciousness present also when we dream.

To get these ancient and modern ideas right, however, we need to understand dreaming as imagination, not false perception. Wakefulness isn't a hallucinatory state constrained by sensory inputs; it's an imaginative state fed by sense perception. Dreaming isn't a hallucinatory state cut off from sensory inputs; it's an imaginative state fed by memory and emotion. As we've seen, imagination permeates waking and dreaming consciousness; in both states it underlies how we experience the ego or the "I-Me-Mine."

Here I part company with Llinás, as well as philosopher Thomas Metzinger. Llinás writes, "Comforting or disturbing, the fact is that we are basically dreaming machines that construct virtual models of the world."[35] Metzinger elaborates: "A fruitful way of looking at the brain is as a system which, even in the ordinary waking states, constantly hallucinates at the world, as a system that constantly lets its internal autonomous simulational dynamics collide with the ongoing flow of sensory input, vigorously dreaming at the world and thereby generating the content of phenomenal experience."[36]

I share Llinás's and Metzinger's conception of the brain as a self-organizing system that participates in bringing forth what we experience as reality. But I think their way of looking at things goes astray in two ways. They assume that waking states and dream states are exactly alike on the inside as experiences and differ only in how they fit into their outside surroundings; and they assume that how an experience fits into its surroundings makes no difference to the kind of experience it is.

First, as we've seen, waking perceptual states and dream states differ experientially in fundamental ways having to do with attention and volition; these differences indicate that dreams aren't false perceptions, but ways of imagining. So it's not the case that waking perceptions and dream images are exactly alike on the inside and differ only in whether or not they match what's on the outside. Rather, they differ on the inside as experiences.

Second, to perceive the world isn't to hallucinate and get things right. To perceive is to explore the world with your sensing and moving body. Perception creates meaning through sensorimotor exploration. You don't get a being with a sense of self by taking a hallucinating brain and tacking on some sensory inputs and motor outputs. You get a being with a sense of self by taking a brain with a capacity for imagination—for imaging its past and future—and embedding it within a sensing and moving body.

Perception, therefore, isn't online hallucination; it's sensorimotor engagement with the world. Dreaming isn't offline hallucination; it's spontaneous imagination during sleep. We aren't dreaming machines but imaginative beings. We don't hallucinate at the world; we imaginatively perceive it.

REALITY CHECK

Are you dreaming or awake right now? How do you know you're not dreaming? Do these words stay the same as you read them? What happens if you look away and then look back again?

Sometimes when I dream I have the impression of reading a text and holding onto its meaning in my mind, yet all the while the letters and words keep shifting. I wake up with the feeling of having just read something important, yet I'm at a loss to say what it was.

Psychologists call trying to answer the question, "Am I dreaming or awake?" state testing or reality testing. One way to become lucid while dreaming is to ask this question and apply some test to determine your state. In order to remember to do this, you need to make a habit of asking the question and applying the test while you're awake. What sort of test should you use? Dreams can seem as vivid as waking experiences, sometimes even more vivid, but unlike waking experiences, they tend to be unstable. If you try to read something in a dream, chances are it won't stay still but will start mutating before you make it to the end of the sentence.

Reality testing focuses on the differences between waking and dreaming. Dream yoga focuses on the similarities. It encourages us to view all waking events as dreamlike, as forms of appearance related to the mind. How the world shows up for us depends on how we imaginatively perceive and conceptualize things. What we experience cannot be separated from how we experience. The thoughts, images, and emotions we bring to whatever we encounter, as well as the meanings we mentally impose, condition what "reality" can mean to us. In this way, the waking world is mind-involving and mind-dependent. Realizing this fact—waking up to our participation in the creation of our world—resembles becoming lucid in a dream. Dream yoga encourages us to cultivate this critical mindfulness in waking life and to carry it into dreaming.

As a way to learn to recognize the dream state, the dream yoga method may have certain advantages over reality testing. After all, doing a reality test doesn't necessarily mean you'll arrive at the right answer:

I'm in a cafeteria trying to get some food for lunch. I notice a woman I obsessed over in college. I'm surprised to see her and exclaim, "What are you doing here?" She's very attractively dressed, and I want to sit with her. I'm older and more confident now, and I want to impress her. I turn back to the counter, but the bread and cheese are gone. I can see she's at the cashier; I have to catch up to her, but I can't find the food I want. It occurs to me I might be dreaming; that's why the bread and cheese have disappeared. I stop for a moment to think about this possibility. I decide it can't be true because everything seems so real. I conclude I'm definitely awake, not dreaming. I run after her because now she's left for the dining room.

What I find funny about this dream is that I do notice certain crucial discrepancies—the woman's unexpected presence, the bread and cheese being there and then being gone—and I do think to question my state, but given my main preoccupation, I come to the completely wrong conclusion. Then, forgetting all about my lunch, I run after her, this time taking no notice of the discontinuity.

If I'd made a habit of telling myself, "I'm dreaming," while I was awake, would I have remembered to do this in this dream? And if I had remembered, would I have been able to shift my attention to the dreamlike quality of the state or wound up dreaming I was dreaming? The second seems more likely here, given the wish governing this dream.

Can these two perspectives—regarding waking life as dreamlike and reality testing—both be true? How can testing for reality be possible if waking reality is like a dream?

Western philosophy has long grappled with the problem of how to tell the difference between the waking and the dreaming states. In Plato's dialogue *Theaetetus*, Socrates asks the young mathematician Theaetetus what evidence he could give to prove either that they're awake and talking to each other or that they're asleep and all their thoughts are a dream. Theaetetus admits he doesn't know how to prove the one any more than the other, for the two states seem to match; even dreaming that you're telling someone a dream seems strangely similar to really telling someone a dream. "So you see," Socrates concludes, "it is disputable even whether we are awake or asleep."

Two thousand years later, René Descartes (1596–1650) asked the same question.[37] He argued that there are no "sure signs" to differentiate with complete certainty waking sense experiences from dream experiences, though we can point to a feature that reliably distinguishes between the two—memory during dreams doesn't connect dream experiences to the rest of one's life, but waking memory does.

Some philosophers have responded to this problem with the following line of thought.[38] We know we're awake when we're awake; the problem is that we don't know we're dreaming when we dream. In other words, the problem isn't that we can't know we're awake and not dreaming in the same way we can't know we're dreaming and not awake. On the contrary, we know we're awake because we can reflect on being awake in the waking state, but we can't reflect this way in the dream state; we know only afterward that we were dreaming, thanks to waking memory.

Sartre takes this line of thought one step further. We can always reflect on waking consciousness, he argues, but we can never reflect on dream consciousness. Reflection is impossible in the dream state because any real reflection (as opposed to an imaginary or dreamed one) entails a momentary awakening and hence a departure from the dream state. Reflection confirms and reinforces waking consciousness, but "destroys the dream." We can't assert, "I'm dreaming"; we can only say retrospectively from the waking state, "I was just now dreaming."

I expect by now you can see that this line of thought won't work—indeed, it gets things backward. On the one hand, we can know we're dreaming while we dream: we can be aware of our imaging during sleep and attend to it as dream imaging. Contrary to Sartre, this kind of meta-awareness doesn't destroy the dream and does enable us to assert, "I'm dreaming." So meta-awareness of dreaming can establish that we're dreaming. This is precisely what happens in a strong lucid dream.

On the other hand, as we've seen, we can't know we're awake simply by reflecting on our state and concluding we're awake. Here's another dream that shows why:

I've lost my seven-year-old son, Max. I'm in a network of corridors and passageways, and I can't find him anywhere. I'm frantic. I turn the corner—it's a dead end, but someone is sitting on a chair at the end of the hall. I decide to go ask him whether he's seen my son. As I get closer, I see he has brilliant green eyes and a flaming red beard; he's smoking a pipe, with one leg crossed over the other. He's a tall leprechaun. He looks me straight in the eye and says, "Don't you know, silly, . . . this . . . is . . . a . . . DREAM!" The realization hits me like a wave. I feel pulled into the undertow; everything around me now seems liquid and lacks solidity. I try to stabilize myself by sitting down and centering my awareness. I succeed for a few seconds, no longer afraid and excited to know I'm dreaming. Then I wake up. It's morning, so I get up, go to the bathroom, head downstairs, check on Max and Gareth still asleep in their room, and start to make tea to take with me out on the deck while I do my morning tai chi. I wake up again. It's 4 a.m. I lie there, not sure whether I've woken up or am still dreaming. When I get up later in the morning, I wonder again whether I'm awake or dreaming. I also wonder whether I really woke up in the middle of the night or that too was part of my dream.

As this dream illustrates, while we're dreaming—that is, without ever leaving the dream state—it can seem that we've woken up, are reflecting on being awake, and remember having just been dreaming. This kind of dream experience, known as "false awakening," shows that we can't confirm we're awake simply by thinking reflectively that we are.

Some philosophers might try appealing to the imagination conception of dreaming at this point. In a false awakening, they'd argue, we don't really reflect on our seeming to be awake; we only imagine that we're reflecting this way. Although the presence of imaginary reflection can't show we're awake, the presence of real reflection can.

But this line of thought just pushes the problem back. How do we know for sure that we're really reflecting and not dreaming we're reflecting?

As philosophers Jennifer Windt and Thomas Metzinger discuss, false awakenings lie midway along the continuum of metacognitive insight we can have when we're in the dream state.[39] On the one hand, a false awakening following a nonlucid dream amounts to a cognitive gain, for now we know we were just dreaming, even though we're mistaken to think we're now awake. On the other hand, a false awakening following a lucid dream (like the one in my dream above) amounts to a cognitive loss, for now we think we're awake when we're actually still dreaming.

As Frederik van Eeden reported in his 1913 article, waking from a lucid dream into a false awakening and then eventually waking up into the real world (maybe after more false awakenings) makes the lucid dream seem the deepest of these states. In his words: "The impression is as if I had been rising through spheres of different depths, of which the lucid dream was the deepest."[40] Francisco Varela describes a similar feeling in this dream report, which I found among some of his notes after his death (the title is his):

Levels for a Cat

This dream happened last night. I am in the countryside, alone, except for my tiger-colored cat. I go to bed late, quite concerned because he has not come back home to sleep, as is his unchangeable habit. My sleep is deep and eventless until suddenly a dream begins. The scene happens in my bedroom that is full of other people. In particular an old acquaintance, who is behaving in a most aggressive manner to the weaker persons present there. I feel ill at ease, and seek to protect a young child, whom I hug next to me. At that moment I have the intuition or sense that my cat has returned. I raise myself up through layers of dream and sleep, until emerging from this side of lucidity. Almost in the same gesture of motion, I pull away my covers and look by the moonlight through the glass pane of the bedroom door. There he is, indeed, the cat standing up on his two hind legs, scratching the door to be let in, as is his habit. Half in the dream, and half in my body, with a sigh of relief, I get up, and open the door. Outside the night is fresh and full of spring smells. I wake up further only to realize there is no cat. His sight has disappeared into the thin air of the night. I put myself back in bed, fall asleep lightly. A few minutes later a scratch comes, and I see exactly the same scene of the mirage, with the animal at the door. This time I get up and let him in for real. I return to bed with the strong sense of having collapsed several levels of the so-called reality.

Francisco J. Varela

Menerbes, Provence, France

Easter Weekend 1999.

Although false awakenings often are nonlucid, we can also have lucid false awakenings, either by recognizing one when it happens or by regaining lucidity immediately afterward. As Windt and Metzinger explain, "When lucidity is attained in a false awakening, especially following a lucid dream, the subject is aware of both the ongoing dream state and of previously having experienced a dream. In this sense, one might say that a lucid false awakening presents a higher degree of lucid awareness than an ordinary lucid dream, because it combines both concurrent and retrospective insight into the dream state."[41]

Indian literature contains numerous narratives of what we would call lucid dreams and false awakenings. As scholar Wendy Doniger discusses in her classic book, *Dreams, Illusion and Other Realities*,[42] these narratives illustrate the following thought: we can *verify* the hypothesis that we're *dreaming* by waking up—either from the dream (normal waking) or within the dream (lucid dreaming). And (what amounts to the same thing) we can *falsify* the hypothesis that we're *awake* by waking up—again either from the dream or within the dream. But we can't *falsify* the hypothesis that we're *dreaming* or (what amounts to the same thing) *verify* the hypothesis that we're *awake*. The reason is that for any experience we choose—specifically, any experience we take to be a waking one—it seems conceivable that we could wake up from *that* experience.

Doniger calls Indian myths based on this thought "myths of the receding frame." They present the dreamer as never being certain that he has awakened from the last of a series of dreams within dreams or false awakenings. The Indian stories of dreams within dreams within dreams go on for pages, but the following contemporary poem, by the Indonesian poet Sukasah Syadan, captures the image in a few words:

Dream

last night
I dreamed
that I dreamed
that I awoke
a sleepless man
posting
on what I dreamed
last night[43]

Indian philosophers also addressed the question of how we can distinguish between waking and dreaming if life is like a dream. One answer I like comes from Śaṅkara (788–820 C.E.), one of India's great Hindu philosophers of the Advaita Vedānta school.[44] ("Advaita" means nondual; "Vedānta" refers to the end of the Vedas, that is, the texts of the *Upanishads*.) He answered that waking life includes recognition of the dream state in a way the dream state doesn't include recognition of waking life.

Gauḍapāda (c. eighth century C.E.)—the teacher of Śaṅkara's teacher—had proposed that from the standpoint of knowledge of the true self (*ātman*) and its union with ultimate nondual reality (*brahman*), the whole of our waking experience is no more than an insubstantial and illusory dream. Śaṅkara accepted that profound similarities link waking and dreaming, but he was unwilling to blur the distinction between them this way. Although waking and dreaming both take their objects to be real in ways they're not, waking experience has greater "reality" than dream experience.[45]

By "reality" Śaṅkara didn't mean things as they are in themselves, apart from experience. Little use can be made of this notion because we can never step outside experience to examine how it matches or fails to match reality so conceived. Rather, by "reality" he meant that which is revealed by an experience that hasn't been canceled, revoked, or contradicted by another experience. Mere appearance, however, can be canceled, revoked, or contradicted this way.

For example, as you step outside your house at dusk, you see a snake at your feet and instinctively jump back. Closer inspection proves it's not a snake but an old bit of rope. Your current perception of it as a rope cancels your previous perception of it as a snake, so you now take "rope" to be reality and "snake" to be mere appearance.

According to this criterion, waking experience has greater reality than dream experience because it's less subject to this kind of cancellation. It's a distinctive feature of dream experiences that they are constantly being revoked or contradicted both by other dream experiences and by waking experience:

> The words I'm reading in my dream aren't English because I see now they're really Chinese characters. But no, they're English again.
> I took myself to be awake, but the words keep shifting between English and Chinese, so I know I must be dreaming.
> I must have lost lucidity because I kept trying to find that strange book I was reading, but now that I'm awake I can see I was only dreaming.

Waking experiences aren't constantly revoked or contradicted this way, either by other waking experiences or by dream experiences. Waking experience has more stability than dream experience and allows for greater awareness of one's state and more cognitive control. Lucid dreaming brings these features of the waking state into the dream state.

Waking also encompasses an understanding of the dream state. We weigh the similarities and differences between the two from the standpoint of the waking state. We appreciate our dreams from the waking state, but we don't appreciate our waking life from the dream state—except in lucid dreams when we wake up within the dream and remember waking life.

When Śaṅkara says that waking experience has greater reality than dream experience, he means it includes an understanding of the dream state in a way the dream state doesn't include an understanding of the waking state. Using his technical philosophical term (in one of its English translations), we can say that the waking state "sublates" the dreaming state—it cancels the dream state's mistaken conception of itself and reevaluates the dream state from a more inclusive vantage point.

Lucid dreams don't contradict this point; they reinforce it. In a lucid dream, we wake up inside the dream. Hence waking and dreaming can't be opposed states. Waking life includes lucid dreaming; it includes recognition of the dream state from within the dream, not just retrospectively. The dream state, however, doesn't include recognition of waking life unless it's a lucid dream—a dream in which we're awake. Hence "waking" becomes the higher-order concept that subsumes both ordinary waking and lucid dreaming.

In short, "waking" in this sense isn't a state opposed to dreaming; it's a quality of awareness that can be present in any conscious state, including dreaming.

A DAOIST AFTERWORD: THE BUTTERFLY DREAM

Are lucid dreams better than nonlucid ones? If recognizing the dream state or waking up within the dream is a metaphor for enlightenment, for seeing things as they truly are, while the dream state is a metaphor for the unenlightened and deluded mind, is it always better to be lucid when we dream?

In March 2010 I went to a four-day dream yoga meditation retreat with Alan Wallace at the Upaya Zen Center in Santa Fe, New Mexico. In one of his talks, Alan said, "When you're awake, it's always better to be mindful than not mindful; when you dream, it's always better to be lucid than not lucid. Not being lucid means being both ignorant—not knowing you're dreaming—and deluded—believing you're awake. When you recognize the dream state in a lucid dream, you replace not knowing with knowing, and delusion with true comprehension."

This way of looking at things—its strong prescriptive or normative placement of lucid dreams above nonlucid ones—troubled me. During the question period, I tried to explain why.

Nonlucid dreams, I said, may have their own value, so we shouldn't think of them as being inferior to what they would be if we were lucid in them. There seem to be different kinds of dreams— dreams where you replay recent memories (what Freud called "day residue"), and dreams with a wholly different quality of clarity and intensity and creative imagination. In the first kind, being nonlucid might be crucial for learning and memory consolidation—for acquiring skills and strengthening the ability to recall important information. In the second kind, being nonlucid might be essential to the dream's power and immediacy, and hence its imaginative and creative value. Even mere lucidity—simply knowing you're dreaming without in any way trying to control the dream—might inhibit or obstruct these features. So being lucid shouldn't be given absolute value over not being lucid.

Alan wasn't convinced. He thought it would still be possible to have meta-awareness of these kinds of dreams and to be attentive to their qualities without impeding or interfering with them. And he also thought that no matter how valuable certain kinds of dreams might be, they nonetheless remain essentially ignorant and delusional in the absence of lucidity.

In my view, whether it's possible to be lucid without affecting other important functions of dreams remains an open question, given how little we know at the interface between dream yoga and dream science. In any case, the question seems an empirical one that dream yoga and dream science could try to settle.

The view that lucid dreams are better than nonlucid dreams, however, isn't empirical; it's evaluative or normative. It places different values on different kinds of human experience. I find myself unable to accept this. It seems to diminish the worth of something natural in our experience; it seems to go against the grain.

The classical Chinese Daoist (Taoist) philosopher Zhuang Zi (or Chuang Tzu, ca. 369–286 B.C.E.) offers a different viewpoint in his famous parable known as "The Butterfly Dream":

Once Zhuang Zhou dreamt—and then he was a butterfly, a fluttering butterfly, self-content and in accord with its intentions. The butterfly did not know about Zhou. Suddenly it awoke—and then it was fully and completely Zhou. One does not know whether there is a Zhou becoming a butterfly in a dream or whether there is a butterfly becoming a Zhou in a dream. There is a Zhou and there is a butterfly, so there is necessarily a distinction between them. This is called: the changing of things.[46]

In this parable there's no lucidity—no recognition of the dream state. Zhuang Zhou dreams and is completely immersed in being a butterfly, happy to be flying about and doing what butterflies naturally do, with no knowledge or memory of Zhuang Zhou. Then this dream state ends and there's no more butterfly; there's Zhuang Zhou, who doesn't remember the dream. Guo Xiang (232–312 C.E.), the most important commentator on Zhuang Zi, makes this mutual forgetfulness and ignorance clear: "Now Zhuang Zhou is just as ignorant about the butterfly as the butterfly was ignorant about Zhuang Zhou during the dream."[47] Instead of lucidity and memory, there's nonlucidity and forgetting.

Yet nonlucidity and forgetting here aren't equivalent to mindlessness as the opposite of mindfulness. Rather, they're expressions of a kind of selflessness, radical acceptance, and full presence in the here and now.

The absence of any ego or "I" in the story shows the selflessness. There's no one ego who is first awake, then dreaming, and then awake again. Instead, the story unfolds through three successive phases, one transforming into the other, with no overarching memory bridge connecting them—Zhuang Zhou awake, the dream butterfly, and another Zhuang Zhou awake. This second Zhuang Zhou isn't the same as the first one. He's continuous with him, in the way a river downstream is continuous with the river upstream but isn't identical to it.

Given that there's no one underlying ego, no phase in this transformation—human or butterfly—counts as more real or authentic than the others. Each phase experiences itself fully and completely as a human being or as a butterfly, with no knowledge of the others. Hence the narrator says: "One does not know whether there is a Zhou becoming a butterfly in a dream or whether there is a butterfly becoming a Zhou in a dream." As commentator Wang Xianqian (1842–1917) says, "You can say that it is Zhuang Zhou being the butterfly, or that it is the butterfly being Zhuang Zhou. Either is acceptable—and just this is their oneness, their transformation into each other."[48] Or in the words of another commentator, Lu Zhi (1527–1602): "we cannot say . . . that this one is awake and that one is dreaming, that Zhou is right and the butterfly is wrong."[49] One phase isn't reality and the other merely a dream; one isn't true and the other false. Each phase is authentic and fully accepted as it is. Thus selflessness or the absence of ego leads to radical acceptance of all phases equally.

In this Daoist vision, accepting each phase as equally real, along with accepting the natural distinction between waking and dreaming, is what enables one to be fully present in the here and now. Being aware of the butterfly dream as only a dream would prevent its being an experience of being fully and completely a butterfly, "self-content and in accord with its intentions." Remembering the dream—thinking back to it—would prevent the waking state from being an experience of being fully and completely human. The fullness of each experience requires not violating the natural borders between dreaming and waking, not supplanting forgetting with remembering. Transgressing these borders means fighting against change ("the change of things")—a losing battle that detracts from reality rather than bringing about a higher reality. Detracting from reality inevitably leads to suffering, not happiness. Thus the Daoist, in a gesture of radical acceptance, embraces dreaming and forgetfulness without judgment or qualification.

You might now be wondering whether this line of thought leads to a repudiation of lucid dreaming and dream yoga. But that's not the moral we should take from this parable. Lucid dreaming belongs to human experience, so for the Daoist, it's something we should embrace too. Instead, the moral is that our experience necessarily includes forgetting as well as remembering, dreaming as well as lucid dreaming, and so we devalue our experience and distort our nature if we set one kind of experience above the other.

The danger in cultivating lucid dreaming is that we become attached to it as something intrinsically more valuable than nonlucid dreaming. We need to be able to let go of lucidity and release ourselves to the full presence of the dream. If we can't, then we deprive ourselves of certain natural and valuable experiences. The point isn't to make a forced choice between lucidity or nonlucidity according to some imposed standard; it's to be able to move flexibly between them without attachment.

The Daoist sage embodies this flexibility. The sage doesn't go against the grain and make distinctions between high and low; he responds fluidly and effortlessly according to the circumstances and what's appropriate.

It's the sage who narrates the butterfly parable; he's the one telling us the story. Zhuang Zi, the author and presumably the narrator of the parable, speaks for the sage. He is "Master Zhuang," not Zhuang Zhou ("Zi" is the honorific title "Master"; "Zhou" is the family name). Master Zhuang doesn't tell the story in the first person about himself in the past; there's no "I" narrating the events, for there's no single, enduring ego underlying Zhuang Zhou, the butterfly, and Master Zhuang. He tells a story about something that happened in the past when once there was a dreaming Zhuang Zhou, and a butterfly dream, and then a Zhuang Zhou awake. He tells us that we don't know whether there was a man who became a butterfly in a dream or a butterfly who became a man in a dream. Master Zhuang, speaking for the Daoist sage, embraces both the butterfly and the man and affirms them equally, but identifies with neither one.

How can Master Zhuang know about both Zhuang Zhou and the butterfly while they know nothing of each other? The parable gives no answer to this question.[50] But implied in the parable's structure—in its narrative frame—is that the sage witnesses all three phases, embracing them with equal acceptance without identifying with any of them. In this kind of all-embracing, witnessing awareness, we rejoin the dispassionate meta-awareness that dream yoga cultivates.

Guo Xiang, in his commentary, reads the parable as an allegory of life and death: "The distinction between dreaming and waking is no different from the differentiation, the debate, between life and death."[51] Dreaming is no less real than waking; death is no less real than life. The "changing of things" includes waking and dreaming, as well as life and death. When living, we're like Zhuang Zhou with no knowledge of the butterfly; when dead, we're like the butterfly with no knowledge of Zhuang Zhou. The sage doesn't fear death but accepts life and death equally, not identifying with either one.

7
FLOATING

Where Am I?

I'm ten years old, sleeping in the top bunk bed in my room.
I know I'm asleep because I'm looking down at myself lying
in the bed below. I'm floating above the bed, just beneath
the ceiling.

The next morning I tell my mother my strange dream.
I felt light, almost weightless. A yellow glow brightened the
area around me while the rest of the room was dark. At
one moment I could see, from the vantage point of my bed,
my face looking down at me. It almost seemed like I could
see myself from two places at once and in two places at
the same time. Or maybe the two views were just flipping
back and forth. I'm not sure. Then I was up at the ceiling
again, with a clear view of myself under the covers in the
bed below. Everything seemed real; it didn't feel like any
dream I'd had before.

She listens and asks me how I felt in the dream. I tell her
I was a bit scared, though not like in a nightmare. Nothing
in particular frightened me; I just felt strange. Telling her
the dream reassures me.

Later that day, after dinner, my father calls me up to his
study. The second-floor room in the old Ontario farmhouse
is self-contained, with its own stairway off the kitchen by
the front entrance to the house. Sometimes he calls me up

to see whether I want to meditate with him before I go to bed. I try to sit cross-legged with my back straight, eyes closed, attention focused at the point between my eyebrows, while mentally saying *"Hong"* (rhymes with *song*) with the in-breath and *"Sau"* (pronounced *saw*) with the out-breath. I find it hard to sit still for more than about five minutes, but I like it anyway.

I hear you had a strange dream last night, he says. After I tell him the dream, he takes down one of his many books, a blue hardcover volume, and opens it to a drawing of two bodies, one lying in bed and the other floating above it, with what looks like some kind of thread connecting them.

We have more than one body, he explains. Besides our physical body, we have other kinds of bodies. One of these is called the astral body; it belongs to the astral plane, a subtler plane of existence than the physical plane. When you fall asleep, your astral body separates from your physical body so it can travel on its own and learn important lessons. Many dreams are the garbled memories of these experiences. Sometimes you can wake up in the dream state and become aware of your astral body, as you did. Advanced yogis can consciously project their astral bodies while they sleep and then accurately remember their astral travels. Your dream wasn't an ordinary dream; it was a kind of out-of-body experience called astral projection. It's good you remember it and nothing to be afraid of.

When I got a bit older my father told me the full, syncretic Yoga-Theosophy cosmology of the body. In Yoga, the physical body is called the "food sheath" (*annamayakośa*); it's the outermost "sheath" or "vesture" (*kośa*) of our full being. The "life-energy sheath" (*prāṇamayakośa*), which Theosophy calls the "etheric body," makes up our vital being. The "mental sheath" (*manomayakośa*) and the "higher intelligence sheath" (*vijñānamayakośa*) make up the astral body or our mental being. Subtler than the astral body is what Yoga calls the "bliss sheath" (*anandamayakośa*) and Theosophy calls the "causal body." The physical body is the body of ordinary waking life, the astral body is the body of dreams, and the causal body is the body of deep and dreamless sleep.

For years this cosmology of the body gave me a way to understand my out-of-body experience and other experiences. But as I learned more about the brain and studied philosophy of mind, I began to wonder. Do out-of-body experiences really show that one's consciousness can separate from the body? Are these experiences truly a form of disembodied consciousness, or some kind of illusion or hallucination?

What I've come to think is that out-of-body experiences are a specific kind of altered state of consciousness that's dependent on the brain and body. They aren't so much experiences of disembodiment as experiences of altered embodiment. Far from showing the separability of the self from the body, out-of-body experiences reinforce the intimate link between our body and our sense of self. Such experiences have specific neural correlates that overlap with the neural correlates of switching between first-person and third-person perspectives when we imagine and when we dream. Like dreams, out-of-body experiences are mental simulations or creations of the imagination, but like lucid dreams, they're subject to voluntary control, and you can know when you're having one.

DOUBLING YOURSELF

Where are you right now? Since you're reading these words, maybe you're at home sitting in a chair, or at a café, or browsing in a bookstore. Such answers, however, are partial—they report the place where you happen to be located, but such locales can change. Is there someplace where you always are—or at least where you always seem to be—that doesn't change this way?

Most of us would probably answer that we're located wherever our body is located. Or, more precisely, we feel ourselves to be located wherever we feel our body to be located, regardless of our surroundings. We feel that we are or own our body, and that we see the world through its eyes. In these ways, our sense of who we are here and now—our sense of self in the present moment—is an embodied or bodily sense of self.

Philosophers call this sense of self "bodily self-awareness." Bodily self-awareness includes feelings of ownership ("this body is mine"), agency ("I'm the one making this movement"), self-location ("I'm in my body"), and egocentric perception ("I'm seeing the world through my eyes"). Notice that these aspects of bodily self-awareness normally coincide or have a certain unity—we feel that the body through whose eyes we see the world (egocentric perception) is our own body (ownership) where we're located (self-location) and that we control from within (agency).

So-called "autoscopic" phenomena alter and disrupt these aspects of bodily self-awareness.[1] "Autoscopy" means seeing one's body from an outside perspective. Autoscopic phenomena can occur spontaneously while falling asleep or waking up, after severe accidents, during surgery when conscious awareness intermittently returns, and as a result of damage to certain areas of the brain.[2]

In an "autoscopic hallucination," you see a double or duplicate of yourself in your personal space. In other words, you see a body that looks like you, but you don't experience any ownership of this body or control over it, nor do you locate yourself in it; it's your double's body, not your own body.

In "heautoscopy," your egocentric visual perspective switches back and forth between your normal body and your illusory double body. Sometimes your visual perspective can even seem to be at these two positions at the same time. You feel ownership of both bodies and you locate yourself in both bodies. There's a feeling of doubling or of having two selves.

In out-of-body experiences (OBEs), you see yourself from a position outside your physical body and locate yourself at that vantage point. Your experience is thus autoscopic—you see your body from the outside—but unlike an autoscopic hallucination, you experience this body as yours, not as belonging to a double. In this way, your experience is like heautoscopy, except that you locate yourself fully at the outside vantage point.

The childhood experience I described above had all the characteristics of an out-of-body experience, but it also included moments of heautoscopy. I felt as if I were outside my physical body with an elevated center of awareness from which I could view my body down below me in the bed. But at certain points my experience also switched back and forth between looking down at my sleeping body and seeing my face up above looking down at me in the bed.

OUT-OF-BODY BUT NOT DISEMBODIED

People often describe out-of-body experiences as experiences of disembodiment. Some people believe you're literally outside your physical body and are conscious without it. Yet this hardly follows from the experience itself. As psychologist and writer Susan Blackmore notes, describing the first time she had an out-of-body experience: "I was not functioning *without* my physical body. I seemed to be in a different place from that body, but there is no doubt that it was functioning quite well. It may have been tired, but it was not dead. It was sitting up, moving and talking. It is therefore unjustified to conclude that such an experience could take place without a functioning body."[3]

In fact, the state and position of one's body are important contributing factors to out-of-body experiences. Most of them occur when the physical body is relaxed and inactive. Blackmore's first, as she tells us in her book, *Beyond the Body: An Investigation of Out-of-the-Body Experiences*, happened after she had smoked some hash and was feeling sleepy while listening to music with friends.[4] And, although she was sitting, most out-of-body experiences happen to people when they're lying down. The techniques for inducing an out-of-body experience also usually require that you lie on your back and enter a relaxed and sleepy state.[5] Finally, when neurological patients experience autoscopic hallucinations and heautoscopy they're either sitting or standing, but when they have out-of-body experiences they're in a supine position. Thus, "OBE and autoscopic hallucination depend differently on the patient's position prior to the experience, suggesting that proprioceptive and tactile mechanisms influence both phenomena."[6]

Whether there's some aspect of consciousness that transcends the physical body during an out-of-body experience is a question I'll come back to later. The point I want to make now concerns the experience itself, that is, what it's like to have an out-of-body experience. Scientists and philosophers who reject the idea that we have another astral or spiritual body that leaves the physical body nevertheless often describe these experiences as ones where you seem to yourself to be disembodied. But this description is inaccurate.

Out-of-body experiences feel as if they're happening in a bodily space, that is, a space that's perceived and felt in a bodily way. Throughout, you experience yourself as having a visual-spatial perspective (egocentric perception), as being located at the origin of that perspective (self-location), and as being able to move deliberately through space (agency). These features are always present, even though some people report experiencing themselves not as having a body but as being a blob or a point of light.[7] Yet no matter what form you take, to have a visual-spatial perspective and be able to move through space mean that you experience yourself as occupying space, and any sense of self that's spatial in these ways is also bodily, at least in a minimal sense. As philosopher Thomas Metzinger writes, "It is interesting to note how OBEs, phenomenologically, are *not* states of disembodiment. On the contrary, there always seems to be a spatially located phenomenal self, even if its embodiment is reduced to a pure spatial point of visuo-attentional agency."[8] In other words, even if you experience yourself as a passive observational point of view, you still experience yourself as able to look in this or that direction, that is, as being able to direct your visual attention within phenomenal space, and this means that you're experiencing yourself as spatially located, not as a disembodied self that lacks all spatial properties (like René Descartes's conception of the disembodied thinking self).

So far, we've been thinking of the spatial perspective or frame of reference in out-of-body experiences as an egocentric visual perspective, but it's also a gravity-centered or geocentric spatial reference frame with vertical "up" and "down" directions. In general in ordinary experience and out-of-body experiences, we have a constant knowledge of the vertical orientation and of which way is up. Sensory receptors in the inner ear, called otolith vestibular receptors, are sensitive to gravity, linear acceleration (the rate of change in velocity as you move in one direction), and horizontal movement. In out-of-body experiences, people usually report that they seem to be above their physical bodies, and they typically describe feelings of rising rapidly, floating, and flying horizontally. Such sensations all involve the vestibular system of the inner ear (the semicircular canals indicating rotational movements, and the otoliths indicating linear accelerations). In short, the sense of self in out-of-body experiences includes not just a visual egocentric perspective but also a vestibular geocentric perspective with a bodily sense of balance, movement, and up-down orientation.

For these reasons, instead of describing out-of-body experiences as experiences of disembodiment, we should describe them as experiences of altered embodiment. You see your body from the outside as being in a location that doesn't coincide with the felt location of your awareness. In other words, you see your body as an object at a place that doesn't coincide with the felt location of your visual and vestibular awareness as a subject.

In this way, out-of-body experiences bring us back to a crucial distinction seen in earlier chapters examining memory and dreaming—between the sense of self as subject and the sense of self as object. We need to take a closer look at this distinction in the context of out-of-body experiences.

BODY AS SUBJECT, BODY AS OBJECT

As we've seen, when you're conscious of something, you grasp or apprehend it in a certain way—you perceive the pattern of light and color as a sunset, you take the dream image to be a real person chasing you, you pay attention to the dream as a dream, and so on. In these and many other ways, your consciousness is "about" or "directed toward" the objects of your experience. What happens when the object of your experience is you?

One way to experience yourself is as the direct object of your awareness. When you look down at your hands, see your face in a mirror, or direct your attention to how you feel on the inside, you immediately recognize the hands, the mirror image, and the internal sensations as you or yours. In such cases, your awareness is transitive—it takes an object—and you experience that object immediately as you or yours.

Phenomenologists call this way of experiencing yourself the "self-as-object." We can map this onto our earlier discussion of bodily self-awareness by saying that when you experience yourself as an object of outer or inner perception, you're also having an experience of your "body-as-object."

As we know from earlier chapters, you also experience yourself as a subject. When you look down at your hands and recognize them as your own, you know and feel that you're the one having this experience from your unique perspective. Your felt awareness includes not just your hands but also your seeing of them. In this way, you know and are aware of this visual perception as your own experience.

Phenomenologists call this way of experiencing yourself the "self-as-subject." We can also map this kind of self-experience onto bodily self-awareness by saying that whenever you perceive the world, you also experience your "body-as-subject." In particular, whenever you perceive your body as an object in the world, you also experience your body as you yourself undergoing this perception.

In sum, bodily self-awareness includes both transitive awareness of the body as an object of perception and feeling, and intransitive bodily self-awareness as a perceiving and feeling subject.

Out-of-body experiences illustrate the importance of distinguishing between the body-as-subject and the body-as-object.[9] Your body-as-object is the body you see from the outside lying in bed, whereas your body-as-subject is you the perceiver. To put the point another way, your body-as-object is the external body image you identify as your body, whereas your body-as-subject is the felt origin of the visual (egocentric) and vestibular (geocentric) perspective from which you make that identification.

We can now say in more precise terms what makes an out-of-body experience an experience of altered embodiment rather than of disembodiment: there's a dissociation between your body-as-object and your body-as-subject. Normally you experience them as being in the same place. In an out-of-body experience, however, this unity comes apart, so that your body-as-object and your body-as-subject have different locations. For example, your body-as-object lies below on the bed while your body-as-subject floats above near the ceiling.

Out-of-body experiences reveal something crucial about the sense of self: *You locate yourself as an experiential subject wherever your attentional perspective feels located.* Thus, as Metzinger remarks, "attentional agency . . . is one of the essential core properties underlying the conscious experience of selfhood."[10] Although attentional agency and the body-as-object are normally unified, out-of-body experiences show they can come apart. These experiences also show that when they do, your sense of self adheres to your sense of attentional agency, and your sense of self-location adheres to your sense of visual-spatial perspective. In other words, your sense of who you are and where you're located goes with your self-as-subject and not your body-as-object.

IN THE BRAIN AND OUT-OF-BODY

What's going on in the brain during an out-of-body experience? Although we still can't give a definitive answer to this question, evidence from neuroscience suggests that out-of-body experiences depend on a particular area of the brain where the temporal and parietal lobes meet, called the temporoparietal junction.[11] This area is crucial for multisensory integration—integrating signals from the different sensory modalities of sight, sound, touch, and self-movement—and for being able to switch between taking a first-person perspective and a third-person perspective in mental imagery.

In 2002, neuroscientist Olaf Blanke and his colleagues in Geneva and Lausanne reported in the journal *Nature* that they had repeatedly induced out-of-body experiences by electrically stimulating the brain of a patient being treated for a drug-resistant epilepsy.[12] The doctors had failed to find any brain lesion using brain imaging methods, so they implanted electrodes on the surface of the cortex in order to locate the region where the seizures began. The doctors also used focal electrical stimulation to map the relationship between specific brain areas and specific functions. For example, stimulating different areas of the motor cortex evokes different body movements. During stimulation of the right angular gyrus—a structure belonging to the area of the temporoparietal junction—the patient reported having experiences resembling out-of-body experiences. The first stimulations produced vestibular feelings that the patient described as "sinking into the bed" or "falling from a height." When the doctors increased the amplitude of the electrical current, the patient reported, "I see myself lying in bed, from above, but I only see my legs and lower trunk." Two further stimulations produced the same experience, including an instantaneous feeling of "lightness" and "floating" about two meters above the bed, near the ceiling.

In a subsequent study, Blanke and his colleagues recorded detailed phenomenological information about spontaneous out-of-body experiences in this patient and five other neurological patients. Here's a description of another patient's out-of-body experience just prior to surgery:

> The patient was lying in bed and awakened from sleep, and the first thing she remembered was "the feeling of being at the ceiling of the room." She ". . . had the impression that I was dreaming that I would float above [under the ceiling] of the room. . . ." The patient also saw herself in bed (in front view) and gave the description that "the bed was seen from above" and that "there was a man and that she was very frightened." The scene was in colour, and was visually clear and very realistic.[13]

Blanke and his colleagues found that five of these patients had brain damage in the temporoparietal junction of the right hemisphere. In a later study, the same brain region was found to be activated within half a second when healthy individuals were asked to imagine seeing things from an out-of-body perspective. Furthermore, interfering with this region by magnetically stimulating it impaired the ability to imagine this transformation of body position.[14]

These findings, along with others, have helped neuroscientists to build up a picture of the temporoparietal junction as a key neural site for the integration of information related to our bodily sense of self, including how we recognize others on the basis of their bodies and our sense of how our own body looks from the outside to other people.[15] The right temporoparietal junction includes the core region of the vestibular cortex, which is responsible for our sense of balance and spatial orientation. Other regions belonging to the temporoparietal junction coordinate proprioceptive, tactile, and visual information about the body. In addition, the temporoparietal junction is involved in the perception of the human body, imagining one's own body, switching between first-person and third-person perspectives, and being able to distinguish between oneself and others.

In out-of-body experiences, vision, proprioception, and vestibular awareness come apart. You see yourself as being at a location that doesn't coincide with the source of your egocentric visual perspective and with the source of your vestibular awareness. Since the temporoparietal junction is involved in integrating these different kinds of sensory information, it makes sense to suppose out-of-body experiences depend on some kind of disruption to multisensory integration there.

Blanke proposes that out-of-body experiences happen when the temporoparietal junction suddenly fails to integrate sensory signals from the body in the normal way.[16] Specifically, he proposes that two disturbances to multisensory integration combine to produce out-of-body experiences. On the one hand, proprioceptive, tactile, and visual signals about one's own body are not properly matched. On the other hand, the vestibular frame of reference for one's personal space is not properly matched to the visual frame of reference for external space. These two disruptions combine to create the experience of seeing one's body in a position that doesn't coincide with its felt location, together with the experience of floating and seeing things from an elevated visual-spatial perspective.

In Blanke's view, these sensory-integration disruptions can occur spontaneously in healthy individuals as well as through experimental manipulations such as direct cortical stimulation. They can also be made to occur in healthy individuals through computer-generated virtual reality techniques.

VIRTUALLY OUT-OF-BODY

In 2007 two neuroscience teams—one in Lausanne led by Bigna Lenggenhager and Olaf Blanke and including philosopher Thomas Metzinger, and the other in Stockholm led by Henrik Ehrsson—reported in the journal *Science* that they had independently induced elements of out-of-body experiences using head-mounted video displays to give people an outside perspective on their own bodies.[17] These experiments all relied on the power of vision to influence or control the other senses.

In Lenggenhager's experiment, the participants stood in front of a camera and saw their own back presented from the camera's view through video-display goggles. A computer enhanced the image to create a three-dimensional "virtual own body." When the participants' backs were stroked with a pen at the same time they saw their virtual back being stroked—but not when the tactile and visual stroking were asynchronous—the participants reported that the touch sensation seemed to be located at their virtual back.

One way to describe this result is to say that given the conflict between vision and touch—between the visual cues and the tactile cues—vision dominated. Vision "captured" the touch sensations so they were referred to the seen body. In this way, the participants self-identified with the seen virtual body as their own body.

But there was more to the experiment. When the scientists turned off the camera, guided the participants back a few steps, blindfolded them, and then asked them to walk back to their original position, they returned to a position that was closer to the apparent location of their virtual body, as if they had been located in front of the position where they had been standing during the back stroking. Thus the participants' sense of self-location was shifted or biased toward the virtual body.

In Ehrsson's experiment, the participants sat in a chair and viewed themselves from behind through a video headset. Ehrsson used one hand to stroke the participant's chest with a plastic stick, while his other hand held another plastic stick and stroked a location just behind the participant and in front of the cameras. Thus, the participants felt the stroking on their real chest, which was out of view, but saw the other arm touching the apparent location of their virtual chest just below the cameras' field of view. When the strokes were synchronous—so that the participants felt the real chest stroke at the same time as they saw the virtual chest stroke—they reported feeling that the stick they saw was the one touching their real chest, and that they were sitting behind their physical body. By referring the felt touch to the visually perceived location of the virtual chest, the participants were in this way self-identifying with the virtual chest as their own body. Ehrsson also swung a hammer in front of the camera so that it seemed to hit the area of the virtual chest. The participants reported feeling anxious even though they knew the hammer posed no threat, and the change in their skin conductance response indicated that the hammer swinging increased their level of physiological arousal. Thus both the participants' descriptions of how they felt and their skin conductance response indicated that they experienced being at the position of the cameras. Their sense of self-location was shifted or biased toward the visual-spatial perspective of the cameras outside their physical bodies, which now appeared to them from an external perspective, as in an out-of-body experience.

In both experiments, the participants' sense of where they were located was shifted or biased toward the location where they saw the touch occurring (the seen location of the stroking stick). Thus, when conflicting visual signals and tactile signals occur in synchrony, vision dominates and strongly influences the sense of self-location.

Yet there were differences between the two experiments. In the first one, the participants referred the touch sensation on their back to the virtual body appearing in front of them, whereas in the second, the participants referred the touch sensation on their chest to the visual-spatial perspective of the camera behind them.

In an effort to harmonize these findings, Lenggenhager and Blanke devised a new experiment that used identical body positions and identical measures of self-location.[18] In this study, the participants were lying down in a prone position. They were again fitted with a head-mounted video display that showed an image of their body from the perspective of a camera above. In order to measure their sense of self-location, the scientists asked the participants to imagine dropping a ball from their hand to the ground, and to estimate its falling time. The experimenters reasoned that the estimated falling time would be greater when the participants perceived themselves to be higher from the ground, and smaller when they perceived themselves to be closer to the ground. The experimenters also predicted that when the participants felt their back being stroked at the same as they saw it being stroked ("synchronous back stroking"), they would locate themselves "downward," closer to the virtual body, whereas when they felt their chest being stroked at the same time as they saw it being stroked ("synchronous chest stroking"), they would locate themselves "upward," away from the virtual body. As predicted, the "mental ball dropping" time estimates (quantifying self-location) were lower for synchronous back stroking than for synchronous chest stroking. In other words, when the participants were stroked on the back, their sense of self-location shifted toward the virtual body appearing below, whereas when they were stroked on the chest, their sense of self-location shifted upward toward the visual-spatial perspective of the camera above. In addition, the participants reported that during the chest stroking and upward shift they experienced vestibular sensations of elevation and floating, like the sensations reported in out-of-body experiences.

How do these results relate to the idea that out-of-body experiences depend on the way the temporoparietal junction integrates sensory signals from the body? Blanke's team, led this time by Silvio

Ionta, next addressed this question in another study that adapted the horizontal body position and mental ball-dropping task for use in a brain scanning experiment.[19]

The participants were inside a magnetic resonance imaging (MRI) scanner, so they had to lie on their back and look upward at the virtual body. Once again, only during synchronous stroking—when the participants felt their back being stroked at the same time as they saw the virtual back being stroked—did they report feeling as though the virtual body was their own. The mental ball-dropping time estimates also showed that the participants perceived their physical body drifting toward the virtual one during synchronous stroking. Yet they also reported striking differences in the direction of their visual perspective. Half of them had the impression of being below the virtual body and looking up at it (the Up group); the other half had the impression of being above the virtual body and looking down at it (the Down group). In other words, the visual-spatial perspective of the Up group was consistent with their actual supine physical position and first-person perspective in the scanner, whereas these were inconsistent for the Down group, who experienced an elevated visual-spatial perspective and sensations of floating, as in out-of-body experiences.

When Ionta and Blanke examined the brain activity, they found that the illusory changes in self-location correlated specifically with changes in activity at the temporoparietal junction. They also found there were different patterns of activity in this area for the Up and Down groups. Ionta and Blanke interpret these differences as showing that activity in the temporoparietal junction reflects not only how the integration of visual and tactile signals influences one's sense of self-location, but also how the integration of visual and vestibular information about self-location and the orientation of one's visual-spatial perspective influences one's first-person perspective.

These results of combining brain imaging and virtual reality methods of manipulating bodily self-awareness support the idea that out-of-body experiences depend on how the temporoparietal junction deals with visual, tactile, and proprioceptive cues about one's own body, and with visual and vestibular cues about one's bodily orientation in space.

More generally, what these and other virtual reality experiments have done is to use the power of vision over other senses to manipulate bodily self-awareness in systematic ways that reveal different aspects of our bodily sense of self. We self-identify with our body, but we can be made to self-identify with a virtual body. We locate ourselves within our body, but we can be made to feel that we're located at places outside the borders of our body. We experience the world from the visual-spatial perspective of our body, but we can be made to experience the world from perspectives outside our body with different "up" or "down" orientations. We feel that we are or own our bodies, but manipulating both our visual-spatial perspective and the sensory cues we get from vision, touch, and proprioception can make us feel ownership for an artificial body or a purely virtual body.[20]

In sum, bodily self-awareness includes feelings of self-identification, self-location, having a first-person perspective, and body ownership, all of which depend on the way our body's sensory and motor systems converse with our brain. Using virtual reality, we can alter this conversation in systematic ways on the sensory and motor side, and thereby create corresponding alterations to all these aspects of our bodily sense of self.

ASTRAL BODY OR MENTAL SIMULATION?

If out-of-body experiences have definite neural correlates and can be partly induced through direct stimulation of the brain and computer-generated virtual reality environments, what should we make of the idea that there's an astral body that leaves the physical body in out-of-body experiences?

Before tackling this question, we need to realize that so far, none of the virtual reality experiments has managed to induce the kind of full-blown out-of-body experiences that enthusiasts of "astral travel" report they've had and have been able to bring about voluntarily.[21] These subjectively compelling experiences include the sensation of separating from your physical body, seeing your physical body from an outside position, having some kind of ethereal second body, locating yourself at the origin of that ethereal body's visual-spatial perspective, being able to travel through space by mentally directing your ethereal body, and being able to pass through physical barriers, such as the walls and ceiling of your house.

Nor do the out-of-body experiences induced in epileptic patients through direct brain stimulation have these features, though they do seem closer to full-blown out-of-body experiences than the ones induced in healthy people using virtual reality.

So we need to evaluate two possible ways of thinking about experimentally induced out-of-body experiences in relation to our opening question. On the one hand, those who believe in the astral body and astral travel would argue that what virtual reality and direct brain stimulation do is create a partial illusory version of what happens for real in astral projection or genuine out-of-body experiences.

This is the line that cardiologist and near-death experience researcher Pim van Lommel takes.[22] He argues that Blanke's patients suffered from out-of-body illusions, whereas the out-of-body experiences that have occurred as part of near-death experiences—where patients say they saw themselves from an outside perspective while in the operating room or while doctors and nurses were trying to resuscitate them—aren't illusory because the patients had accurate and verifiable perceptions of themselves and their surroundings.

On the other hand, those who believe that out-of-body experiences are illusions, hallucinations, or creations of the imagination would argue that although virtual reality and direct brain stimulation so far have managed to bring about only partial aspects of full-blown out-of-body experiences, someday it may be possible to induce full experiences, especially if scientists draw on the methods for voluntarily bringing them about and work with individuals skilled in using these methods.

This is the line taken by researchers such as Olaf Blanke, Susan Blackmore, and Thomas Metzinger. In Metzinger's words:

Rigorous and systematic research on OBEs can also support a research strategy I would like to call "neurophenomenological reduction of paranormal belief systems." . . . The conscious brain is an "ontology engine," it creates a model of reality constructed from assumptions about what exists and what doesn't. . . . It seems plausible that many reports about "paranormal" events and experiences are absolutely sincere reports about specific and highly realistic phenomenology—e.g., of moving outside one's body—which can now be explained in a more parsimonious manner.[23]

I'm inclined to agree with this assessment, with one qualification. I think there are important phenomenological differences between the partial out-of-body experiences induced through virtual reality or direct brain stimulation and the self-induced and full-blown out-of-body experiences that people in different times and cultures have trained themselves to have.[24] Blanke's neuropsychological model of out-of-body experiences as arising from disruptions to sensory integration at the temporoparietal junction doesn't yet account for how these experiences can be self-induced and manipulated voluntarily in ways that apparently produce vivid and detailed impressions of viewing the world while moving outside one's body. So neurophenomenological accounts need further elaboration and refinement.

Like Metzinger, who describes his own out-of-body experiences in his book, *The Ego Tunnel*,[25] and Blackmore, who describes hers in *Beyond the Body*, I used to believe that out-of-body experiences show we have an astral body that can journey on its own in a subtler plane of reality than the physical plane. Gradually, however, I've come to the view that out-of-body experiences don't offer convincing evidence for an astral body. Although there are numerous accounts of subjectively compelling out-of-body experiences from many times and cultures,[26] it seems pretty clear that there's little, if any reliable evidence for the existence of an astral body that departs from the physical body. More parsimonious psychological and physiological explanations are now available.

What makes this assessment a bitter pill to swallow—at least for those of us who've had these experiences—is that the experiences themselves can be so realistic and one's mental state in them so lucid that it's very difficult not to believe that you've separated from your physical body and are truly perceiving physical features of yourself and your environment. To quote Metzinger once more:

> For anyone who actually had that type of experience it is almost impossible not to become an ontological dualist afterwards. In all their realism, cognitive clarity and general coherence, these phenomenal experiences almost inevitably lead the experiencing subject to conclude that conscious experience can, as a matter of fact, take place *independently* of the brain and the body: what is phenomenally possible in such a clear and vivid manner must also be metaphysically possible or actually the case. Although many OBE reports are certainly colored by the interpretational schemes offered by the metaphysical ideologies available to experiencing subjects in their time and culture, the experiences as such must be taken seriously. Although their conceptual and ontological interpretations are often seriously misguided, the truthfulness of centuries of reports about ecstatic states, soul-travel and second bodies as such can hardly be doubted.[27]

At this point in the argument, some people would claim that veridical perception happens at least in some out-of-body experiences. As we saw above, this is how van Lommel differentiates the out-of-body experiences that occur in near-death experiences from the illusory experiences induced through virtual reality and direct brain stimulation. Since this view is bound up with the question of how to understand near-death experiences, I'm going to postpone discussion of it until chapter 9. What we need to ask now is whether there's any evidence for veridical perception in the out-of-body experiences that occur in healthy individuals.

SEEING OR THINKING YOU'RE SEEING?

A few months after my first out-of-body experience, I had a vivid flying dream.

I'm above the fields behind our house, beyond the large green barn and gully, near where our neighbors live. Looking below, I can see they're walking our dog, Bridey. I'm surprised to see her on a leash because we've always let her run free. I wake up and notice the clock says 6 a.m.

We're staying with my grandparents in Los Angeles, and later that day my mother tells me she spoke on the phone to our neighbors back home. Bridey is in heat for the first time, so they have to keep her inside and walk her on a leash. They took her for a walk this morning around 9 a.m. I tell my parents about my dream, which I notice must have happened around the same time they were walking Bridey, given the three-hour time difference between California and Ontario. Another astral projection, my dad says, and you can tell this one wasn't just a dream because you saw something that was really happening.

Later my father referred to this out-of-body experience (as we had come to think of it) in an argument with our friend Gregory Bateson, the well-known anthropologist, cybernetic theorist, and author of the 1970s counterculture bible, *Steps to an Ecology of Mind*.[28] Gregory denied there was any evidence for "extrasensory perception," a concept he thought was incoherent. My father objected, citing Charles Tart's parapsychological research on verifiable perception in out-of-body experiences.[29] Gregory was unimpressed, so my father told him about my out-of-body journey from California to Ontario. "Hummph," Gregory snorted, which was what he always did when he didn't believe a word and had nothing more to say.

The out-of-body literature is full of stories like mine, in which people think they've truly perceived real events outside them, either at the place where their physical body is resting or at faraway locales. In the case of my experience, however, although the dream was subjectively compelling and strengthened my felt connection to my dog—making the dream a meaningful and valuable experience for me—it was a singular and unrepeatable event, not a repeatable and testable one. Having a dream about my dog wasn't unlikely given the circumstances: I was probably missing her, wondering when she would first go into heat, hoping she was being well cared for and wouldn't forget me, and so on. As for her being walked on a leash at roughly the same time I was dreaming about her being walked this way—that I can't explain, but I also see no way to rule out its being a coincidence.

When we examine accounts of alleged out-of-body perception, a striking feature stands out: they always seem to contain a mixture of right and wrong information, of accurate but also highly inaccurate descriptions of the environment supposedly being perceived.[30]

Take the case of Pam Reynolds—famous in the near-death experience literature.[31] She had an out-of-body experience during a long and complicated medical operation to treat an aneurysm. She accurately described that a small area of hair had been shaved from her scalp and that she saw a bone saw that looked like an electric toothbrush, but her description of the bone saw didn't fit the model actually used in her operation. And although her out-of-body experience took place from a visual perspective above the surgeon's shoulder, she didn't report seeing that her head had been turned to one side and held there by a mechanical head-holder—a detail that would have been in plain sight had she really been able to see herself.[32]

Another case is that of Robert Monroe, author of the popular book *Journeys Out of the Body*. When Charles Tart in California asked Monroe in Virginia to visit him in an out-of-body experience at an unspecified time, the time of his out-of-body experience seemed to match the time when Tart and his wife were concentrating on encouraging him to visit, but the description he gave of Tart's home and what they were doing was inaccurate—he saw too many people in the room, and he saw Tart and his wife doing things they didn't do.[33]

At the time of writing this book, attempts to verify out-of-body perception in controlled laboratory conditions—or in hospital rooms where patients have near-death experiences—have not provided any reliable evidence. Although it's possible such evidence could be forthcoming some day, the previous research on this issue doesn't give me reason to think it very likely.

In my view, the impression of seeing things in an out-of-body experience is like the impression of seeing things in a dream; in both cases, what's happening isn't perception but the mental simulation of perception. I agree with Blackmore:[34] in an out-of-body experience, you mentally simulate seeing your body from a third-person perspective, as you do when you remember yourself from the outside or see yourself from the outside in a dream. Using memory and your internal "cognitive maps" of the layout of the outside world, you mentally simulate seeing the environment. As Metzinger notes in support of this idea, when you move in an out-of-body experience, you don't move in a smooth, continuous path, but in discrete shifts or "jumps" as you "think" yourself from one place to the next—places that correspond to salient points in your cognitive map.[35] Notice that this is also how you move in dreams.

In short, the world of out-of-body experiences seems to be the world of the imagination.

There is one interesting way I can think of to push back against this conclusion. Consider first that in ordinary perception we often get all sorts of details wrong and miss things in plain sight, and that the world as we see it in waking perception is also a cognitive construction. Consider next that out-of-body experiences can convey accurate information about certain aspects of the environment while still being mental simulations. After all, on the basis of subliminal sensory cues, you do accurately represent yourself as lying in bed, and you do accurately represent some of the features of the room on the basis of memory. Now consider that if some of the information gained in out-of-body experiences came through (unknown) extrasensory channels, and so wasn't strictly speaking perceptual, your mind would still need to incorporate this information into its ongoing mental simulation. The result would be a mental simulation that gets some things right and some things wrong. So although out-of-body experiences would be mental simulations, they wouldn't be mere hallucinations.

Blackmore classifies this type of theory as parapsychological, which she describes as any theory that combines mental simulation and ESP (extrasensory perception) in order to account for out-of-body experiences. Like Blackmore, I think "ESP" is a catch-all term that explains nothing. Indeed, I'm inclined to go further and agree with Bateson that the concept is nonsensical. Nevertheless, it still seems possible to me that we could possess "extraordinary" ways of knowing that we do not at present understand or know how to conceptualize properly.[36] Moreover, if some out-of-body experiences were shown to provide verifiable and veridical information about aspects of the environment, then the idea of "extraordinary knowing" incorporated into the running of a mental simulation would seem a better model than one based on the reification of an astral body traveling in an astral plane.

Metzinger comments that if any scientific headway is to be made in this area, then we need to pursue fine-grained neurophenomenological analyses of out-of-body experiences "until we are able to make the target phenomenon *repeatable*, an object of investigation that can be reliably reproduced in a rigorously controlled experimental setting. Then we can directly investigate claims with regard to extrasensory perception [or extraordinary knowing] during the OBE state in a systematic manner."[37]

Although I have a great deal of sympathy with this assessment, I also wonder whether certain out-of-body experiences, as well as other experiences purportedly involving "extraordinary" or "anomalous knowing," are singular and intensely personal experiences occurring in exceptional circumstances, and therefore aren't repeatable and controllable in standard laboratory ways. If this were true, then we'd need a proper way to investigate them, where the singular, personal, and exceptional were recognized as essential features of the phenomena, instead of eliminated as meaningless noise or controlled in such a way that they disappear.

ARE OUT-OF-BODY EXPERIENCES LUCID DREAMS?

The last point I want to make in this chapter is that out-of-body experiences share many features of lucid dreams. Out-of-body experiences tend to occur during sleep in the same conditions as lucid dreams. You can know you're having an out-of-body experience while it's occurring and can voluntarily control the experience to varying degrees. Lucid dreams and out-of-body experiences also both tend to be highly realistic or hyperrealistic. These similarities suggest that out-of-body experiences could be either a subtype of lucid dream or a closely related sleep phenomenon.

Many out-of-body experiences, as we've seen, occur during a relaxed and prone state. They also seem to start from either the hypnagogic state leading into sleep or the hypnopompic state between dreaming and waking. Consider Charles Tart's description of "Miss Z," one of the individuals he studied who had many out-of-body experiences (and who, on one occasion, according to Tart, reported correctly the random target number 25132 that Tart had hidden near the ceiling above her bed):

> While chatting about various things with a young woman who babysat for our children, I found out that, ever since early childhood, it was an ordinary part of her sleep experience to occasionally feel as if she had awakened from sleep mentally but was floating near the ceiling, looking down on her physical body, which was still asleep in bed. This experience was clearly different to her from her dreams, and usually lasted only a few seconds. As a child, not knowing better, she thought this was a normal part of sleeping. You go to sleep, dream a bit, float near the ceiling for a bit, dream a bit, wake up, get dressed, eat breakfast, and go to school. After mentioning it to friends once or twice as a teenager, she found it wasn't "normal" and that she shouldn't talk about it anymore if she didn't want to be considered weird![38]

This story reminds me of a remarkable person I interviewed about lucid dreaming. She was a vivid lucid dreamer who not only experienced grapheme-color synesthesia (the consistent association of letters and numbers with colors) but also a rare kind of synesthesia in which numbers, letters, simple shapes, and even objects such as furniture have rich and detailed personalities. She had attracted the interest of psychologists who showed in a study published in the *Journal of Cognitive Neuroscience* that her object-personality pairings were extremely detailed and stable over time.[39] She participated in this study when she was a seventeen-year-old high school student; when we spoke, she was a third-year undergraduate student. T.E., as she was called in the published study, told me that when her older brother (also a synesthete) described lucid dreaming to her and asked her whether she'd ever had a lucid dream, she was surprised to learn that there was any other kind. All her dreams were and always had been lucid. Whereas I was interested to hear all about her lucid dreaming, she seemed much more interested to hear from me what it was like to dream and have no idea you were dreaming!

Besides showing that what counts as "ordinary" versus "unusual" in experience is highly relative to the individual, these cases show that you can experience various states of consciousness in reliable and robust ways, as well as self-induce and voluntarily guide them, without knowing exactly what those states are and how they fit into the full spectrum of consciousness. Starting from a relaxed state while lying down, you can enter a hypnagogic state and then have a lucid out-of-body experience, where you know you're having an out-of-body experience without knowing you're really in a dream state (you could falsely believe you're really traveling outside your body). If you knew you were dreaming, then your out-of-body experience would also be a lucid dream. Or you can have a lucid dream without knowing there's any other kind of dream state, and vice versa. Or you can have a lucid out-of-body experience and then think you've exited the state and are awake, but really you've slipped into a dream state by way of a false awakening. And so on.

These examples suggest that first-person phenomenology may not be enough to tell us what kind of experience we're having or what kind of conscious state we're in. We may also need the outside perspective of neuroscience and psychology. From both perspectives—the inside one of phenomenology, and the outside one of neuroscience and psychology—there are strong resemblances between out-of-body experiences and lucid dreams.

Many of the out-of-body experiences that Sylvan Muldoon and Hereward Carrington describe in their classic 1929 book, *The Projection of the Astral Body*—the book with the pictures my father showed me when I was ten years old—start from sleep states where lucid dreaming is also likely to occur—the hypnagogic and hypnopompic states, states of sleep paralysis, and flying dreams.

Although I've never been able to bring about deliberately an out-of-body experience, the times I've come the closest have been during afternoon naps when I'm in a state of sleep paralysis and know I'm dreaming. On several occasions, I've managed to produce the feeling of a second body by imagining my body moving. Instead of fighting the paralysis and trying to move my real body, I try to visualize my body getting up and looking back at my body in the bed. I've never been completely successful, but I have managed to create the strong sensation of the kind of second body that Frederik van Eeden describes in his classic article on lucid dreaming:

> In the night of January 19 to 20, I dreamt that I was lying in the garden before the windows of my study, and saw the eyes of my dog through the glass pane. I was lying on my chest and was observing the dog very keenly. At the same time, however, I knew with perfect certainty that I was dreaming and lying on my back in my bed. And then I resolved to wake up slowly and carefully and observe how my sensation of lying on my chest would change to the sensation of lying on my back. And so I did, slowly and deliberately, and the transition—which I have since undergone many times—is most wonderful. It is like the feeling of slipping from one body into another, and there is distinctly a double recollection of the two bodies. I remembered what I felt in my dream, lying on my chest; but returning into the day-life, I remembered also that my physical body had been quietly lying on its back all the while. This observation of a double memory I've had many times since. It is so indubitable that it leads almost unavoidably to the conception of a dream-body.[40]

Other experiential similarities between out-of-body experiences and lucid dreams are that both tend to be highly realistic, even hyper-realistic, and both often contain vivid impressions of light illuminating the scene. These can be startling when you see yourself asleep in the bed below and know that the room is dark.

From the outside perspective of neuroscience, the little information we have points to a close relation between the out-of-body experiences occurring during sleep and lucid dreams. Charles Tart, in his EEG studies of Miss Z and Robert Monroe, found that their out-of-body experiences seemed to happen during stage 1 sleep, when hypnagogic experiences usually occur.[41] And Lynne Levitan and Stephen LaBerge, in a study of lucid dreams, REM sleep, and out-of-body experiences, found that lucid dreams occurring after brief awakenings from REM sleep (so-called wake-initiated lucid dreams) were significantly more likely to be judged as out-of-body experiences than lucid dreams occurring during uninterrupted REM sleep.[42] Finally, among dream variables, the occurrence of lucid dreams is the most consistent predictor of out-of-body experiences.[43]

So, are out-of-body experiences really lucid dreams or a subset of lucid dreams? I'm inclined to think they are, or at least that the ones occurring during sleep or in the liminal zones between waking and sleeping are a subset of lucid dreams. But we still don't know enough about these states. We need more information—neurophenomenological information—to say for sure.

Believers in the astral body and astral travel will be disappointed in me for this conclusion. But I'd reply that to be "out-of-body" in the world of imagination, with the knowledge that this is where you are, allows for far more freedom and creativity than being adrift in an astral state you falsely take for a reality outside your mind.

8
SLEEPING

Are We Conscious in Deep Sleep?

lie down on the bed to rest, tired from an overnight flight to Europe followed by a morning train ride. All of a sudden I'm awake and the phone is ringing. I have no idea where I am and what time it is. Without thinking, I pick up the phone. A familiar but unknown voice is asking, "Did I wake you up?" "Yes," I hear myself saying, "but that's okay." I can't find any memories to tell me who I am and how I got here. Then I realize it's Francisco's voice, and this one recognition slowly brings the other memories back to life.

Experiences of deep and dreamless sleep, like this one, are puzzling. On the one hand, if consciousness is entirely absent in deep sleep, then how does the life of consciousness hold together across the gaps between being asleep and being awake? On the other hand, if some kind of consciousness is present in deep sleep, then why do we seem to have been unaware of being asleep, and not to remember anything from deep sleep when we wake up?

The obvious way to deal with these puzzles is by appealing to memory. When you wake up, you remember—eventually, if not always right away—your life before you fell asleep. Memory is how consciousness bridges the chasm of deep sleep. But invoking memory raises new questions. When you wake up, do you remember being asleep? Or do you infer that you were asleep from your memory of getting into bed and your experience of waking up? How, exactly, do you know you were asleep?

We can sharpen these questions by thinking about our consciousness of the passage of time. According to Edmund Husserl (1859–1938), the founder of the Western philosophical movement called Phenomenology, our conscious awareness of whatever is happening now includes a function that he calls "retention."[1] Retention holds onto the just-past so that what we experience as happening now, we also experience at the same time as receding into the past. His example is listening to a melody. As I write these words, I'm listening to a recording of an Indian morning *rāga* played on the sarod. Each note figures in a phrase I hear all at once as rising and subsiding while new notes are arriving at a faster and faster pace. Retention is that aspect of my consciousness that presents the notes as becoming past while I hear them sound. The just-past notes trail off, but I retain them in the phrase I'm currently hearing. Yet I retain them not as present—I don't hear them as a chord—but precisely in their mode of becoming past, so that they form a flowing sequence. In this way, I'm perceiving the just-past notes as they flow away, not recollecting them. Later, once the music is over, the melody may return spontaneously to my mind or I may deliberately recall and mentally replay it. Such experiences of passive remembering and active recollection depend on the retentional function that belongs essentially to my earlier listening consciousness. For this reason, Husserl sometimes calls retention "primary memory," and remembering or recollection "secondary memory." The crucial point is that retention or primary memory is an essential ingredient of every conscious experience; it's not a separate mental act in the way that remembering or secondary memory is.

This is precisely what enables us to sharpen our questions about deep sleep. If we're aware, upon awakening, of having just been asleep, or if we remember being asleep, then we must undergo some kind of experience in dreamless sleep. Specifically, we must have some kind of flowing retentional consciousness in the deep sleep state if the later remembering of this state is to be possible.

But now we face a puzzle. During deep and dreamless sleep, it seems that the sense of "I" disappears. Yet retentional consciousness—at least in its familiar waking forms—includes a minimal kind of self-awareness. I'm aware of the notes of the *rāga* as slipping into the past through my awareness of the notes as *having just been heard by me*. Notice that this kind of self-awareness isn't reflective or introspective. I'm not reflecting on my hearing or introspectively paying attention to it; I'm paying attention to the notes. Nevertheless, I retain them by retaining my experience of hearing them, and this retention of my own auditory experience enables me to remember the melody later once it's gone. In this way, my consciousness is reflexive; it retains itself as it flows away—and it retains itself precisely as flowing away. Western phenomenologists call this kind of reflexivity or minimal self-awareness "prereflective self-awareness." The question is whether it makes sense to suppose that this kind of minimal self-awareness is present during deep and dreamless sleep.

If we don't experience anything in deep sleep, then when we wake up we can't be retentionally aware of having just been asleep; we can only infer that we were asleep. On this view, retrospective inference is what fills in the gap created by deep sleep.

But we face a puzzle here too. What exactly is the basis for this inference? Wouldn't it require as evidence at least some kind of awareness of entering sleep and emerging from sleep? And wouldn't this requirement mean that sleep isn't a simple void but rather a felt absence with its own peculiar phenomenal qualities?

These kinds of questions about sleep occupied India's contemplative philosophers, especially in the Yoga and Vedānta traditions. Tibetan Buddhism also has much to say about deep and dreamless sleep. According to these traditions, deep and dreamless sleep is a mode of consciousness, not a condition where consciousness is absent. This way of thinking raises questions about the standard conception of deep sleep in contemporary Western philosophy of mind and casts new light on the neuroscience of sleep and consciousness.

THE FEELING OF BEING ALIVE

Before we examine the Indian views of sleep, it's worth taking a look at some Western treatments of the moment of awakening from a deep and dreamless sleep. One of my favorite descriptions of the kind of disorientation we sometimes feel comes from Marcel Proust. In a long passage at the beginning of the first volume of his novel, *In Search of Lost Time*, the unnamed narrator describes awakening from sleep:

> A sleeping man holds in a circle around him the sequence of the hours, the order of the years and world. He consults them instinctively as he wakes and reads in them in a second the point on the earth he occupies, the time that has elapsed up to his waking; but their ranks can be mixed up, broken. If towards morning, after a bout of insomnia, sleep overcomes him as he is reading, in a position too different from the one in which he usually sleeps, his raised arm alone is enough to stop the sun and make it retreat, and, in the first minute of his waking, he will no longer know what time it is, he will think he has only just gone to bed. If he dozes off in a position still more displaced and divergent, for instance after dinner sitting in an armchair, then the confusion among the disordered worlds will be complete, the magic armchair will send him travelling at top speed through time and space, and, at the moment of opening his eyelids, he will believe he went to bed several months earlier in another country. But it was enough if, in my own bed, my sleep was deep and allowed my mind to relax entirely; then it would let go of the map of the place where I had fallen asleep and, when I woke in the middle of the night, since I did not know where I was, I did not even understand in the first moment who I was; all I had, in its original simplicity, was the sense of existence as it may quiver in the depths of an animal; I was more bereft than a caveman; but then the memory—not yet of the place where I was, but of several of those where I had lived and where I might have been—would come to me like help from on high to pull me out of the void from which I could not have got out on my own; I passed over centuries of civilization in one second, and the

image confusedly glimpsed of oil lamps, then of wing-collar shirts, gradually recomposed my self's original features.[2]

Proust depicts the moment of awakening from deep sleep as one where we've lost all sense of the self derived from memories of the episodes of our lives. Instead of the autobiographical or narrative sense of self as a person with a story line through time, there remains only the sensation of existing at that moment. What marks the first instant of awakening isn't the self of memory but the feeling of being alive, what Proust calls "the sense of existence as it may quiver in the depths of an animal."

Tomas Transtömer, in his prose poem "The Name," also describes the feeling of being alive and being stripped of all autobiographical memory in the instant of awakening.[3] Drowsy while driving, he pulls over and goes to sleep in the backseat of the car. Hours go by, and all of a sudden he wakes up in darkness with no idea where he is or who he is. Eventually the memory of his life comes back to him.

Proust and Tranströmer both portray memory as seeming to come back from the outside, from someplace that doesn't coincide with present consciousness. For Proust, memory returns like help from on high to pull him up out of the void. Tranströmer describes memory as coming to him "like an angel" or like footsteps that "come quickly quickly down the long staircase" to save him.[4] In Proust's words, memory doesn't so much recover the self as "gradually recompose" it. Memory re-creates the autobiographical self anew in the process of awakening, so that the remembered self is a result and not a cause of waking up.

Literary scholar Daniel Heller-Roazen elaborates on this thought in his discussion of Proust by quoting the French poet-philosopher Paul Valéry: "One should not say I wake . . . but There is waking—for the I is the result, the end, the ultimate Q.E.D. of the congruence-superimposition of what one finds on what one must have been expecting to find."[5]

If what one finds is what's given in the first instant of awakening—the feeling of being alive—then what one must have been expecting to find is the world supplied by memory, especially the remembered self.

The instant of awakening thus reveals two kinds of self-experience, or ways of experiencing who and what we are—the bodily self-experience of being alive in the present moment, of being sentient, and the autobiographical self-experience of being a person with a story line, a thinking being who travels mentally in time. The first kind of self-awareness we experience immediately upon awakening, but as we reach automatically for the second, sometimes it goes missing.

This loss of self—not finding what we expect to find, and instead, an absence where memory is supposed to be—feels different to different people. For Proust's narrator, the absence is a "void" he can't get out of on his own; for Tranströmer, it's "the hell of nothingness."[6] But others take enjoyment in the loss, as American poet Jane Hirshfield describes in her poem, "Moment," which she wrote in response to Tranströmer.[7] Some panic, the poem tells us; others feel pleasure in the moment. "How each kind later envies the other, who must so love their lives."[8]

Whether it's pleasure or panic, the enjoyable or distressing feeling that arises upon awakening from dreamless sleep suggests that some kind of awareness may be present during sleep, and that this awareness contributes to how you feel the moment you wake up.

Consider that although deep sleep creates a gap or rupture in our consciousness, we feel the gap from within upon awakening. Our waking sense that we were just asleep and unaware isn't outside knowledge, it's inside, firsthand experience. We're aware of the gap in our consciousness from within our consciousness. Although we may forget many things about ourselves when we first wake up—where we are, how we got there, maybe even our name—we never have to turn around to see who it was who was just asleep and unknowing, if by "who" we mean the self as the subject of present-moment experience in contrast to the self as the mentally represented object of autobiographical memory. This intimate and immediate self-awareness as we emerge from sleep into waking life suggests that there may be some kind of deep sleep awareness, a taste of which we retain in the waking state, despite there being no specific memory content to recall. But if so, then there must be something it's like to be deeply asleep, in which case consciousness can't be entirely absent.[9] This line of thought, as we'll see, lies behind the Yoga and Vedānta view that dreamless sleep is a mode of consciousness.

"I SLEPT PEACEFULLY AND I DID NOT KNOW ANYTHING"

Already in the earliest texts of the *Upanishads*, dating from the seventh century B.C.E., dreamless sleep is singled out as one of the principal states of the self (see chapter 1). Various descriptions are given. Some texts describe dreamless sleep as a state of oblivion, while others characterize it as a mode of unknowing or noncognitive consciousness that lacks both the outer sensory objects of the waking state and the inner mental images of the dream state.[10] It's this second characterization that we find in the later texts of the Yoga and Vedānta schools. These texts also present a basic form of argument for dreamless sleep being a mode of consciousness. When you wake up from a dreamless sleep, you're aware of having had a peaceful sleep. You know this directly from memory, so the argument says, not from inference. In other words, you don't need to reason, "I feel well rested now, so I must have had a peaceful sleep." Rather, you're immediately aware of having been happily asleep. Memory, however, presupposes the existence of traces that are caused by previous experiences, so in remembering you slept peacefully, the peaceful feeling must have been experienced. To put the thought another way, the memory report, "I slept peacefully," would not be possible if consciousness were altogether absent from deep sleep, but to say consciousness is present in deep sleep is to say that deep sleep is a mode of consciousness.

The earliest version of this argument comes from the Yoga tradition, specifically from the fourth-century C.E. author Vyāsa, who wrote the first and primary commentary on Patañjali's *Yoga Sūtras*, a text codified sometime between the second and fourth centuries C.E.[11]

Patañjali defines yoga as the stilling or restraining of the "fluctuations" of consciousness (*Yoga Sūtras* I:2). When this stilling is accomplished, the "seer" or "witness" can abide in its true form; otherwise the seer identifies with the fluctuations of consciousness—with the movements of thought and emotion (I:3–4). Patañjali identifies five kinds of fluctuations or changing states of consciousness— correct cognition, error, imagining, sleep, and memory (I:5–6)—and he defines deep sleep as a state of consciousness that's based on an "absence" (I:10).

As the traditional commentaries indicate, "absence" doesn't mean absence of consciousness; it means absence of an object presented to consciousness.[12] Deep sleep is a kind of consciousness without an object. When we're awake we cognize outer objects, and when we're dreaming we cognize mental images. When we're deeply asleep, however, we don't cognize anything—there's no object being cognized and no awareness of the "I" as knower. Nevertheless, according to Yoga, we feel this absence while we sleep and remember it upon awakening, as evidenced by our saying, "I slept peacefully and I did not know anything."

The traditional commentaries describe the absence experienced during sleep as a kind of "darkness" that completely overwhelms and envelops consciousness. This image or metaphor of darkness is suggestive. In waking perception, complete darkness means there's no visual object to be seen. Yet darkness is a visual quality with its own phenomenal presence. Similarly, in the "darkness" of deep and dreamless sleep, there's nothing to be cognized or known, yet this absence itself is said to be subliminally experienced and remembered upon awakening. So the absence is a felt absence, not a simple nonexistence.

Given this conception, we might ask why Yoga classifies deep sleep as a "fluctuation" of consciousness, instead of as a state where consciousness is stilled. And given that one of the goals of yoga is to still consciousness, we might also ask why deep sleep isn't a way to achieve this goal.

These questions bring us to Vyāsa's commentary on the *Yoga Sūtras*. Vyāsa explains the classification of deep sleep as a fluctuation of consciousness in the following way:

Since we can remember when we wake up that we had been sleeping, sleep is called a mental modification, as indicated in the feelings by phrases such as 'I slept well, I am feeling cheerful, it has cleared my brain' or 'I slept poorly; on account of disturbed sleep, my mind has become restless, and is wandering unsteadily,' or 'I was in deep sleep as if in a stupor, my limbs are heavy, my brain is tired and languid, as if it has been stolen by somebody else and lying dormant.' If during sleep there was no cognition of the inert state, then on waking, one would not have remembered that experience. There would not also have been recollection of the state in which the mind was in sleep. That is why sleep is regarded as a particular kind of mental state, and should be shut out like other cognitions when concentration is practiced.[13]

Here we see the original statement of the Indian philosophical argument for considering deep sleep to be a subliminal mode of awareness: if awareness were absent, then you wouldn't be able to recollect the quality of sleep, because a recollection is the recalling of an earlier experience based on the subliminal traces it leaves in the mind. But since you do remember how you slept, you must be aware in deep sleep.

Vyāsa distinguishes three types of sleep that are remembered upon awakening—peaceful and refreshing, disturbed and restless, and dull and heavy. The subliminal experience of these qualitatively different kinds of sleep leaves mental impressions that produce memories.

Although object-directed cognition ceases and deep sleep can be peaceful and refreshing, the deep sleep state isn't one where consciousness has achieved the stillness that yoga cultivates. In the words of yoga scholar Edwin Bryant, whereas the meditative stilling of consciousness "occurs in full vibrant wakefulness and in complete lucidity as to the nature of reality; in deep sleep, awareness is simply aware of the dense motionless darkness . . . in which it is enveloped."[14]

Furthermore, in deep sleep the seer identifies with this enveloping darkness, instead of abiding in its true form as pure witnessing awareness. For this reason, Vyāsa says that sleep—or more precisely, the identification with sleep as the true form of the seer—should be brought under control, that is, brought to stillness.

MEMORY OR INFERENCE?

When you wake up and say, "I slept well and I didn't know a thing," are you really remembering what you experienced during sleep? Or are you inferring how you slept based on how you feel when you wake up?

The question is important because if you're making an inference and not remembering, then the traditional Indian yogic argument for deep sleep being a mode of consciousness won't work. Although memory of specific events does imply previous experience of them, if you're not remembering but inferring how you slept based on how you feel upon awakening, then the crucial reason for thinking that you were subliminally conscious during sleep is lost. Why not say instead that consciousness is "switched off," and the absence of consciousness explains why you're not aware of cognizing anything, including yourself, while you're in deep sleep?

In Indian philosophy, two of the main Hindu schools take opposite positions on this issue. On one side stands the Nyāya school, known for its writings on logic and the theory of knowledge. The Nyāya philosophers—known as Naiyāyikas—argue that the statement, "I slept peacefully and I did not know anything," is a case of inference, not memory. The self continues to exist in deep sleep but loses the property of being conscious, so consciousness isn't an essential property of the self.

On the other side stands the Vedānta school, especially its most prominent subschool, known as Advaita Vedānta (Nondual Vedānta). Advaita Vedānta follows the Yoga tradition in holding that deep sleep is a mode of consciousness. Against the Nyāya school, the Advaita Vedānta philosophers or Advaitins try to show that the statement, "I slept peacefully and I did not know anything," can't be an inference but has to be a memory report. The self essentially is consciousness, or more precisely, pure witnessing awareness, which we ignorantly confuse with our mental and bodily sense of ego or "I-Me-Mine." Although the ego sense comes and goes—it's present in the waking and dreaming states, but not in dreamless sleep—the "witness consciousness" remains constant through waking, dreaming, and dreamless sleep.

The issue before us is how we know, when we wake up, that we slept well and didn't know anything. As Rāmānuja (c. 1017–1137 C.E.), a philosopher who belonged to the Viśiṣṭādvaita Vedānta (Qualified Nondual Vedānta) school, pointed out, when we wake up we don't think, "As I now feel pleasure, so I slept then also." Instead, we think simply, "I slept well."[15] In other words, the waking thought refers only to the past sleep state, not to the present moment. In this way, the thought purports to be a memory, not an inference.

The debate between the Naiyāyikas and the Advaitins focuses on the ignorance or absence of knowledge in deep sleep, specifically on how we know or establish the waking report, "I knew nothing." The not knowing occurs during deep sleep, but while we're asleep we know nothing of this ignorance; we come to know it only upon waking up. Given that we don't remain ignorant of our own ignorance, how is this knowing of not knowing possible? The Naiyāyikas claim that we infer we were ignorant because we don't remember anything, but the Advaitins argue that retrospective forgetting is no proof of a prior lack of consciousness; after all, you can experience something and forget it afterward, as often happens in the case of dreams. Moreover, when we wake up we have the feeling of having been asleep and having not known anything. This feeling, the Advaitins claim, is better regarded as a kind of memory brought about by the traces of previous experience. So in some sense we must experience our ignorance—the unknowing stillness of our mind—in dreamless sleep.

In reply, the Naiyāyikas claim that we have no consciousness in dreamless sleep and that when we wake up we make an inference by reasoning in the following way: "While I was in deep sleep, I knew nothing, because I was in a special state (I wasn't awake) and I lacked the necessary means for knowledge (my senses and mental faculties were shut down)." Of course, they aren't saying that we explicitly make this inference when we wake up from a deep and dreamless sleep, but that what looks like memory is really a case of implicit reasoning having this inferential form.[16]

In order to understand the kind of inference that the Naiyāyikas think we make, as well as why the Advaitins reject the Nyāya position, it will be helpful to state the inference in the form of the standard Nyāya syllogism, an important part of the Nyāya theory of inferential knowledge.

Suppose we're looking at a hill and you say to me, "There's fire on the hill." I doubt what you say, however, so you need to convince me. You point to the hill and say, "There's smoke on the hill." I see the smoke and I'm convinced. According to Nyāya, if we want to unpack how perception and inference have worked together to convince me that you're right, we need to formulate the inferential cognition in the following five steps:

1. There is fire on the hill.
   [This is the proposition to be proved. It's what you think when you look at the hill, and it's what you want to convince me is the case.]
2. Because there is smoke on the hill.
   [This is the reason you give to support what you say.]
3. Wherever there is smoke there is fire.
   [This step states the universal concomitance of the presence of smoke and the presence of fire.]
4. As in the case of the kitchen.
   [This step provides an example or actual case of the concomitance, to which we both agree.]
5. There is fire on the hill.
   [This step states the conclusion, which is the proposition with which we began, but now stated as established and generated by the preceding inferential process.]

Let's now apply this five-step syllogism to dreamless sleep. The Nyāya view is that our knowledge that we knew nothing in dreamless sleep is based on the following sort of inference:

1. While I was in dreamless sleep, I knew nothing (there was an absence of knowledge in my self).
2. Because (i) I (my self) was in a special state (that is, not awake) or (ii) I (my self) lacked the necessary means for knowledge (that is, my senses and mental faculties were shut down).
3. Whenever (i) I (my self) am in a special state (whenever I'm not awake) or (ii) I (my self) lacks the necessary means for knowledge (my senses and mental faculties are shut down), I know nothing (there is an absence of knowledge in my self).
4. As in the case of fainting or a blow to the head.
5. While I was in dreamless sleep, I knew nothing (there was an absence of knowledge in my self).

Notice the parallel between the previous inference concerning fire and the present inference concerning dreamless sleep. In the previous case, our concern is to establish the presence of fire on the hill. In the present case, our concern is to establish the absence of knowledge in the self during dreamless sleep. Nevertheless, the form of reasoning is the same.

Again, the Naiyāyikas aren't saying that we explicitly go through this inference step by step when we wake up, but that we know by inference that we were ignorant during dreamless sleep, and that our inference can be shown to be correct when we make explicit all the steps that it contains. So there's no need to suppose there's any kind of consciousness during dreamless sleep.

We're now ready to go back to the Advaita Vedānta view. The Advaitins respond that this inference is faulty and can't be how we know that we knew nothing during deep sleep. The problem is that I need some way to know or establish the reasons for inferring that I knew nothing (step 3 above), and there seems no way to do this without relying on the kind of memory these reasons are supposed to obviate.

The first reason the Naiyāyikas give for me to infer that I knew nothing is that I was in a special state, different from the normal waking state. But how do I know that I was in this special state? If I say, "Because I knew nothing in this state," then I'm reasoning in a circle.

The second reason the Naiyāyikas give for me to infer that I knew nothing is that the necessary means for knowledge were lacking, that is, my senses and mental faculties were shut down. But here too we need to ask, how do I know that these means were lacking, my senses and mental faculties were inactive?

Suppose I say, "I infer my senses were shut down because they felt refreshed when I woke up." But here the same basic problem repeats itself. How do I know or establish that there's a relationship between my senses feeling refreshed and their previously having been inactive? Wouldn't I need to have some experience of knowing that my senses were inoperative, together with an experience of knowing I felt refreshed, in order to establish a relationship between the two? But while I'm asleep I don't have any experience of knowing my senses are inactive; I know this only upon awakening. So how do I establish this relationship? If I appeal to yet another inference, then it looks like I'm headed off on an infinite regress.

More generally, the only way I can know that the means for knowledge were absent in deep sleep is by knowing that there was no knowledge present in this state. Only by knowing the effect—my not knowing anything—can I infer the cause—the absence of the means for knowledge. So unless I already know what the inference is trying to establish—that I knew nothing—I can't establish the reason on which the inference relies.

The Advaita Vedānta conclusion is that I know on the basis of memory, not inference, that I knew nothing in deep sleep. In other words, I remember having not known anything. But a memory is of something previously experienced, so the not-knowing must be experiential.

Underlying this debate is a metaphysical dispute about the nature of the self. For the Naiyāyikas, the self is a nonphysical substance. Unlike Descartes, however, who also held that the self is a nonphysical substance, the Naiyāyikas don't think that consciousness is the essence of the self. Instead, they maintain that the self is the substratum of consciousness and that consciousness is an adventitious quality present only given the appropriate conditions (namely, when the sensory and mental faculties are functioning to cognize objects). For the Advaitins, however, the self is pure consciousness, that is, sheer witnessing awareness distinct from any changing cognitive state. Thus, unlike the Naiyāyikas, the Advaitins cannot allow that consciousness disappears in dreamless sleep, since they also think that it's one and the same self who goes to sleep, wakes up, and remembers having gone to sleep.

It seems possible, however, to extract the phenomenological core of the Advaita Vedānta conception of dreamless sleep from the Vedānta metaphysics. The core insight is that when I wake up from a dreamless sleep, I can knowingly say I've just emerged from a dreamless sleep, and my saying so is arguably a reporting of my awareness, not a product of having to reason things out.[17] This insight provides the basis of the Advaita Vedānta view that consciousness continues in dreamless sleep, and is logically distinct from the Vedānta belief that the self is essentially pure consciousness.

From a contemporary, cross-cultural philosophical perspective, we can see in this core Vedānta insight the basis for what Western philosophers since Immanuel Kant (1724–1804) have called a "transcendental argument." Transcendental arguments aim to deduce what must be the case in order for some aspect of our experience to be possible. Here, the aspect of experience with which we're concerned is not simply that we sleep but that we know we sleep. What are the necessary conditions of possibility for this kind of self-knowledge? To put the question in a more phenomenological way, how is it possible for you as a conscious subject to experience yourself as one and the same being who falls asleep, does not actively know anything while asleep, and emerges from sleep into waking life? The Vedānta view is that a retrospective inference across the gap of a complete absence of consciousness can't suffice to make this kind of unified self-experience possible. Rather, you must have some kind of experiential acquaintance with dreamless sleep as a mode of your conscious being.

Western phenomenology in the tradition of Kant and Husserl can also help us to see that the Vedānta argument is not, in the first instance, about remembrance or recollection (secondary memory), but rather about retentional awareness (primary memory). The point is that, at the moment of waking up, I'm aware of having just been asleep, that is, I experience by retentional awareness my having just been asleep and not knowing anything.[18] What the Naiyāyikas fail to see, according to the Advaitins, is that I need this retentional awareness in order to know that I slept and to ground any retrospective inference I may subsequently make.

Although the logical refutation of the Nyāya position makes up an important part of the Advaita Vedānta argument for deep sleep consciousness, it's not the most important part. The core of the argument is an appeal to the evidence of direct experience. When we wake up from a deep sleep, it seems that we're aware of having slept well and having known nothing, and this awareness feels like retentional consciousness, not inference. Ruling out the Nyāya view on logical grounds serves to justify this feeling.

Nevertheless, justifying the feeling of retention doesn't explain how the retention and subsequent secondary memory come about. How can we remember not knowing anything if we have no knowledge of our ignorance during deep sleep? Answering this question takes us to the heart of the Vedānta conception of deep sleep.

WITNESSING NOT KNOWING

According to Advaita Vedānta, although we don't *know*, while we're deeply asleep, that we know nothing, we are *aware* of our ignorance. More precisely, consciousness witnesses the not-knowing, and this witnessing awareness enables us to recall the not-knowing later in the waking state.

To understand this conception of deep sleep consciousness, we need some familiarity with the Vedānta conceptions of knowledge and ignorance. Knowledge, in this context, means true cognition, that is, a specific cognitive act that has the property of being true and is based on a valid means of knowing, such as perception or inference. I know there's a fire on the hill because I see it, or because I infer its presence from my perception of the rising smoke. In general, no cognitive acts of knowing occur in deep sleep, so in particular there's no knowing that one does not know.

Ignorance, however, according to Vedānta, isn't the mere nonexistence of knowledge; it's an experience of not-knowing that "conceals" the nature of things. Suppose I mistake a shell on the beach for a coin. My ignorance here involves nonapprehension—I don't see the shell as a shell—and misapprehension—I see the shell as a coin. There is both ignorance of the real thing and the mistaken perception of it as something other than what it is. In Vedānta language, not-knowing "conceals" the shell by "superimposing" the illusion of a coin. In general, when I'm ignorant of a thing, according to Vedānta, my awareness presents some appearance that conceals the nature or the existence of that thing.

The Advaita Vedānta philosophers Gauḍapāda (c. eighth century C.E.) and Śaṅkara (788–820 C.E.) apply this conception of ignorance to dreams and dreamless sleep in the following way.[19] In nonlucid dreams, we don't see the dream as a dream (nonapprehension of the dream), and we mistake our mental images for real things outside us (misapprehension of the dream as reality). We don't know that we do not know, that is, we don't know that our dreaming experience is one of not-knowing. We gain this knowledge only when we wake up, unless the dream becomes a lucid dream.

In deep and dreamless sleep, we experience a kind of blankness or nothingness. In other words, deep sleep isn't a *nothingness of experience* but rather an *experience of nothingness*. Here our ignorance is an experience of pure nonapprehension without misapprehension. Since there's no object of awareness, there's nothing for us to mistake for anything else. As total darkness in waking life conceals everything, leaving only not-seeing with no way to misperceive one thing as another, so in deep sleep there is only not-knowing. We have no knowledge of this ignorance when we sleep, but the nonapprehension is a kind of awareness, and it's this ignorant awareness that is retained in the moment of waking up and that waking memory recalls.

If we project some terminology from contemporary Western philosophy of mind onto Yoga and Vedānta, we can say that deep sleep counts as a "phenomenal" state or state of "phenomenal consciousness"—a state for which there's something it's like to be in that state. Yoga and Vedānta describe it as peaceful, one undifferentiated awareness not divided up into a feeling of being a subject aware of a distinct object, and blissfully unknowing. Yet deep sleep doesn't normally count as a state of "access consciousness"—a state we can mentally access and use to guide our attention and thinking. We have no cognitive access to being asleep during sleep; we gain access retrospectively in the waking state.

As we'll see at the end of this chapter, however, this way of describing things will ultimately need to be revised in order to make room for a central commitment of Yoga and Vedānta, as well as Indo-Tibetan Buddhism, which is that we can gain access to deep and dreamless sleep through meditative mental training. On this view, certain qualities of awareness in dreamless sleep can be made cognitively accessible, though normally we have no cognitive access to them while we're in the state of dreamless sleep.

Moreover, strictly speaking, according to Advaita Vedānta, the awareness in dreamless sleep isn't "my" awareness, if "my" means belonging to my ego, because the mental and bodily sense of ego or "I-Me-Mine" shuts down in deep sleep. Instead, the awareness consists in what Vedānta calls "witness consciousness." Witness consciousness isn't an active cognitive knower; it's a passive nonconceptual observer, pure awareness distinct from the changing states of mind. Pure awareness watches the carousel of sleeping, dreaming, and waking, but without participating in this mental whirling.

We can now state the Advaita Vedānta answer to the puzzle about whether some kind of minimal self-awareness is present in dreamless sleep. The Advaita Vedānta answer is yes, but this kind of self-awareness is the self-luminous witness consciousness, not any cognitive or bodily sense of ego or "I-Me-Mine."

THE INNER SENSE IN SLEEP

Although Advaita Vedānta follows closely the Yoga tradition's view of deep sleep, the two schools also differ in ways that raise important issues about how the mind functions in deep sleep—issues we'll look at later from the point of view of neuroscience.

According to Yoga, deep sleep is one of the "fluctuations" or changing states of consciousness. More precisely, it is a state of the "inner sense," which includes mental or cognitive awareness (as distinct from sensory awareness) and the sense of ego or "I-Me-Mine." The mind continues to function in deep sleep, though it's much less active than in the dreaming or waking states, and in the deepest dreamless sleep it comes to rest entirely.

According to Advaita Vedānta, however, deep sleep isn't a state of the inner sense, for that completely shuts down. In other words, mental or cognitive awareness and the sense of ego cease to function in deep sleep, and only the witness consciousness and ignorance remain.

Yoga also describes different qualities of deep sleep—peaceful, disturbed, or dull—whereas Advaita Vedānta singles out peace or bliss, and attributes it to the cessation of the workings of the inner sense.

But now a difficult question arises for the Advaita Vedānta view. If the inner sense stops functioning in deep sleep, then how does the waking memory, "I slept peacefully and I did not know anything," get formed? Memory implies previous experience, so if there's no bodily and mental sense of "I" in deep sleep, then how can I remember that *I* slept well?

The Advaita Vedānta answer is ingenious. In deep sleep, ignorance completely envelops the mind. Since the ego sense is inoperative, it doesn't appropriate this ignorance to itself, so there's no feeling of the ignorance belonging to an "I." At the moment of awakening, however, the ego sense, grounded on the felt presence of the body, reactivates, and the mind starts up its cognitive workings. Immediately, the ego sense appropriates the lingering impression or retention of not-knowing and associates this retention with itself, thereby generating the retrospective thought, "I did not know anything."[20]

Given this account, Advaita Vedānta can agree with Paul Valéry's remark, quoted earlier: "One should not say *I* wake . . . but There is waking—for the *I* is the result, the end, the ultimate Q.E.D. of the congruence-superimposition of what one finds on what one must have been expecting to find." If "I" is taken to refer to the sense of ego or "I-Me-Mine," then Advaita Vedānta would concur: this feeling of "I" doesn't function in deep sleep, but reconfigures itself as a result of waking up. However, this "I" isn't the true self; the true self is the egoless witness consciousness, which endures throughout waking, dreaming, and deep sleep.

The Advaitin takes this witness consciousness to be transcendental—meaning not fundamentally embodied. It's open to us today, however, to think that the egoless consciousness in dreamless sleep is a fundamentally embodied consciousness, by which I mean a consciousness that's contingent on the brain and other systems of the body. Here is another place where it may be possible to remove the Advaita Vedānta conception of dreamless sleep from its original metaphysical framework and graft it onto a contemporary conception of the embodied mind.

EAST MEETS WEST

Let me summarize things up to this point. Our guiding question is whether some kind of consciousness is present in deep and dreamless sleep. Related to this question is whether, at the moment of awakening, we have some kind of memory of the deep sleep state or make a retrospective inference about having been asleep and unaware. Further related questions are what happens to the self and whether there's some kind of self-experience in deep sleep.

The Indian philosophical treatments of these questions about deep and dreamless sleep give careful attention to a part of human life that contemporary Western philosophy of mind basically neglects.[21] Although philosophers of mind have written about dreaming, they've said almost nothing about dreamless sleep. Even phenomenology in the tradition of Edmund Husserl—the philosophical tradition that reigns supreme in the Western investigation of consciousness—says little about deep and dreamless sleep compared to the richness of the Indian philosophical discussions.[22]

What about neuroscience? What does it have to say about deep and dreamless sleep?

According to the standard neuroscience way of thinking, deep and dreamless sleep is a state where consciousness fades and sometimes disappears completely. Indeed, neuroscientists often try to define consciousness as that which disappears in deep sleep. As neuroscientists Giulio Tononi and Christof Koch write, "When you fall asleep . . . the level of consciousness decreases to the point that you become virtually unconscious—the degree to which you are conscious (of anything) becomes progressively less and less."[23] Elsewhere Tononi writes: "Everybody knows what consciousness is: it is what vanishes every night when we fall into a dreamless sleep and reappears when we wake up or when we dream."[24] Philosopher John Searle agrees: "Consciousness consists of inner, qualitative, subjective states and processes of sentience or awareness. Consciousness, so defined, begins when we wake in the morning from a dreamless sleep and continues until we fall asleep again, die, go into a coma, or otherwise become 'unconscious.'"[25]

From the Indian and Tibetan contemplative perspectives, however, these descriptions are inaccurate. Although object-directed consciousness becomes progressively less and less as we move from waking or dreaming into deep and dreamless sleep, awareness or sentience continues. For Yoga and Vedānta, whereas dreaming is a form of object-directed consciousness—the objects in dreams being mental images—dreamless sleep is a mode of consciousness without an object. Similarly, according to Tibetan Buddhism, deep sleep is a state of "subtle consciousness" without sensory or cognitive content, and it's the basis upon which dreaming and waking consciousness arise.[26]

The Indian and Tibetan conceptions of deep and dreamless sleep bring a new perspective to the neuroscience of consciousness, especially to experimental investigations of brain activity during sleep. At the same time, findings from the neuroscience of sleep are relevant to the Indian debates, especially to the differences between the Yoga and Vedānta views of mental functioning in deep and dreamless sleep.

WHAT WERE YOU THINKING?

Why have neuroscientists thought that consciousness fades or disappears during deep and dreamless sleep? One reason comes from the reports people give when they're woken up from non-REM or NREM sleep, especially when the EEG shows slow waves in the delta frequency range (0.5–4 Hertz) during sleep stages 3 and 4 (so-called slow-wave sleep). When given the instruction, "report anything that was going through your mind just before waking up," people tend to report short and fragmentary thoughts or not being able to remember anything at all.[27] On that basis, scientists conclude that the sleepers were aware of little or nothing at all prior to being woken up, and hence that slow-wave sleep is a state of reduced or absent consciousness.

Yet we should be cautious here. The fact that you have no memory of some period of time doesn't necessarily imply that you lacked all consciousness during that time. You might have been conscious—in the sense of undergoing qualitative states or processes of sentience or awareness—but for one reason or another not been able to form the kind of memories that later you can retrieve and verbally report.

This point is familiar to scientists who study the effects of anesthetics. At certain doses, some anesthetics prevent memory formation while sparing awareness. As neuroscientists Michael Alkire, Giulio Tononi, and their colleagues state in an article on consciousness and anesthesia:

> At doses near the unconsciousness threshold, some anesthetics block working memory. Thus, patients may fail to respond because they immediately forget what to do. At much lower doses, anesthetics cause profound amnesia. Studies with the isolated forearm technique, in which a tourniquet is applied to the arm before paralysis is induced (to allow the hand to move while the rest of the body is paralyzed), show that patients under general anesthesia can sometimes carry on a conversation using hand signals, but postoperatively

deny ever being awake. Thus, retrospective oblivion is no proof of unconsciousness.[28]

Although dreamless sleep and anesthesia aren't the same condition, the general point that retrospective oblivion doesn't prove a prior lack of consciousness must be kept in mind whenever we're tempted to infer that consciousness is absent in deep sleep because people report not being able to remember anything when they're woken up.

If consciousness continues in deep sleep, there may be various reasons people report not being able to remember anything when they're woken up. One reason commonly given in Yoga and Tibetan Buddhism is that deeper aspects of consciousness unfamiliar to ordinary waking awareness can't be cognitively accessed and reported without a high degree of meditative mental training.

We also need to think about the kinds of verbal reports that people are asked to make when they're woken up in the sleep lab. The instruction to report "anything going through your mind just before waking up" encourages you to direct your attention and memory to the *objects* of your awareness—to anything you might have been thinking *about*. But what about the felt qualities of awareness itself? A different instruction, to report "anything you were feeling just before waking up," would encourage you to direct your attention and memory to the felt quality of your sleep. Did you have any feeling of being aware or in some kind of sentient state? Was your sleep peaceful and clear or agitated, restless, or sluggish? Or do you have no impression of any feeling or quality of awareness? The point is to guide people away from focusing exclusively on the objects of consciousness, which may be absent in deep sleep, and to orient them toward the felt qualities of awareness itself.

The connection between this point and the earlier one about meditative mental training is that individuals with such training, especially in lucid dream yoga and sleep yoga, may be able to give more detailed reports about qualities of awareness during sleep than untrained individuals can. We'll return to this idea at the end of this chapter.

SLOW-WAVE SLEEP AND THE BRAIN

Another reason neuroscientists think that consciousness fades or ceases in deep sleep comes from comparing brain activity during slow-wave sleep with brain activity during waking consciousness.

For example, Marcello Massimini, Giulio Tononi, and their colleagues at the University of Wisconsin, Madison, studied how the brain responds to being stimulated by a brief pulse of electricity at a small and precisely chosen region when subjects are awake versus when they're in deep sleep.[29] During wakefulness, the pulse triggers a sustained EEG response that lasts for 300 milliseconds and is made up of rapidly changing waves that propagate in specific directions over long distances in the cortex. During deep sleep, however, although the initial response is stronger than during wakefulness, it remains localized to the stimulated brain region and lasts only 150 milliseconds. In short, whereas the waking brain responds to stimulation with a complex pattern of large-scale activity across many interconnected regions, the deeply sleeping brain responds with localized and short-lived activity.

Tononi and his colleagues interpret these findings as showing that "effective connectivity"—the ability of neural systems to influence each other—breaks down in deep sleep. As a result, "large-scale integration" in the brain can't happen, that is, the brain can't generate the kinds of dynamically changing large-scale patterns of activity—such as the neural synchrony patterns discussed in chapter 2—that characterize moment-to-moment awareness in the waking state.

What causes these losses in effective connectivity and large-scale integration in deep sleep? Part of the answer has to do with what are called "up" and "down" states in slow-wave sleep. During deep sleep, virtually all cortical neurons alternate between being active—the up state—and being completely inactive—the down state. In the up state, neurons fire at their waking rates for about a second; in the down state, they're silent. The synchronous occurrence of the up state in numerous neuronal populations is what generates the large-amplitude slow waves of the EEG measured at the scalp. The more active the neurons and the longer they stay in the up state, the more likely they are to fall into the down state, after which they revert to another up state. Think of a light bulb that's more likely to go off depending on how brightly it burns and how long it's on, but then turns back on again after being off. Instead of having one stable state, the bulb is bistable, going on and off. Similarly, the up state in cortical neurons during slow-wave sleep isn't stable the way it is in wakefulness or REM sleep; rather, slow-wave sleep is inherently bistable, with the up state precipitating the down state, followed by a rebound to the up state, and so on. Any local activation—any turning on of the neurons at a particular region—will eventually trigger a down state that prevents those neurons from communicating with more distant ones. In this way, the effective connectivity between regions breaks down and large-scale integration across selective regions cannot happen.

But what is it about the loss of effective connectivity and large-scale integration that makes neuroscientists think that consciousness disappears in deep sleep? To put the question another way, what's the connection between the presence of consciousness and the presence of effective connectivity and large-scale integration?

To answer this question, neuroscientists usually rely on the idea that a content of consciousness is reportable, and that reportable contents can be attentionally selected, held in working memory, and used to guide thought and action. These cognitive processes— selective attention, working memory, sequential thought, and action guidance—require the large-scale integration of brain activity.

One of the more principled versions of this idea is Giulio Tononi's "integrated information theory" of consciousness.[30] According to this theory, any typical conscious experience has two crucial properties. First, it's highly "informative," in the technical sense that it rules out a huge number of alternative experiences. Even an apparently simple conscious experience, such as lying on your back and seeing the clear blue sky throughout your whole visual field, is richly informative in the sense that it rules out a vast number of other experiences you could have had at that moment. You could have seen the sky as red or some other color, or your eyes could have been closed, or you could have experienced a flock of birds flying overhead, or you could have been focusing attentively on a nearby conversation, and so on. Second, the experience is highly "integrated," in the sense that it can't be subdivided into parts that you experience on their own, such as the top and bottom portions of your visual field, or the color and the space of the sky.

Given this model of consciousness as "integrated information," Tononi proposes that the level of consciousness of a system at a given time is a matter of how many possible states (information) are available to the system as a whole (integration). In the waking state, many possible states are available to the whole system (the system is rich in integrated information), whereas in deep sleep this repertoire drastically shrinks to just a few states (the system is poor in integrated information). Transposed onto the brain, the idea is that during slow-wave sleep there's a massive loss of integrated information in the brain. Effective connectivity breaks down, leaving isolated islands that can't talk to each other (loss of integration), while the repertoire of possible states contracts to a few largely uniform states (loss of information). Hence, according to the integrated information model, deep sleep is a state where consciousness reduces to a very low level or disappears entirely.

Although the integrated information theory offers a useful way to think about the qualitative richness and coherence of consciousness in informational terms, it has a serious limitation as a theory of phenomenal consciousness, so it would be a mistake to use it to rule out the possibility of consciousness during dreamless sleep. Despite Tononi's bold claim that "consciousness is one and the same thing as integrated information,"[31] integrated information doesn't seem even sufficient for consciousness. Computers can possess a high amount of integrated information, but they aren't conscious. More generally, as philosopher Ned Block points out, the integrated information theory doesn't distinguish between intelligence, in the sense of being able to solve complex problems by integrating multiple sources of information, and consciousness, in the sense of sentience or felt awareness (phenomenal consciousness).[32] Since integrated information doesn't seem sufficient for consciousness—let alone identical to it—its presence or absence shouldn't be taken as the definitive mark of whether a state is conscious or not conscious.

We also need to keep in mind the distinction between phenomenal consciousness and access consciousness. To be phenomenally conscious means to be in a state of felt awareness. For example, you're phenomenally conscious when you dream. To be access conscious means to be in a state where there is cognitive access to the contents of awareness. Most people dream throughout the night but have little cognitive access to their dreams—they don't remember them, so they can't report them. When you're access conscious you're able to hold the contents of awareness in memory long enough to report them and use them in your subsequent thinking. Although large-scale integration in the cortex is crucial for cognitively accessible conscious experience, it may not be crucial for every kind of phenomenal consciousness, for example, the kind of cognitively unaccessed consciousness without an object that Yoga and Vedānta believe happens in deep and dreamless sleep.

Of course, Yoga and Vedānta, as well as Tibetan Buddhism, also say that deep sleep consciousness can become cognitively accessible through meditative mental training. We'll come back to this idea shortly.

REMEMBERING IN SLEEP

Although Yoga and Vedānta share the view that deep sleep is a state of consciousness, they differ in their conceptions of what happens to the mind in deep and dreamless sleep. According to Yoga, deep sleep is one of the changing states of the inner mental sense, so cognitive activity, particularly the formation of memories, continues. According to Vedānta, however, the inner mental sense shuts down entirely in deep sleep and reactivates upon awakening. How does this difference between Yoga and Vedānta look from the perspective of Western sleep science?

If we set aside the question of consciousness in deep sleep and restrict the question to whether memory processes are active, the answer from science is unequivocal: memory processes are highly active in slow-wave sleep. Evidence from psychology and neuroscience clearly shows that slow-wave sleep promotes the formation of stable memories of events that were consciously experienced earlier when awake.[33]

One recent experiment took advantage of the strong effect of smell on memory—the way that particular smells can trigger vivid memories, the most famous example being Proust's description of the way the smell and taste of a madeleine dipped in tea brought back to life his narrator's long-forgotten childhood world in the French village of Combray.[34] In the experiment, the subjects learned locations in a spatial memory task while being exposed to the scent of roses. The scent was presented again while the subjects were in slow-wave sleep that night. Compared to the control condition where the scent wasn't presented again during sleep, the presentation during slow-wave sleep resulted in significantly improved recall of the locations in the task on the following day. In addition, the presentation of the scent during sleep resulted in significant activation in the hippocampus, a subcortical structure known to be crucial for the formation and recall of memories for experienced events.

This study built on other ones showing that the same neural networks in the hippocampus that are activated in the acquisition of new memories during waking life are reactivated in slow-wave sleep. For example, studies in rats have shown that when they learn their way in a maze, neurons in the hippocampus that fire in response to specific places—so-called hippocampal place cells—fire in the same order during subsequent slow-wave sleep, a phenomenon known as "hippocampal replay."[35] It's as if the rats rerun the maze offline. More precisely, the neural networks the rats need to run the maze are repeating their waking activity patterns and thereby solidifying those patterns for future use. Hippocampal replay is also found in humans: areas of the hippocampus that are activated when people learn a route through a virtual town in a computer game are reactivated during slow-wave sleep; in addition, the stronger the hippocampal reactivation during sleep, the better people are at remembering the route the next day.[36]

These and other studies tell us that slow-wave sleep strengthens newly acquired memories and integrates them with older ones. Psychologists call this process "memory consolidation."

Hippocampal replay is key in one of the main models of how memory consolidation happens in slow-wave sleep. According to the model, the hippocampus and the neocortex (the outer layer and uniquely mammalian part of the cortex) engage in a dialogue that serves to transform new memories, which the hippocampus and neocortex hold together, into long-term memories, which the neocortex holds alone (or, according to another model, that the hippocampus and neocortex hold together in a strengthened way).

Here are the basics of how the dialogue works. There's a flow of replay activity from the hippocampus to the neocortex, but the neocortex orchestrates the flow. To be more specific, hippocampal replay triggers a similar replay in the neocortex, so that the same neocortical networks that were active in the acquiring of new memories are reactivated in slow-wave sleep. In this way, the hippocampus tunes the neocortex, so that new memories there are preferentially strengthened and integrated into the preexisting network of long-term memory. At the same time, the neocortex organizes the flow of replay activity into successive "frames." This framing happens through the slow oscillation between up and down states in cortical neurons. Recall that in the up state, the neurons fire at their waking rates, whereas in the down state, they're completely silent. Neocortical up states trigger hippocampal up states and thereby determine the successive moments at which the hippocampal replay occurs, which in turn drives the neocortical replay.

This "framing" of memory consolidation into successive momentary pulses is reminiscent of the successive "frames" in the flow of conscious waking perception discussed in chapter 2. In both cases—active perceiving and the subsequent laying down of memories—what at first sight might have looked like one continuous process turns out on closer inspection to have a discrete and periodic structure.

The so-called "hippocampal-neocortical dialogue" is an example of what neuroscientists call "active system consolidation," the strengthening of memories by replaying them during sleep. At the neuronal level, active consolidation consists in selectively reactivating groups of neurons and thereby strengthening the synaptic connections between them.

According to Yoga, deep sleep is a state where memories are put together from subtle and subliminal mental impressions. Active memory consolidation during slow-wave sleep is a neuroscience counterpart to the Yoga view.

SEED SLEEP

We can also find in neuroscience a counterpart to the Vedānta view that deep sleep contains the "seed" of dreaming and waking consciousness. The Advaita Vedānta philosophers Gauḍapāda and Śaṅkara describe deep sleep as "seed sleep."[37] By this they mean that deep sleep is the causal source of waking and dreaming consciousness. Thus another word they use to describe it is "causal." Deep sleep is the causal state immediately prior to dreaming or waking, and it strongly shapes how dreams and waking experiences arise. In the Vedānta framework, whereas consciousness identifies with the physical body as the self in the waking state and with the mental dream body as the self in the dream state, it identifies with a subtle "causal body" in deep and dreamless sleep.

At one level, this view of deep sleep differs considerably from the neuroscience view. For neuroscience, waking sense experience is the basis for all consciousness, which disappears in deep sleep. For Vedānta, waking and dreaming consciousness arise out of deep sleep, and the progression from deep sleep to dreaming to waking is a progression from subtler to grosser levels of consciousness and embodiment. To use an analogy from physics, for Vedānta deep sleep is a kind of "ground state" of consciousness, a lowest-energy state from which the "excited states" of dreaming and waking arise.

At another level, however, the idea that deep sleep is the ground for future experience has a strong analogue in neuroscience. It's now well established that sleep actively promotes the ability to learn and acquire new memories in the waking state.[38] In addition, slow-wave sleep may strongly affect subsequent REM sleep, the sleep stage when dreaming is most likely to occur.[39] According to this idea, slow-wave sleep not only consolidates memories by replaying them, it primes memory networks for further consolidation during subsequent REM sleep, which always follows slow-wave sleep. In this way, memory replay in slow-wave sleep may shape the kinds of dreams we have as the proportion of REM to NREM sleep increases throughout the night.

Neuroscientist György Buzsáki, one of the leading researchers in the study of self-organizing brain activity, calls sleep the brain's "default state" (not to be confused with the "default mode of brain function" discussed in chapter 10).[40] By this he means that sleep is a self-organized state—one that emerges spontaneously without being managed or directed from outside—to which the brain always naturally returns. On the one hand, waking experience influences the way we sleep and rest; on the other hand, "After each day's experience . . . the brain falls back to the default pattern to rerun and intertwine the immediate and past experiences of the brain's owner."[41]

Buzsáki proposes that the self-organized processes of sleep strongly affect how the waking brain responds to the outside world. For example, every mental illness is associated with some kind of change in sleep. The sleep disorder is usually taken to result from the daily environmental interactions of the waking brain, but as Buzsáki points out, the causation probably goes the other way too: the symptoms displayed by the waking brain may result from disruptions to the brain's default state of sleep.[42]

In these newly emerging ideas from neuroscience we find a parallel to the older Yoga and Vedānta view that deep sleep provides the causal source for waking life and the ground in which waking life plants the seeds.

CONTEMPLATIVE SLEEP

In juxtaposing the Indian and neuroscience conceptions of deep sleep, I've proceeded so far as if the Indian notion of dreamless sleep corresponds to NREM slow-wave sleep. But this correspondence is actually too simplistic. What the Indian conception of deep sleep suggests is that we need a finer taxonomy of sleep states—a taxonomy that's not just physiological but also phenomenological, and that accommodates the ways that sleep may be culturally variable, as well as flexible and trainable through contemplative practices.

Recall that Vyāsa, in his commentary on the *Yoga Sūtras*, distinguishes three types of sleep—peaceful sleep, disturbed sleep, and heavy sleep. According to the cosmology that informs Yoga, these three types result from whichever of three "strands" or tendencies of material nature—called the three *guṇas*—predominates in the mind-body complex. Overall, the quality of dullness or the tendency to inactivity (*tamas*) dominates the mind in ordinary sleep. Sleep is heavy or stupefying when this quality isn't modified by either of the two other qualities or tendencies. Sleep is disturbed and restless when the quality of excitation or tendency to activity (*rajas*) is present. And sleep is peaceful and refreshing when the quality of lightness or tendency to clarity (*sattva*) is present. When the Vedānta philosophers describe deep and dreamless sleep as blissful, they have deep sleep with this quality of clarity in mind.

When people are roused from NREM sleep, however, they sometimes report they've been thinking while they were asleep, and often they describe going around in a repetitive loop of rumination. Although this kind of thinking probably occurs mainly in stage 2 NREM sleep, it's also reported during awakenings from slow-wave sleep.

Philosopher Owen Flanagan, in his book *Dreaming Souls: Sleep, Dreams, and the Evolution of the Conscious Mind,* appeals to this finding in order to argue that there's no such thing as dreamless sleep and hence no sleep completely lacking in consciousness.[43] Contrary to the standard neuroscience view, Flanagan thinks we're always conscious while asleep because we're always dreaming. Dreaming, he proposes, is any conscious mental activity occurring during sleep, not just mental activity involving sensory imagery. If ruminative thinking occurring in NREM sleep counts as dreaming, and if this kind of mental activity can happen during slow-wave sleep, then all sleep stages involve dreaming and at least some degree of consciousness.

From the Indian yogic perspective, however, we need to distinguish clearly whether there's such a thing as dreamless sleep, and whether we're always conscious while we sleep. Yoga and Vedānta agree that consciousness is always present when we're asleep, but this isn't because we're always dreaming, even if we define "dreaming" widely to mean any kind of thinking during sleep. On the contrary, what Yoga and Vedānta mean by "deep sleep" is the sleep state where there are no sensory or mental objects of awareness, that is, no images and no thoughts. Nevertheless, there is awareness, so this is a conscious state; it's a mode of consciousness without an object. In the Yoga framework, reports of ruminative thinking upon awakening indicate a coarser or shallower sleep state—closer to the surface of thinking consciousness—with a strong quality of excitation or tendency toward movement of the mind (*rajasic* sleep).

Consider now the reasons that sleep and dream scientist J. Allan Hobson gives for doubting the reliability of waking reports of perseverative thinking during slow-wave sleep:

> Reports of antecedent mental activity elicited following awakenings from deep sleep are rendered unreliable by the brain fog through which they must pass. . . . Even if the deeply sleeping brain were capable of the low-level ruminations sometimes implied by experimental reports, it is unlikely that they would survive the inertia of awakening. It may even be that the tumult of the awakening process triggers the chaotic and fragmentary mentation that is reported. And even when deep sleepers are sufficiently aroused to be interviewed, they may still generate huge slow waves in their EEGs, indicating that they are in a semistuporous state quite different from either sleeping or waking. Indeed, they may even hallucinate, become anxious, and confabulate as if they suffered from delirium. This is precisely what happens in the night terrors of children.[44]

Clearly, this too is a far cry from the Indian yogic conception of deep sleep. Neither reports of ruminative thinking nor waking hallucinatory confabulations correspond to the Yoga and Vedānta descriptions of deep sleep as a peaceful or blissful state free of mental activity and from which one awakens feeling alert and refreshed (*sattvic* sleep). From the Yoga perspective, what Hobson describes are sleep states strongly marked by a quality of dullness combined with mental excitation upon awakening.

My point is not at all that sleep science should refine its taxonomy using the ancient Indian notion of the three *guṇas*. It's rather that ultimately we can't map the Indian notion of deep and dreamless sleep using already established scientific categories, especially the physiologically defined sleep stages, which, even from a scientific perspective, are now recognized as too crude to capture the moment-to-moment dynamics of electrical brain activity during sleep, let alone the experiences that may be correlated with them.[45] Not only is the background metaphysics of the Indian view different from that of modern science, the Indian view is phenomenological, not physiological, and it's embedded in a normative framework that understands sleep in contemplative terms. To bridge from sleep science and the neuroscience of consciousness to the Indian conception of deep and dreamless sleep, we need to view sleep as a mode of consciousness that's trainable through meditation.

From the Yoga perspective, entering a state of blissful deep sleep on a regular basis requires leading a calm and peaceful life guided by the fundamental value of nonviolence (*ahiṃsā*), practicing daily meditation, and treating going to sleep and waking up as themselves occasions for meditation—for watching the mind as it enters into and emerges from sleep in order to inhibit the otherwise automatic identification with the changing states of waking, dreaming, and deep sleep as the true form of the "seer" or witnessing awareness. Thus the modern practice of *yoga nidrā* or "yogic sleep," which seems to have emerged in the twentieth-century neo-Vedanta movement but traces its origin to the much older movements known as Tantra, uses breathing methods, concentration, visualization, attention to the body, and emptying the mind of images and thoughts in order to lead the waking mind into a unique state of lucid awareness at the borderland of waking and deep sleep.[46] One long-term effect of this practice is said to be a deep sleep state that's peaceful and refreshing. Another effect is said to be a greater ability to witness lucidly the sleeping process and to remember qualities of sleep upon awakening.

CLEAR LIGHT SLEEP

In this chapter I've focused on Yoga and Vedānta, not Buddhism, because the philosophers belonging to these schools seem to have argued more explicitly for deep sleep being a mode of consciousness than the Indian Buddhist philosophers did. The idea that deep sleep can be lucidly witnessed, however, is central to the Tibetan Buddhist practice of sleep yoga, which is said to derive from the Indian Buddhist teacher Padmasambhava (ca. eighth century C.E.), so this practice needs mention here.[47]

According to the Tibetan Buddhist sleep yoga teachings, as we fall asleep our awareness withdraws from the five senses and the sixth mental sense until we eventually go blank and fall into darkness. After some time, which could be long or short, dreams arise. The state between the moment of falling asleep and the arising of dreams is deep and dreamless sleep.

Deep sleep, however, can happen in more than one way. Besides ordinary deep sleep, there's lucid deep sleep. Ordinary deep sleep is called the "sleep of ignorance"; awareness is void or blank and in total darkness. Lucid deep sleep is called "clear light sleep." "It occurs when the body is sleeping but the practitioner is neither lost in darkness nor in dreams, but instead abides in pure awareness."[48] Being lucid in this way in deep sleep is much harder than being lucid in the dream state; it takes a high degree of meditative realization.

Since we ordinarily identify with the gross levels of our consciousness—the five senses and the sixth mental sense—and these are shut down in deep sleep, deep sleep seems to the ordinary mind to be a state of unconsciousness. But a subtler level of pure awareness, which constitutes a "substrate" or "base consciousness" underlying sensory and mental consciousness, continues from moment to moment throughout waking, dreaming, and deep sleep. Lucid deep sleep affords an opportunity to experience directly pure awareness in its basic nature of clarity or luminosity—an experience that's described, as it is in Yoga and Vedānta, as blissful.

The Tibetan Buddhist and Yoga and Vedānta descriptions of deep sleep coincide in other ways. In Yoga and Vedānta, the blankness or voidness of deep sleep comes from there being no sensory or mental objects of awareness, whereas the darkness is due to ignorance concealing the presence of pure awareness. Being able to witness lucidly deep sleep means that the concealment of ignorance is removed and pure awareness shines forth in its own radiance, like the sun revealed by the clearing of a heavily overcast sky. Tibetan Buddhist descriptions sound similar, as these words by the contemporary teacher Dzogchen Ponlop indicate:

> The essence of deep sleep is, in fact, great luminosity, the true nature of mind. It is utterly bright and utterly vivid. It is a dense clarity, and because its clarity is so dense, it has a blinding effect on the confused mind. When we purify the ignorance of deep sleep, when we transcend that delusion and further penetrate the intense clarity, then we experience the clear, luminous nature of mind.[49]

In the practice of sleep yoga, the moment of falling asleep is said to be the crucial moment when the luminosity or clarity of pure awareness can be recognized:

> When do we meditate on this luminosity? Primarily, we should try to directly experience the true nature of mind at the very moment of the dissolution of the waking state. At that time, we generate bodhicitta [the wish to attain enlightenment in order to benefit all sentient beings] and, without being interrupted by other thoughts, we look with mindfulness and awareness directly at the mind itself with the intention of observing its aspect of clarity. At the very moment of dropping off to sleep, it is taught that pure awareness shines clearly,

full of vivid and bright qualities. This is a very short moment. Although we may miss it the first time, and the second, and so on, if we become accustomed to looking in this way, eventually we will be able to see this luminosity. By not allowing our mindfulness to wander, we will be able to sustain the experience as we transition from a conscious state into the state of sleep.[50]

Another way to enter lucid deep sleep is the practice of seeing through the dream state in a lucid dream. Recall the dream yoga instructions Alan Wallace gave me in Dharamsala, described in chapter 5. After recognizing the dream state and transforming the dream, you can try to see through the dream by dissolving it completely. You release imagery and thoughts and rest in the awareness of being aware. To be aware this way is to experience lucidly the state of deep and dreamless sleep.

It's also said that the complete shutting down of sensory and mental functions at the moment of falling asleep, together with pure awareness shining clearly before any dreams arise, closely resembles what happens when we die. For this reason, sleep yoga is also a practice for working with the inevitable experience of death.

CONTEMPLATIVE SLEEP SCIENCE

Sleep yoga drives home the point that we can't map the Indian and Tibetan yogic conceptions of deep and dreamless sleep using already established categories from sleep science. These Indian and Tibetan conceptions, besides being closely tied to first-person observations of what happens to consciousness during sleep, are embedded in contemplative frameworks that aim to bring about and promote certain kinds of sleep states. Instead of trying to fit these states into a physiological scheme derived from studying the way twentieth-century Americans and Europeans sleep in the sleep lab, we need to enlarge sleep science to include contemplative ways of understanding and training the mind in sleep. This project will require that sleep scientists, sleep yogis, and contemplative scholars of the Indian and Tibetan traditions work together to map the sleeping mind. In short, we need a new kind of sleep science—a contemplative sleep science.

Consider, for example, the Tibetan Buddhist practice of seeing through the dream state, described above. In this practice, you enter the state of lucid deep sleep from the lucid dream state. Lucid dreaming, however, seems to occur mainly during "phasic" REM sleep, where there are brief and rapid increases to the already high level of cortical activity in REM sleep; in addition, lucid dreaming has been shown to be correlated with large-scale coherent gamma oscillations in the EEG.[51] What happens to these neural patterns when one enters lucid deep sleep from the lucid dream state? Does lucid sleep in general require this sort of large-scale gamma activity, which we also know is strongly correlated with reportable conscious awareness? Or, to put the question more abstractly, does having experiential access to the deep sleep state require the kind of informational integration that occurs when neural systems can communicate quickly over long distances in the cortex? If so, then it's hard to see how lucid dreamless sleep could be a purely slow-wave NREM state in the canonical sleep-science sense. However, given our limited knowledge of both the neurophysiology of sleep and the neural correlates of consciousness, we can hardly assume that we know what the neural signatures of lucid awareness in deep sleep would look like. To date, there are (to my knowledge) no published scientific studies on Tibetan Buddhist dream and sleep yoga, so these kinds of questions about the relationship between lucid dreaming and lucid deep sleep point toward completely unknown territory.

Here's a speculative thought. Perhaps consciousness in deep sleep somehow correlates with features of the up state in slow-wave sleep when neurons fire at their waking rates for about a second. At both the single-neuron level and the larger level of neuronal populations, up state dynamics strongly resemble the dynamics of the awake and activated brain. For this reason, neuroscientists describe up states in slow-wave sleep as "fragments of wakefulness" or as restoring brief moments of "micro-wake-like activity."[52] According to Tibetan Buddhism, the substrate consciousness that's present in deep sleep is momentary in nature, that is, it exists as a continuum of discrete moments. This momentary character seems not unlike the momentary character of up states. In addition, gamma oscillations at both low (40–80 Hz) and high (80–120 Hz) frequencies occur during up states at roughly the same time over multiple cortical areas.[53] Although the function of gamma oscillations in slow-wave sleep is unknown, they appear to support memory consolidation. Another speculative possibility is that these gamma oscillations also correlate with the presence of a subtle conscious awareness in deep sleep, and can be affected by practicing sleep yoga.

More generally, since different parts of the brain can be in up states and down states at the same time—or to put it another way, since different neuronal networks can be "awake" and "asleep" at the same time[54]—the neural correlates of lucid deep sleep might involve one part of the brain being "awake" while other parts are "asleep."

Two neuroscience studies of meditation and sleep are suggestive in this regard. The first is a recent study from Giulio Tononi's and Richard Davidson's labs.[55] They examined slow-wave sleep in highly experienced Theravada Buddhist and Tibetan Buddhist meditation practitioners, and found that the long-term meditators, compared to nonmeditators, had increased higher EEG gamma activity in a parietal-occipital region of the scalp during NREM sleep. The higher activity was positively correlated with the length of meditation training. This finding is notable because gamma-frequency electrical brain activity is a well-known neural marker of conscious cognitive processes and has also been shown to distinguish lucid dreaming from nonlucid dreaming in REM sleep.[56] During NREM sleep, however, gamma activity tends to decrease, so the higher gamma activity in the meditators could reflect a capacity to maintain some level of awareness during sleep.

The second is an older study of sleep in long-term practitioners of Transcendental Meditation (TM) who reported the subjective experience of "witnessing" during sleep.[57] In TM this experience is conceptualized as a "higher state of consciousness" where you feel a quiet and peaceful awareness during sleep and wake up feeling refreshed. In this study, three groups were compared—long-term practitioners, short-term practitioners, and individuals with no TM experience. The main finding was that the long-term practitioners showed a unique EEG pattern during slow-wave sleep, with theta and alpha activity present during stages 3 and 4, as well as decreased skeletal muscle activity as measured by the electromyograph (EMG). Although we can't draw clear conclusions about what these distinctive physiological patterns mean, including whether they're due to TM practice or some other cause, the authors of the study interpret them as supporting the presence of a different kind of slow-wave sleep state in individuals who report witnessing of sleep.

A few other studies have examined whether meditation practices are associated with altered sleep patterns in long-term practitioners compared to nonpractitioners.[58] One study found that experienced practitioners of TM and other forms of yoga meditation showed significantly higher levels of the hormone melatonin, which regulates the sleep-wake cycle and is produced by the pineal gland, immediately following a nighttime period of meditation, compared to the same time after not meditating.[59] Although the physiological pathway by which this increase happens is unknown, the finding suggests that these types of meditation practices can affect basic physiological processes underlying the sleep-wake cycle.

Two other studies found that both experienced Theravada Vipassanā meditation practitioners and experienced practitioners of a yogic breathing method called Sudarshan Kriya Yoga showed a significantly larger amount of slow-wave sleep in their sleep cycles compared to the amount in control subjects of the same age across all age groups from thirty to sixty years old.[60] Although the amount of time spent in slow-wave sleep decreases considerably with age, the middle-aged group of experienced meditators showed the same amount of slow-wave sleep as the younger age group. In addition, the Vipassanā practitioners showed significantly more REM sleep as well as a higher number of sleep cycles through all five stages (NREM stages 1–4 and REM sleep) than did the nonpractitioners across all age groups.

In short, yoga and Vipassanā meditation practices seem to be associated with a host of changes to sleep physiology, so it's not unreasonable to speculate that these meditation practices may also be associated with changes to other sleep-related phenomena, such as learning and memory consolidation, as well as health.

Although these last three studies focused on the potential effects of meditation on sleep physiology, they didn't use contemplative mental training as a way to investigate consciousness and its physiological correlates in sleep. This further step is needed to create a contemplative sleep science and to connect Western science to the Indian and Tibetan contemplative frameworks for understanding sleep.

One benefit of a contemplative sleep science is that it could offer a new approach to the guiding question of this chapter—whether deep and dreamless sleep qualifies as a mode of consciousness. Consider the following testable working hypothesis: in highly experienced practitioners of sleep and dream yoga, we should observe a closer relation between subjective reports of phenomenal qualities of sleep and various objective physiological measures (not just of the brain but also of the rest of the body). If highly experienced sleep yogis were able to provide reports upon awakening about their experience of the state they call deep and dreamless sleep, and if sleep scientists were able to relate these reports to fine-grained features of sleep physiology and to familiar aspects of the neural correlates of consciousness, then we would have new evidence from experimental science that deep and dreamless sleep—at least in certain individuals—is a mode of phenomenal consciousness, some of whose qualities can be made accessible to verbal report.

9
DYING

What Happens When We Die?

've heard anthropologist and Zen teacher Joan Halifax tell the story many times, but it never seems to lose its force. Here it is, in her own words, from her book, *Being with Dying*:

When a group of people gathers together for a meditation retreat, important shifts in one's mind and life may unfold. I often think of one retreat in particular, because what happened one day illustrates with fierce clarity the fragility of these human bodies we inhabit, and the gravity of what Buddhists call "the great matter of life and death."

The retreat took place sometime in the seventies at a quiet center on Cortez Island in Canada, a place then called Cold Mountain Institute. It was the beginning morning of the program, and we had just finished the first period of silent sitting meditation. The bell rang softly to announce the end of the period, and we all stretched our legs and stood up to do walking practice—but one man remained seated.

I remember feeling concern as I turned to look at him: why was he not getting up? He was sitting in the full-lotus position, his legs perfectly folded and his feet resting on his thighs. Then, as I watched in shock, his body tilted over to one side, slumped and sagging, and he fell to the floor. He died on the spot. There were several doctors and nurses participating in the retreat who helped perform CPR and administer oxygen, but it was too late. Later we learned that his aorta had burst while we were all sitting.

This man was healthy enough—perhaps in his late thirties. He almost certainly had not imagined when he came to this retreat that he would die during it. And yet, that day, sixty people sat down to meditate—and only fifty-nine stood up.[1]

All of us will die, sooner or later. This fact is like no other. It's not just that every one of us will die; it's that I, myself, am going to die. How or when, I don't know, but that it will happen, I can be absolutely sure of. And yet, despite this certainty, that I'm going to die is the hardest fact to grasp and the easiest one to turn away from. "What is the most wondrous thing in the world?" asks Yama, the Lord of Death. His son, Yudhiṣthira, answers, "The most wondrous thing in the world is that all around us people can be dying and we don't believe it can happen to us."[2]

Modern Western society, like no other society in human history, reinforces our blindness to the inevitability of our own death. It does this in countless ways, one of which is by hiding death from our view. As I write these words, I'm in my forty-ninth year and I've never seen a dead body. I've been in the close presence of someone near death only twice. At no other time in human history could I have lived so long and seen so little of death.

You might think that we see less of death than our forebears did because biomedicine enables us to live longer. But it also hides death from us in a peculiar and powerful way. Biomedicine talks about death as if it were essentially an objective and impersonal event instead of a subjective and personal one. From a purely biomedical perspective, death consists in the breakdown of the functions of the living body along with the disappearance of all outer signs of consciousness. Missing from this perspective is the subjective experience of this breakdown and the significance of the inevitable fact of one's own death. Biomedicine hides the inner experience of dying and the existential meaning of death.

Materialist scientists and philosophers may argue that there's not much we can say about the experience of death because death is annihilation and the extinction of one's consciousness. Yet even if we set aside the issue of whether science gives us good reason to believe that death entails the complete cessation of all consciousness, this conception is totally inadequate because it says nothing about the experience of *dying*.

In contrast, the Indian and Tibetan yogic traditions claim to provide detailed accounts of the transformations of consciousness during the dying process. Tibetan Buddhism, in particular, as we've seen in earlier chapters, offers a rich contemplative perspective on death, including meditations to prepare for death and to practice as one dies. This kind of experiential view of dying and death is missing from the biomedical perspective.

Nevertheless, we might wonder exactly how these yogic traditions, rooted in foreign cultures and belief systems, can help us to recover an experiential approach to death in our modern Western context. From the other side, we might wonder whether Western science and medicine can help to advance and enrich contemplative ways of approaching death. These were some of the questions I talked about with neuroscientist and psychologist Rebecca Todd—whom I've known since we were kids in the 1970s and whose love I've treasured for over twenty years—as we traveled to the Upaya Institute and Zen Center in Santa Fe, New Mexico, to attend its Being with Dying training program in contemplative end-of-life care.

BEING WITH DYING

It's May 2011, and Rebecca and I have left behind our work lives in Toronto in order to immerse ourselves in the Being with Dying program, which Joan Halifax—or Roshi Joan, as she's known in the Buddhist world—created in 1994 in order to train health-care professionals and dying people in the psychological, spiritual, and social aspects of dying.[3] The program lasts for eight days and goes from seven o'clock in the morning until nine o'clock at night. Out of a group of sixty-five participants, we're the only ones, along with a community leader and former mayor from a small town in Montana, who aren't health-care or other professionals who work with the dying and their families. We're looking forward to learning with and from an extraordinary group of people, many of whom have made care for the dying their life's work—palliative care nurses and doctors, psychiatrists and social workers, chaplains and hospice workers.[4] They've come from all over the world. Many say they're suffering from burnout, not so much from their clinical work, which they love, but from having to deal with the stress and administrative demands of the modern hospital and the effect it has on the dying and those who care for them.

The Being with Dying training combines lectures on science and clinical practice with experiential learning in small groups as well as meditation and yoga. At the heart of the training are reflective and meditative practices for learning to face death—whether one's own or another's—with mindful awareness and compassion.

On the morning of the first day, Roshi Joan asks us to write freely without interruption for five minutes our answer to the question, "What is your worst-case scenario of how you will die?" After the five minutes are up, we note the feelings and emotional reactions we're experiencing. Then we repeat the whole exercise for the question, "How do you really want to die?" It's striking that no one wants to die in a hospital and that many of our worst-case scenarios include dying in a hospital.

The second day begins with a meditation called the "Nine Contemplations," derived from the eleventh-century Indian Buddhist teacher Atiśa. Roshi Joan presents the meditation as a reflective practice for facing the truth of our mortality. Make sure your body is relaxed and calm, she tells us, as we settle ourselves onto the cushions and chairs in the zendo (meditation hall). Close your eyes or leave them lightly open and unfocused. Let your mind settle. Bring your attention to your breath. The nine contemplations are meant to remind you about the nature of life and death. Please consider these truths:

The First Contemplation
All of us will die, sooner or later.
Death is inevitable; no one is exempt.
Holding this thought in mind, I abide in the breath.

The Second Contemplation
My life-span is ever-decreasing.
The human life-span is ever-decreasing; each breath brings us closer to
    death.
Holding this thought in mind, I delve deeply into its truth.

The Third Contemplation
Death comes whether or not I am prepared.
Death will indeed come, whether or not we are prepared.
Holding this thought in mind, I enter fully into the body of life.

The Fourth Contemplation
My life-span is not fixed.
Human life expectancy is uncertain; death can come at any time.
Holding this thought in mind, I am attentive to each moment.

The Fifth Contemplation
Death has many causes.
There are many causes of death—even habits and desires are
    precipitants.
Holding this thought in mind, I consider the endless possibilities.

The Sixth Contemplation
My body is fragile and vulnerable.
The human body is fragile and vulnerable; my life hangs by a breath.
Holding this thought in mind, I attend as I inhale and exhale.

The Seventh Contemplation
My material resources will be of no use to me.
At the time of death, material resources are of no use.
Holding this thought in mind, I invest wholeheartedly in practice.

The Eighth Contemplation
My loved ones cannot save me.
Our loved ones cannot keep us from death; there is no delaying its
advent.
Holding this thought in mind, I exercise non-grasping.

The Ninth Contemplation
My own body cannot help me when death comes.
The body cannot help us at death; it, too, will be lost at that moment.
Holding this thought in mind, I learn to let go.[5]

Before we break up into small groups to talk about these con-templations, Roshi Joan explains that they're like a weather report warning us of a storm. The storm is inevitable, though we can't say exactly when or how it will hit. Accepting the truth of death is how we can begin to prepare for the storm without fear while arousing our awareness of life. Continued practice of the nine contemplations helps to transform fear of death into acceptance of death and grati-tude for life. And this kind of openness, she says, is what's really needed in order to be fully present when caring for dying people and their needs.

What turns out to be the high point of the Being with Dying train-ing for Rebecca and me arrives early in the morning on the last full day of the program. As we enter the zendo after the three-mile walk up to Upaya from our hotel in Santa Fe, we see that all the chairs have been removed; we're supposed to lie down on the cushions. Roshi Joan is going to guide us through a meditation called "Dissolution of the Body," in which you visualize going through the dying process.[6] The meditation is her nonsectarian and modern adaptation of the classical Tibetan Buddhist meditation practice called "Dissolution of the Elements After Death," whose traditional monastic version the Dalai Lama had described at our Mind and Life dialogue in 2007 (see chapter 3).[7]

In the Tibetan Buddhist view, our living body, at a subtle level, consists of a vital energy that's inseparable from consciousness. The Tibetan term for this vital energy is *lüng*, which translates the

Sanskrit word *prāṇa* and literally means "wind." In Tibetan medicine, all forms of outer and inner motion in the body are controlled by "winds" or energies, which also serve as the support for our sensory and mental consciousnesses. A traditional image for the relationship between wind and consciousness is that of a horse and its rider.[8] Consciousness rides on the winds or energies of the body, and when these begin to transform in the dying process, consciousness undergoes radical changes. Ultimately, as the winds dissolve, consciousness falls apart and death occurs. Tibetan yogis visualize or rehearse in imagination the entire dissolution process as a way to familiarize themselves with dying and to gain control over the mental states that arise in the course of dying. Moreover, advanced yogis, it is said, can go beyond simply visualizing the dissolution process and actually bring it about during meditation.

Roshi Joan recommends that we lie in the traditional posture for the dissolution meditation—the "sleeping lion posture," which is the position the Buddha is said to have taken when he was dying. You lie on your right side with your knees drawn up slightly, and your right hand makes a pillow under your cheek while your left arm rests along your left side. Pressing the little finger against the right nostril completes the posture.

We've been told to lie with our heads toward the altar with the statue of Mañjuśrī, who, wielding a sword in his right hand, symbolizes the wisdom that cuts through ignorance and delusion. But Roshi Joan now tells us that we can imagine the top of our head being directed toward any image we choose—a religious image, a loved person, flowers, a nature scene, or any being who represents awakening and compassion. Immediately I see with my mind's eye an image of the dancing Śiva, whose rhythmic movements embody the creation and destruction of the cosmos, and I hear my dad asking me, when I was a little kid, which I liked better, Śiva dancing in a ring of flames or Gautama Buddha sitting under a tree in meditation. I never could decide.

My mind comes back to the room and I open my eyes to take in the sight of our now close-knit community lying on the zendo floor and preparing to practice dying together. As I close my eyes and bring my attention back to my breathing, I hear Roshi Joan explaining that in this practice you imagine letting go of the elements that make up your mind and body as they dissolve when you die. Traditionally, these elements are specified as earth, water, fire, air or wind, and space, and the process is described as one where they progressively dissolve, each element into the next. The meditation practice is about awakening to the dissolution process and releasing what you take to be your identity into a vast spaciousness and radiance. You release your bodily form, your feelings, your perceptions, your inclinations or mental tendencies, and ultimately your consciousness. Through this practice you familiarize yourself with the experience of dying and train yourself to transform it into an experience of enlightenment and liberation. Even in cases of sudden death, she tells us, Tibetan teachers say that you go through the same dissolution process. So training your mind to recognize the process and to let go during it can help you, should you die suddenly.

There's one more instruction Roshi Joan gives before we start the meditation. Since the process of falling asleep, she says, closely resembles what happens when you die, the dissolution meditation may put you at the edge of sleep. But try not to fall asleep, and if you hear your neighbor snoring, gently nudge him and wake him up.

Imagine you are home in bed, she begins. You are dying and fortunate to be surrounded by loved ones. You feel somewhat agitated and irritable, but you accept this state of mind. Your body is weak; you have no energy to do anything. You feel a growing heaviness, going right to the core of your body. You're so weighed down that there's no distinction between you and the bed.

I feel heavy, too heavy for the zendo floor to support me, as if I'm going to fall through it into a void. Vertigo takes over and I open my eyes. The room looks unstable, as if it were about to spin and throw me into the air, and I want to sit up. I wonder whether I'm feeling the effects of the altitude, being at 7,200 feet above sea level, or having some kind of anxiety attack.

Wake up as this body lets go, I hear Roshi Joan saying. I try to return to my breath, notice the anxious feeling, and release it. I calm down a little, but the feeling stays there in the background, rising and falling like a rough sea that threatens to toss a small boat.

Your sight is dim, Roshi Joan says. It's difficult to open and close your eyes. Your sensory grip on the world is loosening. As your body slips away, the outside world slips away too. You sink deeper and deeper into a blurry mental state. Whatever visions you see are like blue mirages.

I've managed to close my eyes again, and pre-sleep images are luring my attention. Using the breath, I try to energize my awareness so I don't lose track of what Roshi Joan is saying.

This is the dissolution of the body, she is telling us—these feelings of heaviness, drowsiness, being weighed down, the loss of control, and your inability to see the form world around you. In this state of mind and body, be awake; be effortlessly present. The mind can be still as you let go. Be present as this body is dying. This body is not you. This is the dissolution of the earth element as form unbinds into feelings.

After a few minutes of silence, the visualization starts up again. Your hearing is diminished, Roshi Joan says, and you've sunk into a blurred and undifferentiated state of mind. Your nose is running, saliva is leaking out of your mouth, a watery discharge is coming out of your eyes, and you can't hold your urine. Your generative fluids are all drying up, and your body is parched. You have a thirst that no amount of water can quench.

Let go fully into this dryness. Release the watery element of your body.

Your mind is hazy, and you've ceased to have experiences of pain, pleasure, or even indifference. Behind your eyes you see a vision of swirling smoke and a haziness that dissolves all differences. This is the dissolution of the water element and the unbinding of feelings. Wake up as you let go into this vision of swirling smoke.

Now the body's fire element begins to dissolve into air, Roshi Joan continues, and your body feels cool. Heat withdraws from your hands and feet into your body's core. You can't smell anything, and you can't drink or swallow. Your ability to perceive anything through the five senses is gone, and your mental discernment alternates between clarity and confusion. Your in-breath is short and weak; your out-breath is long. You can't recognize anyone around you or remember the names of your loved ones.

By this point in the meditation I've managed to settle into my breathing and am feeling less anxious. My awareness seems to be sliding back and forth between inner withdrawal of the senses and outer following of the meditation instructions. I hear snoring nearby, followed by a grunt; someone has fallen asleep and been tapped by his neighbor.

Roshi Joan continues the meditation. You may feel as though you're being consumed in a blaze of fire that rises into space. Let go into this fire. Or you may see sparks of light like fireflies. Wake up in this vision of sparks. This is the dissolution of the body's fire element and the unbinding of your ability to perceive.

You've now given up any sense of volition, she says, and your mind is no longer aware of the outside world. Accept this aimlessness.

As the element of air is dissolved, you're having visions. These visions relate to who you are and how you've lived your life. You may be seeing your family in a peaceful setting or friends welcoming you, or you may be reliving pleasant experiences from your past. Or you may be having hellish visions. If you've hurt others, they may appear to you. Troublesome moments of your life may arise to haunt you. You may even cry out in fear.

It's very important, Roshi Joan says, not to identify with these visions, no matter whether they're beautiful or terrible. Simply let them be. They will pass; you don't have to do anything.

Your body is barely moving now, she tells us. Its last energy is receding into the core, and whatever heat is left now resides in the area of the heart. Your in-breath is barely a sip, and your out-breath is uneven and rattles. Your eyes gaze into emptiness and roll upward.

Your last out-breath is long. Imagine that you're experiencing this last breath and let it go.

Your body lifts slightly to meet the next in-breath, which does not come. Cognitive functions cease altogether. Your breathing has stopped. The perception from the outside is that you are dead. Know this empty state, experience it, and be present to it. This is the dissolution of the air or wind element and the unbinding of mental formations.

At the moment of so-called physical death, Roshi Joan says, you see a small, flickering flame like a candle. Suddenly it's extinguished, and your awareness is gone.

My mind recoils at these words, and the meditation now seems paradoxical. How can I imagine no awareness from the inside? It doesn't make sense; there's nothing inner to visualize. I can visualize myself from the outside, as if I were looking at myself in my deathbed, but this is to visualize an out-of-body experience, where my awareness is still present. Or it's to visualize how I would look to someone else in the room who witnesses my death. Or is the loss of awareness like deep and dreamless sleep? But then there would have to be some kind of awakening from the blackout, in order for the loss to be known. I notice that my conceptual mind is racing, as I've gotten caught up in thinking about the ungraspability of my own death.

After a long silence, Roshi Joan tells us to visualize a tiny white drop descending from the crown of the head into the heart area. As it descends, imagine the energy and experience of anger transforming into fundamental clarity. You experience an immaculate autumn sky filled with brilliant sunlight.

Imagine now, she says, a red drop ascending from the base of the spine toward the heart. As it ascends, desire transforms into profound bliss. You experience a vast and clear copper-red sky, like an autumn sky at dusk.

The white and red drops, she continues, meet in the heart and surround your subtle consciousness. You are now free from the conceptual mind.

A deep, black sky, free of stars or moon, appears. Out of this nothingness, luminescence arises. You are one with a clear dawn sky free of sunlight, moonlight, and darkness. You are bliss and clarity. The clear light of presence is liberated, the mother light of your awareness. Visualize it, be with it, be awake to it; don't turn away from it.

This is your ultimate great perfection. This is the actual moment of death. It is the inner or subtle dissolution, the dissolution of consciousness into space, and the dissolution of space itself into luminosity.

The vertigo I was feeling is gone now and the room is still, its silence held by the hum of mountain life outside the zendo.

When you feel ready, Roshi Joan says, roll onto your back while keeping your eyes closed. Notice how being more physical changes your mental experience. As you move into ending the practice, the tendency is to look for some reference point, physical or mental. But what we want to do is to bring the quality of openness, of courage and presence, from this meditation practice into the whole of our life. Open your eyes, gazing but not seeking a reference point. Let your attention be wide and inclusive. Slowly roll to the side and sit up, while noticing this transition to greater embodiment. Try to sustain the heart-mind of openness. Lift your gaze above the heads of those who are in front of you, and let your breath be deep in your body. Notice again how the mind changes as the body changes. Try to maintain the quality of openness and the taste of luminosity.

When you're ready, she tells us, stand up and gently walk out of the zendo. Bring your gaze into the sky. Let your mind mix with the sky, the luminous New Mexico sky.

A procession forms, and we leave the zendo one by one. I turn around and face the mountains behind Upaya. The blue sky fills my eyes, and in my head I hear these words:

> At the round earth's imagined corners blow
> Your trumpets, angels, and arise, arise
> From death, you numberless infinities
> Of souls, and to your scattered bodies go. . . .
> —John Donne, Holy Sonnet VII

DEATH IN MIND

One of the oldest thoughts about death in the Judaeo-Christian tradition comes from the Book of Ecclesiastes: "For the living know that they shall die: but the dead know not any thing" (9:5). The Greek philosopher Epicurus (341–270 B.C.E.) expressed a similar thought:

"death . . . is nothing to us, because as long as we exist, death is not present, and when death is present we do not exist" (Letter to Menoeceus 124–125). Given this way of thinking about death, if phenomenology is restricted to what we can experience, then although there can be a phenomenology of dying, there cannot be any phenomenology of death, for experience ceases to exist at death.

The Tibetan Buddhist worldview is different.[9] Tibetan Buddhists see death as a transitional moment between two of the six "in-between states" or *bardos* that make up the cycle of existence. The "bardo of this life" begins at the moment of birth and ends once active dying begins. Within the bardo of this life are the "bardo of dream," which begins at the onset of sleep, includes dreaming and deep sleep, and ends with waking up, and the "bardo of meditation," which is the interval of time when the mind rests in meditative absorption. The "bardo of dying" begins with active dying and ends at the moment of death. Immediately following the moment of death is the "luminous bardo of *dharmatā*"; it encompasses the after-death experience of the "clear light" or "ground luminosity," which is the ultimate nature of mind as pure awareness (*dharmatā* means the "ultimate nature of phenomena"). The "bardo of becoming" is the last interval in the cycle; it encompasses the after-death experience of wandering in the form of a subtle "mental body" while seeking a new physical embodiment, and it ends when one enters the womb of one's future mother, at which point the whole cycle begins again.

The guided meditation at Upaya involved visualizing the progression of the bardo of dying from the "outer dissolution" of the senses and elements of the body to the "inner dissolution" of consciousness, culminating in the arising of the clear light or ground luminosity of pure awareness. What would ordinarily be considered the moment of death in a modern clinical setting—the cessation of breathing and other vital signs—corresponds to what Tibetan Buddhists regard as the end of the outer dissolution. What follows next is the inner dissolution, which consists in the gradual dawning of ever more subtle levels of consciousness as the coarser levels of consciousness fall apart. Death occurs only with the arising of the clear light of pure awareness at the end of the inner dissolution, for this is the moment

when the outer sensory consciousnesses and inner mental consciousnesses have completely dissolved back into the ground luminosity or
ultimate nature of the mind.

To be able to recognize this moment when it happens, however,
is said to require having trained oneself to recognize the ground
luminosity of pure awareness in the bardo of this life, including
in the bardo of meditation and the bardo of dream. Furthermore,
immediately before the dawning of the clear light at death, a "blackout" period happens, as if consciousness is lost.[10] The blackout is
described as being like the blackness of an empty sky with no sunlight, starlight, or moonlight. In the words of contemporary Tibetan
teacher Dzogchen Ponlop:

> If we have not trained our mind through practice, then we faint and
> lose all awareness at this point. . . . If we have stabilized our mind
> and developed some insight into its nature, then we will recognize the
> arising in the next moment of the ultimate nature of mind. We will see
> its empty essence, its suchness, which is nothing other than the . . .
> ground luminosity.[11]

Tibetan Buddhists say that the bardo of dying corresponds closely
to what happens when we fall asleep. Our thought processes progressively dissolve during the pre-sleep state. At the moment of dropping
off to sleep, blackness occurs, followed immediately by the emergence of the clear light or ground luminosity of pure awareness,
which we fail to recognize unless we've trained our mind in dream
and sleep yoga. From this deep sleep state, dreams arise as spontaneous manifestations or appearances of the mind's own basic luminosity. Thus falling asleep is akin to the bardo of dying, the emergence
of the clear light in deep sleep is akin to the dawning of the clear
light at death, and dreaming is akin to the bardo of becoming, with
the dream body corresponding to the after-death mental body that
wanders in search of a new physical embodiment.

This comparison between death and sleep is also meant to illustrate how difficult it is to maintain awareness during these in-between
states and hence the importance of training the mind, especially

through the meditation practices of dream and sleep yoga. In the words of another contemporary Tibetan teacher, Sogyal Rinpoche:

> How many of us are aware of the changes in consciousness when we fall asleep? Or of the moment of sleep before dreams begin? How many of us are aware even when we dream that we are dreaming? Imagine, then, how difficult it will be to remain aware during the turmoil of the bardos of death.
>
> How your mind is in the sleep and dream state indicates how your mind will be in the corresponding bardo states; for example, the way in which you react to dreams, nightmares, and difficulties now shows how you might react after you die.
>
> This is why the yoga of sleep and dream plays such an important part in the preparation for death. What a real practitioner seeks to do is to keep, unfailing and unbroken, his or her awareness of the nature of mind throughout day and night, and so use directly the different phases of sleep and dream to recognize and become familiar with what will happen in the bardos during and after death.[12]

The kinship between sleep and death—a kinship found also in Greek mythology, the Old and New Testaments, and the Mesopotamian *Epic of Gilgamesh* (ca. 1800–1300 B.C.E.)[13]—suggests one way we can see the Tibetan Buddhist account of death as involving phenomenology: it is trying to extrapolate from a phenomenology of sleep and dreaming to an understanding of death.

Modern Tibetan Buddhist teachers and their followers, however, often present the Tibetan Buddhist account of death as if it were phenomenological in the sense of being a literal description of what anyone will experience at the moment of death and afterward. I must admit I'm very skeptical of this way of thinking. The main reason is that phenomenology is based on first-person reports of direct experience from human beings with working cognitive abilities, but once someone is at the point of death or has died, there's no way she or he can report anything. Therefore, even if we allow for the sake of argument that consciousness might continue after death, any account of consciousness at the moment of death or afterward can't be based on reports of direct experience but must instead be based on inference or conjecture.

The Dalai Lama seems to admit this point. At our 2007 Mind and Life dialogue, he stated that what happens after death is a mystery and not open to scientific investigation. He did suggest, however, that scientists should investigate what happens to the brain and body of great meditation practitioners who enter the clear light state through meditation when they're dying. It's believed that these practitioners remain in the clear light state for a period of time—in some cases many days—after their respiration and heartbeat cease, because during that time their bodies seem to stay fresh without decay.

The Dalai Lama gave a more philosophical statement at an earlier Mind and Life dialogue in 1992.[14] Tibetan Buddhist philosophy, he explained, distinguishes among three kinds of phenomena—"evident phenomena," which are known through direct perception; "remote or obscure phenomena," which are known through inference; and "extremely remote or extremely obscure phenomena," which are known only through the testimony of a third person. Death and after-death phenomena fall into the third category. The Tibetan Buddhist account of death, therefore, doesn't lend itself to direct experiential or inferential confirmation. Nevertheless, the Dalai Lama went on to argue, one can have a high degree of confidence in this account if there is no evidence contradicting it and if it accords with one's own investigations of what can be known through direct experience and inference. In particular, one can directly experience the dissolution of consciousness and the arising of the clear light nature of awareness in meditation and sleep, and these experiences provide a basis for inferring that a similar, though much more intense, experience happens when one dies. One can also see that the bodies of great meditation practitioners who enter the clear light state as they're dying don't decay in the usual way after physical death. Thus, the Dalai Lama concluded, even though acceptance of the traditional Tibetan Buddhist view of death requires the acceptance of third-person testimony, such acceptance can be reasonable, not a matter of blind faith.

Although this line of thought may be reassuring if one already accepts the Tibetan Buddhist view of death, it's not likely to convince anyone who is skeptical. The reason is that the appeal to third-person testimony just pushes the problem back. Now what we need to know is how there could possibly be third-person testimony about the experience of death or the after-death state. After all, we accept something as testimony when we're confident that it's based on a person reporting what she or he has witnessed, but such reporting seems impossible in the case of death.

At this point, many Buddhists (as well as Hindus, Jains, and devotees of New Age spirituality) will appeal to the cosmology of rebirth or reincarnation and the possibility of memories of past lives, including memories of dying and being reborn. In recent times, people in a variety of settings have reported apparent memories of past lives.[15] If we had good reason to believe that such memory reports were veridical, then we could say that testimony about the intermediate state between death and rebirth might be possible (unless there is no intermediate state and one is born directly into the next life, as some Theravada Buddhists maintain).

Of course, this argument isn't likely to persuade anyone who thinks that the Indian and Tibetan rebirth or reincarnation cosmology grows out of a need to provide a coherent framework for ethics (we all have been in countless different relationships with each other over eons, including being mother to and offspring of every other being), morality (what you do comes back to you, if not in this life, then in the next one), and the possibility of spiritual liberation (it takes many lives to become enlightened and attain liberation), not from being able to remember past lives.[16] Indeed, many of the apparent memory reports (though not all[17]) come from cultures permeated by this worldview.

Modern Western Buddhists and other believers in rebirth often cite the reincarnation research of psychiatrist Ian Stevenson, who over the course of his life documented thousands of reports of purported memories of past lives, particularly in young children.[18] A small number of these cases, Stevenson maintained, were strongly "suggestive" of reincarnation.

Although Stevenson's presentation of these cases often makes for compelling reading, all the evidence is anecdotal and derived from interviews where there is a large amount of room for false memory and after-the-fact reconstruction. The interviews weren't conducted directly with the children when they first reported the memory, but some time later, so there had been plenty of time for the child to assimilate information gotten from family members and to repeat it as if it were his or her experience. And sometimes the children weren't interviewed at all; only family members were. Finally, it's hard to know how to assess whether a memory report about a past life exceeds chance probability, and critics have pointed to a number of serious flaws in Stevenson's statistical reasoning.[19] For these reasons, I don't find Stevenson's evidence convincing, though it does seem possible in principle to investigate claims of past life memories using scientific methods.[20]

With regard to the issue of rebirth, I agree with Stephen Batchelor when he writes:

> It may seem that there are two options: either to believe in rebirth or not. But there is a third alternative: to acknowledge, in all honesty, *I do not know.* We neither have to adopt the literal versions of rebirth presented by religious tradition nor fall into the extreme of regarding death as annihilation. Regardless of what we believe, our actions will reverberate beyond our deaths. Irrespective of our personal survival, the legacy of our thoughts, words, and deeds will continue through the impressions we leave behind in the lives of those we have influenced or touched in any way.[21]

We'll come back to the importance of this attitude of "not knowing" at the end of this chapter.

Another reason not to interpret the Tibetan Buddhist account of death as a literal phenomenological account is that it's steeped in metaphor and symbolism. For example, at the moment of death and in the state immediately after death, the sensory and conceptual mind is said to have dissolved completely, so the experiences must be nonsensory and nonconceptual, and hence ineffable. Yet these experiences are described in sensory and cognitive terms, and in a language that's rich with Buddhist symbolism. For example, it's said that if one fails to "recognize" the ground luminosity of pure awareness, which is also identified with "buddha nature" and the inconceivable and unmanifested "truth body" of buddhahood, then a second phase called the "spontaneously arising luminosity" ensues, in which "an array of dazzling sights and sounds surrounds one."[22] This "luminosity of appearance" is identified with the manifested and subtle "enjoyment body" or "bliss body" of buddhahood, and it appears in the form of one hundred deities, some peaceful and some wrathful, who are nothing other than the nature of one's own mind but who also express various qualities of different celestial "buddha families."[23]

Given this kind of symbolic and religious imagery and language—which permeates the whole bardo cosmology—it's difficult for those of us who stand outside the Tibetan Buddhist worldview to regard its account as a literal description of what it's supposed to be like for consciousness to continue after death. The account doesn't seem phenomenological in that sense. Instead, it seems to be what religious scholars call "soteriological" (pertaining to salvation), for it concerns the Buddhist goal of liberation or enlightenment and how that goal is understood within a religious context of collective ritual and symbolism.

It would be a mistake, however, to think that the Tibetan Buddhist account of death must be either literally true or false. Instead, we can see it as a script for enacting certain states of consciousness as one dies.[24] In this way, it is more performative and prescriptive than descriptive. Looked at from the outside, the Tibetan Buddhist account of death strikes me as a "ritualized phenomenology." The dissolution meditation doesn't so much present a phenomenological description of death as rehearse and enact a phenomenology of death as a ritual performance. Given the power of ritual, it stands to reason that someone who spends his or her whole life practicing the ritual, or even just living in a culture where there are constant reminders of the ritual, would experience its symbolism in a powerful and immediate way at the time of death.

There's another important perspective that does allow us to see the Tibetan Buddhist view as including what some philosophers would think of as a phenomenology of death. Phenomenology, in its philosophical sense, is concerned with describing the essential or necessary structures of consciousness. So if death belongs essentially or necessarily to the structure of consciousness itself, then laying bare this fact would amount to a contribution to a phenomenology of death.

To see what I mean, recall Epicurus's statement, "death . . . is nothing to us, because as long as we exist, death is not present, and when death is present we do not exist." So far I've talked about how Tibetan Buddhists would disagree with the second part of this statement. (Of course, if "we" means our ordinary conceptual mind and gross sensory consciousness, then Tibetan Buddhists would agree with Epicurus, but if "we" includes the subtler levels of our consciousness, they'd reject the statement.) Consider now the first part of what Epicurus says—that as long as we exist, death is not present. Tibetan Buddhists would also disagree with this. To see why, we need to go deeper into what "bardo" means.

"Bardo," as noted, means in-between state. So whenever we're in between two states, no matter what the scale, we're in a bardo state. These two states could be living and dying or being awake and being asleep, but they could also be the just-past moment of thought and the moment about to come. Thus "bardo" includes the gap between the cessation of one moment of thought and the arising of the next moment. To use a mathematical term, "bardo" is a "self-similar" phenomenon, one where the whole is similar to each part of itself, no matter what the scale. For example, each part of a coastline is a scaled-down version of itself, because zooming in on any part of the coastline will reveal a shape similar to the shape of the unmagnified region. Similarly, zooming in on any in-between state or bardo of the cycle of existence will reveal another in-between state or bardo with its own dissolution processes similar to the larger one.

What this self-similarity implies, however, is that, contrary to Epicurus, dying and death are always present and aren't restricted to what happens at the end of the bardo of this life. In other words, dissolution and cessation are always taking place from one moment to the next. In Dzogchen Ponlop's words: "From this point of view, death is taking place in every moment. Every moment ceases, and that is the death of that moment. Another moment arises, and that is the birth of that moment."[25] From the Buddhist perspective, Epicurus's statement misses this way in which death is always present.

The point here isn't that the death taking place in every moment is the same as the death happening at the end of this life. Rather, it's that a proper understanding of the existential meaning of death must include the dissolution of each moment of thought and the gap before the arising of the next one. We hardly ever notice this dissolution, just as we hardly ever notice the moment when we fall asleep. Yet, from the Tibetan Buddhist perspective, noticing the dissolution of each thought and the gap before the arising of the next one—whether in waking life, when we fall asleep, or when we die—gives us the opportunity to experience directly the dissolution that is always present together with luminous pure awareness.

DYING TO MEDITATE

On June 19, 2011, New Zealand Television (TVNZ) reported that a Tibetan Buddhist monk, who had been certified dead by a doctor on May 24, 2011, showed no signs of decay nearly three weeks later at his home where his body had remained in bed.[26] The monk, Jampa Thupten Tulku Rinpoche, was a renowned teacher and the spiritual head of the Dhargyey Buddhist Centre in Dunedin, New Zealand. According to his followers, Thupten Rinpoche had entered a special meditative state in which the practitioner continues to rest in the clear light of death—the intermediate state where the basic nature of the mind or the ground luminosity shines forth after the inner dissolution of conceptual consciousness or the cognitive mind. Only when his meditation was finished would he truly be dead and his body begin to decay. Thupten Rinpoche remained in this state for eighteen days and then was cremated in a ritual fire ceremony on June 13, 2011.

In Tibetan this type of death meditation practice is called *thukdam* (*thugs dam*). *Thukdam* is an honorific term meaning "to be engaged in meditation practice," but it's usually reserved specifically for the meditation practice of abiding in the clear light of pure awareness or the ground luminosity at death. According to the Tibetan Buddhist view, everyone experiences the clear light of death when the gross levels of consciousness dissolve back into the "very subtle consciousness" or ground luminosity, but how long this lasts varies.[27] For most people it's only a few seconds or minutes, but for advanced meditation practitioners, who have learned to control the movement of the subtle consciousness within their body, the experience of abiding in the clear light of death can last for several days or even weeks. As long as the experience of the clear light of death is sustained, the very subtle consciousness remains present in the body, but its energetic aspect withdraws and remains only as a warmth in the area of the heart. As Sogyal Rinpoche writes:

> A realized practitioner continues to abide by the recognition of the nature of mind at the moment of death, and awakens into the Ground Luminosity when it manifests. He or she may even remain in that state for a number of days. Some practitioners and masters die sitting upright in meditation posture, and others in the "posture of the sleeping lion." Besides their perfect poise, there will be other signs that show they are resting in the state of the Ground Luminosity: There is still a certain color and glow in their face, the nose does not sink inward, the skin remains soft and flexible, the body does not become stiff, the eyes are said to keep a soft and compassionate glow, and there is still a warmth at the heart. Great care is taken that the master's body is not touched, and silence is maintained until he or she has arisen from this state of meditation.[28]

Once the clear light experience is over, the very subtle consciousness leaves the body and decay begins. At that point, according to the Tibetan Buddhist tradition, death has occurred.

In the case of Thupten Rinpoche, New Zealand Television interviewed a forensic pathologist, Dr. John Rutherford, who expressed skepticism about the idea that a special death meditation practice was responsible for the lack of decay of Thupten Rinpoche's body. He noted that a dead body wouldn't be expected to show noticeable decay, even over several weeks, given certain conditions, including cool ambient temperature, low humidity, and the intestine being empty of organisms that would normally spread throughout the body and cause putrefaction. The relative lack of such organisms would result from fasting or a wasting disease. Thupten Rinpoche's death seems to fit these conditions: he died of cancer—a wasting disease—and his body was left in a relatively cool, dry room, and wasn't moved.

In recent years there have been reports of a number of cases in which physically dead Tibetan monks have been regarded as abiding in the *thukdam* state because their bodies showed no noticeable decay for periods of time ranging from several days to over two weeks after their breathing stopped and their heartbeat ceased.

Sogyal Rinpoche, in *The Tibetan Book of Living and Dying*, reports the case of the former Gyalwang Karmapa, the leader of one of the four main schools of Tibetan Buddhism, who died in a hospital in the United States in 1981, and whose body was observed by the nursing staff and the chief of surgery not to have decayed thirty-six hours after his death; the body also seemed to have remained warm around the heart area.[29]

Another case is Lama Putse, of Ka-Nying Shedrub Ling Monastery near Boudhanath on the outskirts of Kathmandu, Nepal. He died March 31, 1998, and is reported to have remained in *thukdam* for eleven days, until April 11, 1998. On April 8 and 9, senior physicians from the CIWEC clinic in Kathmandu examined him and noted the lack of decay.[30]

Chogye Trichen Rinpoche died January 22, 2007, in Kathmandu, and is reported to have remained sitting upright in *thukdam* for fifteen days, until February 6, 2007.[31]

Kyabje Tenge Rinpoche, of Benchen Monastery in Kathmandu, died March 30, 2012, and is reported to have remained in *thukdam* for three and a half days, until April 3, 2012. A picture of him sitting upright in the *thukdam* state was widely circulated on the Internet.[32]

Lobsang Nyima, of Drepung Loseling monastery in the south Indian town of Mundgod, died September 14, 2008, and is reported to have remained in *thukdam* for eighteen days, until October 1.[33] It was reported that there was no noticeable decay, odor, or slumping of the body during this period. This case is noteworthy for two reasons.

First, south India is comparatively warm and humid, so some of the conditions conducive to longer preservation of the body—cool ambient temperature and low humidity—may not have been present. The elevation in Mundgod is also considerably lower than the elevation in Dharamsala, Nepal, or Tibet, so high altitude, which is conducive to longer preservation of the body, also wasn't present.

The second reason is that physicians at the hospital in Belgaum, where Lobsang Nyima died, attempted to record his body temperature, EEG, and EKG (electrocardiogram) while he rested in the *thukdam* state. This research was conducted at the request of the Dalai Lama and was supervised by scientists from Richard Davidson's laboratory at the University of Wisconsin-Madison. The scientists had brought a thermal camera and a portable EEG/EKG/respiration monitor to India and had trained people there to use them. At the time of this writing, these data, as well as data from other Tibetan monks that Davidson's team in India has gathered on *thukdam*, haven't yet been published.

From a scientific perspective, the main question such cases raise is whether anything of measurable biological significance is taking place in association with what Tibetan Buddhists identify as *thukdam* and describe as abiding in the clear light of pure awareness at death. If bodies don't noticeably decay for up to three weeks, is this unusual or normal under the conditions of temperature, humidity, altitude, and so on?

After death, the human body decomposes through four main stages when it's left in the open air.[34] In the first stage, from days 1 to 6, cells begin to digest themselves (a process called autolysis) and soft tissue starts to decompose. *Rigor mortis* or stiffening of the muscles peaks in twenty-four hours and then subsides. The body starts to cool, and the skin separates from the muscle and becomes stretchy—a process known as skin slippage. In the second stage, from days 7 to 23, microbes from within the body and the outer environment feed on the corpse, producing numerous gases and causing the body to bloat. The third stage of active decay lasts from days 24 to 50. At this stage, putrefaction gives way to a chemical reaction called saponification that produces so-called corpse or grave wax, also known as adipocere. In the fourth stage, from days 51 to 64, the last traces of tissue decompose, leaving the skeleton.

The Tibetan Buddhist descriptions of *thukdam* certainly seem different from the time scale of this progression. Yet such time scales are approximate and depend on a host of factors, the most important being temperature, humidity, and exposure to oxygen. So we don't really know whether or to what extent the decomposition process in any given case of *thukdam* is different from what we should expect, given the conditions present at the time.

Another issue concerns the observed lack of *rigor mortis* in *thukdam*. At exactly what time after death are these observations being made? It would be unusual and unexpected if no *rigor mortis* were observed during the thirty-six hours immediately after death, but if the observations are being made at a later stage, then the lack of *rigor mortis* would be expected.

These issues came up at a small meeting I attended at the University of Wisconsin-Madison's Center for Investigating Healthy Minds. Richard Davidson, the director of the center, convened the meeting so that scientists could begin to think about the best ways to investigate *thukdam*. The meeting included experts on the molecular biology of cellular metabolism and suspended animation, the neuroscience of coma and brain death, and the forensic anthropology and pathology of human decomposition, as well as experts on Tibetan Buddhist meditation and traditional Tibetan medicine.

The two specialists in Western scientific methods of examining dead bodies, forensic pathologist Vincent Tranchida and forensic anthropologist Daniel Wescott, thought that the apparently slower decomposition phenomena of the four *thukdam* cases that Davidson's team had investigated so far were not necessarily unusual but within the range of what one might expect, given various conditions.

One intriguing possibility that emerged from this meeting came from exchanges between the experts on Tibetan meditation and scientist Mark Roth, who works on the molecular biology of suspended animation at the Fred Hutchinson Cancer Research Center in Seattle. The Tibetan meditation specialists described certain deep meditative states in which the body seems to enter a kind of stasis with no observable breathing. Roth wondered whether these practices are able to reduce oxygen demand and slow or even stop metabolic activity. He speculated that, given lifetime practice of such techniques, meditators who practice them at death might be able to alter the usual physiological course of dying and death by affecting the body's metabolic rate.

In the discussions that followed, it became clear that the best way to test this possibility would be to study practitioners while they're still alive to see whether they can suspend metabolic activity—whether, to use Roth's analogy, they can stop the furnace from burning while keeping the pilot light on—and then see whether the dying and decomposition process is different for them compared to nonpractitioners.

What about the presence or absence of consciousness during *thukdam*? From the Tibetan Buddhist perspective, pure awareness—the very subtle consciousness that constitutes the basic nature of the mind—remains present in the *thukdam* period after physical death and prior to decomposition. Yet finding that the biology of dying and decomposition can be altered as a result of certain meditation practices wouldn't be sufficient to confirm this belief. Although long-term practice of certain types of meditation might affect how the body decays, consciousness could cease to exist at death. More generally, effects resulting from the past presence of consciousness can still occur even when consciousness is gone.

What such a finding would do, however, is dramatically reinforce the need to revise our biomedical model of death in order to include how the mind affects the dying body. Clinicians and end-of-life caregivers already know that one's state of mind strongly influences the dying process. According to the Indian and Tibetan yogic traditions, great contemplatives demonstrate this truth in exemplary ways by being able to disengage from the sense of self or ego as they die.[35] Resting in a more spacious sense of awareness, they can watch the dissolution of their "I" or "Me" consciousness with equanimity. Modern biomedical science has lost sight of the value of such contemplative ways of approaching death. One potential benefit of the collaboration between scientists and Tibetan Buddhists in the examination of *thukdam* is that it may help science to recognize that a full understanding of the biology of death requires understanding how the mind meets death.

NEAR-DEATH EXPERIENCES

Near-death experiences (NDEs) are another case where science confronts the need to understand how the mind meets death. These experiences have been defined in general terms as "profound and sometimes life-changing experiences reported by people who have been physiologically close to death, as in cardiac arrest or other life-threatening conditions, or psychologically close to death, as in accidents or illnesses in which they feared they would die."[36] In addition to what these experiences might be able to tell us about what it's like to be near death, they provide an important case for investigating the relationship between consciousness and the body.

Physician Raymond Moody coined the term "near-death experience" in his 1975 book *Life After Life*.[37] He listed 15 common elements of these experiences, based on a sample of 150 reports. Several of the elements—out-of-body experience, seeing a dark tunnel, experiencing a review of one's life, and seeing a brilliant light—are now central to Western popular culture representations of near-death experiences. Yet no one single near-death experience in Moody's study included all the elements, and none was present in every reported near-death experience.

A few years after Moody's book, psychologist Kenneth Ring developed a scale for determining the "depth" of a near-death experience.[38] The scale is based on assigning different weights to ten features—the awareness of being dead, positive emotions, out-of-body experience, moving through a tunnel, communication with light, observation of colors, observation of a celestial landscape, meeting with deceased persons, experiencing a life review, and the presence of a border or a "point of no return." The report of a given experience is scored according to the presence or absence of each of these weighted features, with the possible score ranging between 0 and 29. A score between 0 and 6 is thought to be too low to merit the label "near-death experience," whereas a score between 7 and 9 indicates a moderately deep near-death experience, and a score between 10 and 29 indicates a deep or very deep near-death experience. Ring also formulated a sequence of five stages that he believed marked what he called the "core near-death experience." These stages are 1) peace and well-being; 2) out-of-body experience; 3) entering a tunnel-like region of darkness; 4) seeing a brilliant light; and 5) passing through the light into another realm.

Finally, psychiatrist Bruce Greyson developed a new scale that includes sixteen items grouped into four components—1) cognitive features (time distortion, thought acceleration, life review, revelation); 2) affective features (peace, joy, cosmic unity, encounter with light); 3) paranormal features (vivid sensory events, apparent extrasensory perception, precognitive visions, out-of-body experiences), and 4) transcendental features (sense of an "otherworldly" environment, sense of a mystical entity, sense of deceased/religious spirits, sense of border/"point of no return").[39] Greyson's "Near-Death Experience Scale" has a maximum score of 32, where a score of 7 or higher indicates a near-death experience.

In 2003 I received an e-mail from Dr. Pim van Lommel, a Dutch cardiologist and prominent near-death experience researcher. He had read my article on neurophenomenology written with neuroscientist Antoine Lutz,[40] and wanted to call our attention to near-death experiences. Many cardiac arrest patients, he told us, could report memories of subjective experiences that had occurred during the period of their clinical death when their blood flow and breathing had stopped. According to the current neuroscience view of consciousness, however, such experiences shouldn't be possible. This information intrigued me, as I hadn't yet examined the near-death experience literature. Van Lommel and his colleagues had recently published a major study of near-death experiences in an important medical journal, *The Lancet*.[41] He attached this article to his e-mail. On first reading, I was excited to think that perhaps these experiences could indeed show that aspects of consciousness transcend the brain. So I dived into a close examination of the near-death experience literature.

Cardiac arrest is a medical emergency where the heart fails to contract effectively and the normal circulation of the blood abruptly stops. Immediately following cardiac arrest, blood pressure drops sharply, resulting in either no blood flow or severely reduced blood flow to the brain. The EEG slows and progresses to a flatline where no electrical brain waves are recorded. As near-death experience researchers Sam Parnia and Peter Fenwick point out, cardiac arrest is the final step in dying, regardless of the cause, so it's the closest physiological model we have of the dying process.[42] For these reasons, the experiences people have during cardiac arrest may offer clues to the experiences that accompany dying, no matter what the cause. Although the majority of cardiac arrest survivors recall nothing from the event, current prospective studies of cardiac arrest patients indicate that approximately 10 percent report having had a near-death experience.[43]

Van Lommel and other prominent near-death experience researchers claim that near-death experiences pose a major challenge to the modern neuroscience view that consciousness is contingent upon the brain.[44] They believe that neuroscience can't account for the occurrence of these experiences during cardiac arrest for two principal reasons. First, patients recall vivid and apparently lucid experiences despite having suffered severe impairment to their cerebral functioning. Second, patients sometimes recall an out-of-body experience in which they appear to have accurately perceived their physical surroundings from an elevated vantage point in the room where efforts are under way to resuscitate them. As Bruce Greyson, Janice Miner Holden, and Pim van Lommel state in a recent exchange: "Current neurophysiological models of NDEs fail to explain lucid experiences that occur during cardiac arrest, when conscious experience should be fragmentary or absent. This problem is exacerbated in resuscitated patients who report perceiving events they should not have been able to perceive, yet are later confirmed."[45]

These statements raise four issues that we need to go through in some detail.

1. The timing of near-death experiences.

At exactly what time do near-death experiences occur? Do they occur during a period of flatline EEG, or immediately before this period, as patients are entering the flatline state, or immediately after, as they recover from this state?

The short answer to these questions is that we still don't know the exact time at which near-death experiences occur in cardiac arrest.

What we know about near-death experiences comes from memory reports given well after the experience is supposed to have occurred. These reports tell us the time when the experience seemed to happen according to the patient, specifically in relation to the patient's memory of losing and regaining consciousness. In other words, the reports give us information about the patient's subjective sense of time, but they don't establish the objective time at which the experience took place.

In addition, there are no documented cases that clearly establish that the near-death experience happened at the precise time that the EEG was flat. Other possibilities are that the experience occurs just before the flatline state or just after this state, when patients are recovering, or that some elements of the experience happen before and others after the flatline state, but memory consolidates them into one remembered episode.

Near-death experience researchers have tried to rule out these possibilities by arguing that the patient becomes unconscious too quickly for the near-death experience to occur before the period of flatline EEG, and that after this period, as the patient is regaining consciousness, the patient's thinking is too confused to support the lucid consciousness that's reported as belonging to the near-death experience.[46]

Excluding these possibilities in this way, however, won't work, for two reasons.[47] First, we don't know how much time it takes to have a near-death experience. Here we need to keep in mind the important difference between time as subjectively experienced and time as objectively measured. For example, you can fall asleep for twenty minutes during an afternoon nap and have a dream in which it seems that hours go by. Or you can sit in meditation for what seems like only a few minutes and be surprised to hear the bell indicating that the forty-minute period has ended. Consider also the life-review element of the near-death experience. Although the person experiences a review of his or her entire life, the review seems to last only a short time. In these and other ways, our sense of the passage of time is very different from time as measured by a clock. Hence it's possible that the near-death experience could occur in the few remaining seconds before the loss of consciousness, even though it seems from within to last much longer.

Second, the attribution of confusion to patients as they regain consciousness is made by an outside observer, but the patients themselves may not feel subjectively confused. One can subjectively feel great clarity even though from the outside one's mental capacities seem diminished. This disparity is known to happen in hypoxia or oxygen deprivation, which occurs in cardiac arrest. So appealing to the subjective sense of clarity, especially as represented in a memory report, isn't enough to rule out the possibility that the near-death experience occurs as the patient emerges from the flatline state.

2. The state of the brain during a near-death experience.

We have virtually no information about what specific brain states are associated with near-death experiences.[48] Cardiac arrest is a medical emergency, so it's not surprising that we have no human EEG data or neuroimaging data about brain activity in the critical period of time when near-death experiences are thought to occur.

In particular, no EEG or neuroimaging studies have been done to compare cardiac arrest patients who report near-death experiences to those who don't report them.

Since no EEG measures are made in most cases of cardiac arrest, the assertion that the EEG is flat is based on inference. Yet this is problematic. The average time for the EEG to start to show changes, such as slowing and attenuation, is over ten seconds after the last heartbeat.[49] How long it takes for the EEG to become flat depends on various factors; in cases where no blood is being delivered to the tissue (zero perfusion), the EEG flatlines after about twenty seconds.[50] Yet efficient external cardiac massage can sometimes restore the EEG after ten to twenty seconds.[51] Defibrillation (electrical stimulation of the heart) can also restore the EEG. Hence, in general in cases of cardiac arrest with successful resuscitation, and in particular in such cases where the patient reports a near-death experience, we can't conclude with certainty that the EEG was flat or determine how long it was flat, unless we have EEG recordings.

Although it has been assumed that brain activity, especially the kind of brain activity associated with consciousness, drops sharply during cardiac arrest, a recent experiment with rats showed that this is far from being the case.[52] During cardiac arrest the rat brain showed a huge surge in the kind of high-frequency and widely integrated activity associated with consciousness in humans. Within the first thirty seconds after cardiac arrest and before the EEG went flat, there was an enormous increase in synchronized gamma frequency oscillations, including a strong increase in the influence of frontal-region oscillations on posterior-region oscillations, as well as an increase in the coupling between the amplitude of the fast gamma oscillations and the phase of slower theta and alpha waves. As we've seen in earlier chapters, this kind of gamma-synchrony brain-wave pattern in humans occurs in moments of conscious perceptual recognition, meditative states of open awareness, and lucid dreaming. As the authors of the study state, "The neural correlates of conscious brain activity identified in this investigation strongly parallel characteristics of human conscious information processing. Predictably, these correlates decreased during general anesthesia.

The return of these neural correlates of conscious brain activity after cardiac arrest at levels exceeding the waking state provides strong evidence for the potential of heightened cognitive processing in the near-death state."[53]

It's also crucial to note that even a flatline EEG doesn't necessarily mean a total loss of brain activity.[54] EEG recordings at the scalp mainly register activity from the cortex. The EEG can fail to pick up activity from deeper subcortical structures, such as the hippocampus and amygdala. Neural discharges in these structures without the involvement of the cortex have been shown to be related to the occurrence of meaningful hallucinations in epileptic patients.[55] Moreover, even at the level of the cortex, the EEG can fail to pick up seizure activity that's visible to fMRI (functional magnetic resonance imaging, a neuroimaging method that can localize activity inside the brain).[56] In addition, a recent study showed that after the EEG went flat as a result of coma, quasi-rhythmic activity in the hippocampus occurred and was transmitted to the cortex.[57] Hence, even if the EEG is flat, we can't conclude that there's no brain activity, or no activity sufficient to support some kind of consciousness, including the near-death experience.

Near-death experience researchers have tried to show that the near-death experience can occur in the absence of brain activity by citing a famous case in the literature, the case of Pam Reynolds, which I mentioned in chapter 7 in my discussion of out-of-body experiences.[58] In 1991, Reynolds, an American singer-songwriter, was diagnosed with a giant aneurysm of the basilar artery (one of the arteries that supplies blood to the brain). In order to have the aneurysm removed, she had to undergo a surgical procedure called "hypothermic cardiac arrest," in which the body temperature is lowered to 15.5 degrees Celsius (60 degrees Fahrenheit) and the heart is stopped. The artificial respirator is turned off, so breathing ceases, and the EEG flatlines, with the brainstem showing no response to auditory stimulation. While apparently unconscious under general anesthesia, Reynolds had a remarkable out-of-body experience and near-death experience. Upon awakening from the operation, she was able to report accurately a number of events that occurred during the surgery. Three years later she came into contact with Michael Sabom, a cardiologist and near-death experience researcher, who published her story in his 1998 book, *Light and Death.*

Upon careful examination, however, the Pam Reynolds case doesn't provide convincing evidence for the presence of consciousness during the complete cessation of brain activity. The crucial issue is whether any part of Pam Reynolds's near-death experience occurred in the approximately 35- to 40-minute period of time when no blood flowed through her brain, her EEG was flat, and her brainstem response was absent.

We know that the out-of-body experience component of her near-death experience probably happened in the period of time before the body-cooling procedure started—that is, when her body temperature and heartbeat were still normal—because the events she reported seeing and hearing took place at this time. She reported seeing the bone saw, hearing a high-pitched sound that she took to be coming from the saw, hearing the high-pitched sound change to a *"Brrrrrrrrrr!"* sound, and hearing someone, whose voice sounded female and whom she believed to be the cardiologist, say that her veins and arteries were very small. These reported observations corresponded to events that happened during the first two hours of the surgery before the body cooling procedure was begun: the neurosurgeon used a pneumatically powered bone saw to remove a large section of the skull, while the female cardiac surgeon tried to introduce the tubing for the cardiac bypass machine into the blood vessels in Pam's right groin; the vessels were too small, so the surgeon switched to the blood vessels in the left groin.

A plausible explanation for Pam Reynolds's ability to perceive these events, as anesthesiologist Gerald Woerlee has argued in his analysis of this case, is that, while Reynolds was sedated, paralyzed, and pain-free, she regained consciousness under anesthesia.[59] This phenomenon is called "anesthesia awareness." It's known to occur in a minority of operations, though anesthesiologists do their best to prevent it from happening.[60] Many cases are horrific—the patient is paralyzed but fully conscious and able to feel pain—but in some cases the patient doesn't feel pain or distress. Out-of-body experiences are also known to occur during anesthesia awareness.[61]

Reynolds's eyes were taped shut, so she wouldn't have been able to see what was going on around her. Although she was wearing fitted earplugs that delivered 40-decibel white noise to one ear and 95-decibel clicks every eleventh of a second to her other ear (in order to monitor her auditory brainstem response), she probably would have been able to hear the sound of the saw through bone conduction (as when you hear inside your head the sound of the dentist's drill). On the basis of hearing the sound, she may have generated a visual image of the saw, which she described as looking like an electric toothbrush. She would have been familiar with the surgical procedure from the surgeon's description and from having read and signed the informed consent form, and she would have seen the layout of the operating room because she was awake when she was wheeled in. So she probably had enough knowledge to create an accurate visual and cognitive map of her surroundings during her out-of-body experience. Reynolds's ability to hear what the cardiac surgeon said may seem less likely, but to my knowledge no one has tried to replicate the auditory stimulus conditions to determine whether speech is comprehensible through those sound levels or during the pauses between the clicks.

Whether Pam Reynolds's out-of-body experience of the operating room provides evidence of veridical out-of-body perception is an issue I will take up shortly. The point I want to emphasize now is that this part of her near-death experience clearly didn't occur at a time when her EEG was flat and her brain was inactive.

Reynolds's near-death experience continued with her feeling herself being pulled into a vortex. She had a sensation of going up very fast, as if in an elevator. She heard her deceased grandmother calling her as she traveled down a dark shaft toward a bright light that got bigger and bigger. She entered the light and met many deceased relatives who were covered with light or had the form of light. It was communicated to her that she couldn't go all the way into the light, because then she'd be unable to return to her body. She felt her relatives were nourishing her and strengthening her. Then her uncle took her back through the tunnel, and when she reached the end she could see her body on the operating table. She felt her body pulling her back and the tunnel pushing her out at the same time. The return to her body hurt, like jumping into a pool of ice water. When she came back, she could hear the Eagles song, "Hotel California," playing in the operating room; the line it had reached was, "You can check out anytime you like, but you can never leave." When she woke up from the operation, she was still on the respirator.

In Sabom's original presentation of this narrative, he weaves together the subjective timeline of Reynolds's experience and the objective timeline of the medical operation, making it seem as if the experience of being pulled into the vortex occurred during the period of hypothermic cardiac arrest when no heartbeat, respiration, or brain activity was present. This representation of Reynolds' near-death experience in relation to the objective timeline of her operation has been widely and uncritically repeated in the near-death experience literature as well as the popular media, notably in a 2002 BBC documentary called "The Day I Died." The supposed correspondence can be found in the form of a chart in the *Wikipedia* article on Pam Reynolds.[62] Yet there's no direct evidence whatsoever for it. Sabom apparently inferred the correspondence by trying to connect Reynolds's story of her near-death experience to the operative report of the surgeons and the neuroanesthesiologist (and it was already three years after the fact when he first examined both accounts). But the inference is completely unwarranted. Nothing in Reynolds's account of her journey through the tunnel to the light and her return to her body can establish that this experience happened during the period of hypothermic cardiac arrest. All we can reasonably infer, based on the events that she was able accurately to report, is that her out-of-body experience of the operating room happened before the cooling of her body began, and that her experience of hearing the Eagles song (which was really playing) happened after her body had been warmed, her heart had been restarted, the artificial respiration had been turned back on, and her brain was active again (it's also possible that at this point, the white noise and clicking sounds presented to her ears had been terminated because her brainstem response no longer needed to be monitored). Although the other elements in her near-death experience narrative occurred between the out-of-body experience of the operating room and her hearing the Eagles song, we have no grounds for supposing that they occurred while her brain was inactive. On the contrary, it's much more reasonable to suppose, based on our medical knowledge of anesthesia awareness, that her experience of the vortex or tunnel, journeying to the light, meeting deceased relatives, and so on occurred either before the cooling of her body or after her body temperature had been restored. It's also more reasonable to suppose that her experience of returning to her body and hearing the Eagles song was a second occurrence of regaining consciousness under anesthesia, after which she lost consciousness again until she awoke in the intensive care unit.[63]

In summary, we still don't have any direct information about what is happening in the human brain during a near-death experience. We also have no compelling evidence for thinking that the brain is inactive or shut down when these experiences occur. The case most widely cited as providing such evidence—the Pam Reynolds case—in fact provides no such evidence. On the contrary, upon careful examination this case actually supports the claim that near-death experiences are contingent on the brain.

3. The quality of the evidence for veridical out-of-body perception.

The Pam Reynolds case is often cited as a case of veridical out-of-body perception. As we've seen, however, her ability to report accurately some of the events in the operating room seems explainable on the basis of normal sensory perception. Moreover, her report also contained inaccuracies and left things out in ways we wouldn't expect if she really could see herself. First, her description of the bone saw didn't fit the model used in the operation. Second, she didn't report seeing that her head had been turned to one side and was being held by a mechanical head-holder, despite her out-of-body visual perspective being just above the neurosurgeon's shoulder. ("I was metaphorically sitting on Dr. Spitzer's shoulder," she says in her report.) Third, although she heard the saw "crank up," she didn't see the surgeon use it on her head. All these details would have been in plain sight had she really been able to see herself.[64] Pam Reynolds's out-of-body experience therefore seems better explained as a case of mental simulation based on ordinary sense perception and memory than as a case of veridical out-of-body perception.

Of course, Reynolds is just one particular case, and what we want to know is whether there's any evidence for veridical out-of-body perception in general.[65] All the evidence that near-death experience researchers cite for veridical or verifiable out-of-body perception in near-death experience is anecdotal.[66] Many of these reports are based on interviews conducted well after the near-death experience occurred. Such reports are subject to memory errors and post-hoc reinterpretations of what happened. Five published studies so far have tried but failed to find evidence of veridical out-of-body perception in rigorously controlled conditions.[67]

In short, at present, there are no documented cases of veridical out-of-body perception in near-death experiences in cardiac arrest patients in rigorously controlled clinical conditions.

4. Are there known neurophysiological processes that plausibly can be said to play a role in giving rise to the constellation of elements that make up near-death experiences in cardiac arrest?

Two general neurophysiological models have been given of the near-death experience—the dying brain model and the reviving brain model.[68] The two models are complementary, not mutually exclusive. Whereas the dying brain model focuses on how brain activity becomes disordered as the brain enters into situations that can lead to death, the reviving brain model focuses on how brain activity becomes disordered as the brain starts to regain its normal functioning after it has been severely compromised. These two models are really two sides of the same coin, for they both relate near-death experiences to what happens to the brain when its activity becomes disordered and unstable in situations such as cardiac arrest.

The central idea of both models is that near-death experiences in life-threatening situations are a consequence of "neural disinhibition." Neurons communicate by exciting or inhibiting each other through electrochemical signals. Neural inhibition is essential for preventing situations in which neurons in a given area excessively excite each other all at the same time—which leads to seizures. Disinhibition happens when the inhibitory signaling between neurons becomes less effective and as a result, the cells start to fire faster and synchronize. Widespread disinhibition leads to the uncoordinated excitation of whole brain areas. Neural disinhibition can be triggered by many psychological and neural factors, including sensory deprivation, epilepsy, migraine, drug use, brain stimulation, and anoxia or the absence of oxygen, which happens in cardiac arrest.[69]

It's well known that reduced oxygen levels (hypoxia) can lead to experiences with many of the elements of near-death experiences. For example, fighter pilots who accelerate too rapidly can lose blood flow to the brain and experience so-called "G-force induced loss of consciousness" or G-LOC. Pilots who undergo G-LOC in both real-world and laboratory conditions report having experiences with many of the elements of near-death experiences, including "tunnel vision and bright lights, floating sensations, automatic movement, autoscopy, out-of-body experiences, not wanting to be disturbed, paralysis, vivid dreamlets of beautiful places, pleasurable sensations, psychological alterations of euphoria and dissociation, inclusion of friends and family, inclusion of prior memories and thoughts, the experience being very memorable (when it can be remembered), confabulation, and a strong urge to understand the experience."[70] Healthy subjects report the same kinds of experiences in medical experiments that use hyperventilation and Valsalva maneuvers (attempting to exhale forcefully with one's mouth closed and nose pinched) to induce transient cerebral hypoxia and fainting.[71]

Pim van Lommel and his colleagues have tried to exclude cerebral anoxia as accounting for near-death experiences in cardiac arrest with the following argument.[72] They reason that if cerebral anoxia were the cause of the experience, then most cardiac arrest survivors should report a near-death experience, since all cardiac arrest patients suffer from cerebral anoxia. Yet in their study, which followed cardiac arrest patients from successful resuscitation to recovery, only 62 of 344 patients (18 percent) reported a near-death experience (and only 41 patients or 12 percent reported a "core" experience).

As other scientists have pointed out, however, this reasoning is faulty.[73] First, van Lommel and his colleagues didn't provide any direct measures of anoxia in their study, so there's no way to know whether the patients had comparable levels of anoxia. Second, the relevant consideration isn't the overall level of anoxia, but rather the rate of change in anoxia or the rate of anoxia onset. If anoxia occurs too fast, then the patient loses consciousness and blacks out. With more prolonged rates of onset, the patient seems dazed. At intermediate levels, intense altered states with near-death experience elements occur. Third, anoxia affects different people in different ways, since there are many structural and functional differences between their individual brains. Finally, individuals differ in their ability to recall events—for example, in their ability to recall their dreams—so individual differences in memory might also account for why some cardiac arrest patients report near-death experiences and other patients don't.

Neural disinhibition resulting from anoxia is only one likely physiological contributor to near-death experiences in cardiac arrest. Another likely contributor is the release of neurotransmitters (chemicals produced within the brain that transmit signals between neurons and other brain cells). For example, endorphins, whose secretion decreases feelings of pain and can lead to feelings of euphoria, may play a role in the generally positive emotional tone of many near-death experiences.

In addition, altered functioning in the brain's temporal lobe has been directly linked to patients who report having had a near-death experience. Psychologists Willoughby Britton and Richard Bootzin found that, compared to a control group, individuals who reported having a near-death experience had more epileptic-like EEG brain waves over the left temporal lobe.[74] These individuals also reported more temporal lobe epileptic symptoms, such as sleepwalking, hypersensitivity to smells, hypergraphia (the overwhelming urge to write), feelings of intense personal significance, and unusual auditory or visual perceptions. Finally, the same individuals had altered sleep patterns; they slept about an hour less than the control group, and they took longer to enter REM sleep and had fewer REM sleep periods than the control group. These findings suggest that altered temporal lobe functioning may contribute to the occurrence of near-death experiences, and that individuals who have had such experiences are physiologically distinct from the general population.

Finally, cognitive neuroscientists Olaf Blanke (whose research on out-of-body experiences we examined in chapter 7) and Sebastian Dieguez have put forward a model of how the distinct brain areas known to be frequently damaged in cardiac arrest patients might contribute to the various elements that make up near-death experiences.[75] Blanke and Dieguez suggest that there may be two types of near-death experience. Both are due to altered functioning or disinhibition in frontal and occipital brain regions, but the first type is due especially to altered functioning in the area of the right hemisphere where the temporal and parietal lobes meet (the right temporoparietal junction, which plays a crucial role in out-of-body experiences), whereas the second type is due especially to altered functioning in the same area of the left hemisphere (the left temporoparietal junction). An out-of-body experience and an altered sense of time, as well as sensations of flying, lightness, and self-motion induced by moving visual appearances, characterize near-death experiences of the first type. The feeling of a presence, meeting and communicating with spirits, seeing glowing bodies, and experiencing voices, sounds, and music, but with no sensation of self-motion, characterize near-death experiences of the second type. Both types include positive emotions and the life review, which Blanke and Dieguez propose are due to altered functioning in the hippocampus and amygdala. And both types include the experience of light and tunnel vision, which they propose results from altered functioning in the occipital cortex.

It also seems possible that a patient could have both types of near-death experience and later link them together into one remembered and reported episode. Pam Reynolds's near-death experience, for example, might have been of this kind.

Although Blanke and Dieguez's model is speculative, as they admit, it serves to illustrate how we can begin to approach near-death experiences from a cognitive neuroscience perspective, instead of supposing, as many near-death experience researchers do, that these experiences pose an insurmountable challenge to neuroscience.[76]

THE UNGRASPABILITY OF DEATH

The true challenge that near-death experiences pose isn't that they contradict contemporary neuroscience. Rather, it's that they call to be investigated and understood in a way that doesn't lose touch with their singular existential meaning and with the ultimate ungraspability of death.[77]

One way to lose touch with the existential meaning of near-death experiences is to argue, on the basis of the kind of cognitive neuroscience perspective just sketched, that these experiences are nothing other than false hallucinations created by a disordered brain. Another way is to argue that these experiences are true presentations of a real, transcendent, spiritual realm to which one's disembodied consciousness will journey after death.

Both these viewpoints fall into the trap of thinking that near-death experiences must be either literally true or literally false. This attitude remains caught in the grip of a purely third-person view of death. In the one case, the experience of drawing near to death is projected onto the plane of a third-person representation of the disordered brain; in the other case, the experience is projected onto the plane of a third-person representation of a transcendent spiritual realm. Both viewpoints turn away from the experience itself and try to translate it into something else or evaluate it according to some outside standard of objective reality.

Ask yourself this question: If you were having a near-death experience, which would matter more—the truth or falsity of the experience according to some outside scientific or religious standard, or your ability to be calm, peaceful, and mindful in the face of what is happening?

Here's another question: As you're dying and at the moment of your death, won't it be your experience that has primacy and not any third-person viewpoint and its criterion of what is real?

From an existential standpoint, death has to be understood in the first person. In Martin Heidegger's famous words: "Insofar as it 'is,' death is always essentially my own."[78] We grieve the death of a loved one; we go to funerals and wakes; and those who devote their lives to caring for the dying witness death every day. Yet none of these ways of encountering death enables us to comprehend the singularity of our own death. In philosopher Todd May's words: "It is not that one's own death matters more in the grand scheme of things than the death of someone else. It is that one's own death cannot be understood by coming to terms with someone else's death. The silencing of one's experience, including the experience of the silencing of another's experience, remains intimately one's own in a way that cannot be understood by analogy with anyone or anything else."[79]

At the same time, my own death is ungraspable. From the relevant standpoint—the first-person standpoint—there is no object "my own death" that one can mentally grasp, precisely because there's no subject for whom this can be an object.

Leo Tolstoy, in *The Death of Ivan Ilyich*, depicts this breakdown of subject-object thinking when he describes how Ivan Ilyich, a lawyer in Czarist Russia, finally comes to the realization that the pain in his gut will not go away and that he is incurable and going to die:

Absorption; the blind gut was curing itself. Then suddenly he could feel the same old dull gnawing pain, quiet, serious, unrelenting. The same nasty taste in his mouth. His heart sank and his head swam. "O God! O God!" he muttered. "It's here again, and it's not going away." And suddenly he saw things from a completely different angle. "The blind gut! The kidney!" he said to himself. "It's got nothing to do with the blind gut or the kidney. It's a matter of living or . . . dying. Yes, I have been alive, and now my life is steadily going away, and I can't stop it. No. There's no point in fooling myself. Can't they all see—everybody but me—that I'm dying? It's only a matter of weeks, or days—maybe any minute now. There has been daylight; now there is darkness. I have been *here*; now I'm going *there*. Where?" A cold shiver ran over him; he stopped breathing. He could hear nothing but the beating of his heart.

"When I'm dead, what happens then? Nothing happens. So where shall I be when I'm no longer here? Is this really death? No, I won't have it!" He jumped up, tried to light the candle, fumbling with trembling hands, dropped the candle and the stick on the floor and flopped back down on to his pillow. "Why bother? It doesn't make any difference," he said to himself, staring into the darkness with his eyes wide open. "Death. Yes, it's death."[80]

Ivan Ilyich, anguished and dying, suddenly realizes that what's happening to him isn't a matter of internal organs conceived in third-person anatomical terms—the blind gut and the kidney. It's a matter of living and dying, of life and death. The blind gut and the kidney do make a difference, but only as contributors to what really matters—the life and imminent death that are Ivan Ilyich's own.[81]

We should say the same thing about the temporal lobe, the temporoparietal junction, or the disordered brain in general in any near-death experience. They make a difference, but only as contributors to what really matters—the living and dying of the person as she experiences them.

Days go by and Ivan Ilyich descends into misery:

Ivan Ilyich could see that he was dying, and he was in constant despair.

In the depths of his soul Ivan Ilyich knew he was dying but, not only could he not get used to the idea, he didn't understand it, couldn't understand it at all.

All his life the syllogism he had learned from Kiesewetter's logic— Julius Caesar is a man, men are mortal, therefore Caesar is mortal— had always seemed to him to be true only when it applied to Caesar, certainly not to him. There was Caesar the man, and man in general, and it was fair enough for them, but he wasn't Caesar the man and he wasn't man in general, he had always been a special being, totally different from all others. . . .

"I've always thought—and all my friends have too—that we're not the same as Caesar. And now look what's happened!" he said to himself. "It can't be, but it is. How can it be? What's it all about?"[82]

Ivan Ilyich cannot grasp the death that is going to be his own. He recognizes the truth of the syllogism stated in the third person but can't comprehend its truth in the first person. We are no different. None of us is Caesar or humanity in general, but each of us is Ivan Ilyich.

The ungraspability of death calls for a radical change of attitude. To borrow Buddhist scholar Robert Thurman's words, "the grasping mind cannot grasp its ultimate inability to grasp; it can only cultivate its tolerance of that inability."[83] Faced with the ungraspability of death, we need to set aside certainties and convictions—whether scientific or religious—and cultivate the qualities that Joan Halifax has put at the heart of the Being with Dying training—"not-knowing," or the tolerance of uncertainty, and "bearing witness" to experience, especially the experience of suffering.[84] This attitude is what Stephen Batchelor called for in the passage quoted earlier, and it's what Francisco Varela had in mind when he talked about "staying with the open question" of what happens to consciousness at death (see the prologue).

We need to take this attitude—the tolerance of uncertainty and bearing witness—into the study of near-death experiences. What this means in pragmatic terms is to stop using accounts of these experiences to justify either neuroreductionist or spiritualist agendas and instead take them seriously for what they truly are—narratives of first-person experience arising from circumstances that we will all in some way face.

This approach calls for a more detailed phenomenological investigation of near-death experiences than has been done so far. Instead of just interviewing patients with questionnaires designed for scoring their experiences according to the Near-Death Experience Scale, we need to use interviewing methods that help individuals to recall their experiences in ways that minimize after-the-fact interpretation and memory reconstruction. One technique—the "explicitation interview"—uses open and undirected questions in order to help individuals recall implicit aspects of their experience to which they may not have immediate cognitive access.[85] This kind of "second-person" method of investigating experience can help to distill what the person directly experienced from how the person may subsequently interpret or evaluate what she experienced. The explicitation interview has been used successfully with epileptic patients to help them gain access to subtle changes in their experience prior to their having a seizure, so that they can detect early warning signs and gain mental control over the seizure's onset and process.[86] In the case of patients who report near-death experiences, the explicitation interview can help to uncover the particularities of each person's experience as well as more subtle common factors across individual experiences that may be missed by the Near-Death Experience Scale. With this richer qualitative information, we can relate each near-death experience more precisely to the person's individual brain and body, as well as culture and life circumstances.

The advantage of this neurophenomenological approach is that it can help us to see near-death experiences from multiple physiological, psychological, cultural, spiritual, and phenomenological perspectives without reducing the experiences to any of these perspectives. It can also help us to remember that only the dying can teach us something about death, and what we're called upon to do is to bear witness to their experience.

CODA

Matsuo Bashō (1644–1694), the great poet and master of the haiku, having fallen seriously ill, was asked by his friends for a death poem. He refused, saying that any of his poems could be his death poem. Yet the next morning he called his friends to his bedside and told them that during the night he had dreamed, and that on waking a poem had come to him. He recited this poem, and died four days later.[87]

On a journey, ill:	*Tabi ni yande*
my dream goes wandering	*yume wa kareno o*
over withered fields.	*kakemeguru*

10

KNOWING

Is the Self an Illusion?

\\ f the self were the same as the conditions on which it depends, it would come to be and pass away as they do; but if the self were different from the conditions on which it depends, it could not have any of their characteristics." These words paraphrase the thoughts expressed in the first and tenth verses of the chapter on the self in the *Fundamental Stanzas on the Middle Way,* one of the most important works of Indian Buddhist philosophy, written by Nāgārjuna (second century C.E.).[1] In this chapter, Nāgārjuna criticizes two extreme views—either the self is a real, independent thing or there is no self at all—and points beyond them to a middle way—the self is "dependently arisen."

According to the first extreme view, the self is a thing or entity with its own inherent being. It's the essence of the person, the one thing whose continued existence is required for the person to continue to exist. Given this view, as Nāgārjuna points out, there are basically only two possibilities: either the self is the same as the body and the mental states of feeling, perception, conscious awareness, and so on, or the self is different from the body and such mental states. But neither possibility will work.

On the one hand, if the self were identical to the psycho-physical states that make up a person, then the self would

be constantly changing, because these states are constantly changing. Mental states especially come and go, arising and ceasing, so
if the self were identical to either the collection of mental states or
some specific mental state, it too would come and go, arising and
ceasing. In other words, the self would not be one real thing that
stays the same from moment to moment and that can be referred to
as "I" or "you."

On the other hand, if the self were different from the body and
mental states, a thing or entity with its own separate and independent being, then the self couldn't have any of their characteristics.
But how could something that didn't have any of your characteristics
be you? How could whatever happens to your mind or your body be
independent of what happens to you? Furthermore, how could you
ever know anything about this kind of self, and why would you care
about it?[2]

Indian Buddhism uses a concept called the "five aggregates" to
describe the bodily and mental states that make up a person. The five
aggregates are the basic components into which the individual person divides upon analysis. They are traditionally listed as material
form, feeling, perception (or cognition), inclination (or volition), and
consciousness. Buddhist scholar Anālayo gives the following example to illustrate the five aggregates:

> During the present act of reading . . . consciousness is aware of each
> word through the physical sense door of the eye. Cognition [per-
> ception] understands the meaning of each word, while feelings are
> responsible for the affective mood: whether one feels positive, nega-
> tive, or neutral about this particular piece of information. Because of
> volition [inclination] one either reads on, or stops to consider a pas-
> sage in more depth, or even refers to a footnote.[3]

The five aggregates make up one of the many early Buddhist taxonomies for describing and classifying experience. Form includes the
physical matter of the body and its sense organs, as well as the various kinds of external material forms to which those organs are sensitive. Feeling is the affective aspect of experience, whereas perception

or cognition is the cognitive aspect. Whenever one of the six senses (the five bodily senses or the sixth mental sense) is stimulated, an associated feeling occurs that is either pleasant, unpleasant, or neutral. In addition, an identification happens, so that the raw sensory or mental data are not only felt but also categorized into something recognizable. This identification both conditions and is conditioned by an inclination, intention, or volitional tendency to act or react mentally and physically in some way. This inclination can be either wholesome, unwholesome, or neutral. Finally, consciousness is aware of the presence of the object, and, as a result of always being conditioned by attention (which, in this classification, belongs to the fourth aggregate), is turned or oriented toward the object.

In his chapter on the self, Nāgārjuna refers to the five aggregates and poses a dilemma for the view that there is an inherently existent self. Here is Jay Garfield's translation of the opening stanza:

> If the self were the aggregates,
> It would have arising and ceasing (as properties).
> If it were different from the aggregates,
> It would not have the characteristics of the aggregates.[4]

Again, if there is an inherently existent self, it must either be identical to the aggregates or be something different from them, but neither alternative will work.

On the one hand, if the self is either one and the same thing as all or some one of the aggregates, then the self will be constantly changing, arising and ceasing, exactly as forms, feelings, cognitions, inclinations, and sensory and mental awarenesses are constantly arising and ceasing. Yet precisely because the aggregates are constantly changing, they cannot be or constitute a self that has its own independent being and remains wholly present from moment to moment.

On the other hand, if the self is something different from the aggregates, then it cannot have any of their characteristics — experiential form, feeling, cognition, inclination, or consciousness. Such a self would be removed from all experience and be thoroughly unknowable.

Faced with the difficulties of identifying an independently real self with the body and mind or with something else entirely, the temptation is to deny the existence of the self altogether. The Buddhist term for this kind of view is "annihilationism" or "nihilism," whereas the view that the self is an independently real entity is called "eternalism" or "reificationism."

Nāgārjuna is not an annihilationist or nihilist about the self. On the contrary, as Garfield explains, "what Nāgārjuna is emphatically *not* doing is arguing that there are no aggregates in any sense or that there are no persons, agents, subjects, and so forth. The hypothesis for reductio is that over and above (or below and beneath) any composite of phenomena collectively denoted by 'I' or by a proper name, there is a single substantial entity that is the referent of such a term."[5] In other words, the pronoun "I," though meaningful, doesn't get its meaning by referring to an independently real or inherently existent self.

To infer from the absence of a single substantial self to the absence of the person, agent, or subject altogether is to fall prey to the nihilist or annihilationist extreme. We find this nihilist extreme today among those neuroscientists and neurophilosophers who, realizing that the brain offers no home for a substantially real self, come to the conclusion that there is no self whatsoever and that our sense of self is a complete illusion. As philosopher Thomas Metzinger states at the beginning of his book *The Ego Tunnel:* "there is no such thing as a self. Contrary to what most people believe, nobody has ever *been* or *had* a self . . . to the best of our current knowledge there is no thing, no indivisible entity, that is *us*, neither in the brain nor in some metaphysical realm beyond this world."[6]

I call this extreme view "neuro-nihilism." It isn't a genuine alternative to the reificationist view of the self; it's simply the negative or shadow version of this view. Neuro-nihilism assumes that were the self to exist, it would have to be an independently real thing or indivisible entity. The problem is that there is no such thing or entity in the brain. So, if it seems to us that we have or are an independently real self, then our sense of self must be an illusion created by our brain. Thus neuro-nihilism agrees with reificationism about the meaning of the concept of the self—both understand it to be the idea of an independently real thing or indivisible entity that is the essence of the person. Neuro-nihilism, however, denies that anything real falls under this concept, so it concludes "no such things as selves exist in the world."[7] In this way, neuro-nihilism is the negative and materialist version of reificationism.

Nāgārjuna, however, and the philosophers of the Madhyamaka or "Middle Way" school derived from his writings, reject the assumption that our ordinary or everyday concept of the self is that of a substantially real thing or entity. Our ordinary or everyday concept of the self is the concept of a subject of experience and an agent of action, not of an inner and substantial essence of the person. Furthermore, when we look carefully at what we apply our ordinary concept of self to in the world of our individual and collective experience, we don't find any inherently existent thing or independent entity; what we find is a collection of interrelated processes, some bodily or physical, some mental or psychological. These processes are all "dependently co-arisen," that is, each one comes to be and ceases to be according to a multitude of interdependent causes and conditions. The proper conclusion to draw from this state of affairs is not the nihilist one that there is no self whatsoever but rather that the self—the everyday subject of experience and agent of action—is a dependently arisen series of events. More simply stated, the self isn't a thing or an entity; it's a process. In Stephen Batchelor's words:

> To have become a person means to have emerged contingently from a matrix of genetic, psychological, social and cultural conditions. You are neither reducible to one or all of them, nor are you separate from them. While a person is *more* than a DNA code, a psychological profile and a social and cultural background, he or she cannot be understood *apart* from such factors. You are unique not because you possess an essential metaphysical quality that differs from the essential metaphysical quality of everyone else, but because you have emerged from a unique and unrepeatable set of conditions.[8]

The problem, however, according to Madhyamaka and Buddhism more generally, is that we don't usually experience our self this way—as a dependently arising process. Instead, we habitually experience our self as if it were a unified agent that functions as the executive controller of what we think and do, and that has a permanent inner essence distinct from our changing mental and physical characteristics. In this way, we're deluded about our nature. Our root error is to mistake something that's dependent and contingent—and hence "empty" of substantiality—for something that's independently existent.

Nāgārjuna compares our fundamental cognitive error with the images in a dream, mirage, or magical performance.[9] Just as these images exist in one way—as illusory—yet appear to exist in another way—as real objects of perception, so the self exists as dependently arisen, yet appears to exist as independently real with its own inner essence. Notice that the point of the analogy isn't that there is no self at all and that any sense of self is completely mistaken. After all, dreams exist as genuine experiential phenomena, yet they lack the independent being that we take them to have when we do not realize we're dreaming. Similarly, we habitually experience our self as if it were an independently real entity when in fact it doesn't exist in that way or have that kind of being. Instead, it exists as dependently arisen and hence as empty of any independent reality. The deep and penetrating experiential realization of this truth is like becoming lucid in a dream or waking up from the dream—the subject of experience still exists but is no longer deluded about its nature.

The Madhyamaka conception of the everyday self as dependently arisen suggests an alternative to neuro-nihilism. The trick is not to assume that the everyday self must be based on some independent entity or essence in the brain or else be nonexistent. The self isn't that kind of thing. Indeed, the self isn't a thing or an entity at all; it is brought forth or enacted in the process of living. This alternative way of thinking about the self isn't just philosophical; it makes a difference for cognitive science, especially the neuroscience of the self. My aim in this final chapter is to show how a conception of the self as dependently arisen can guide us away from neuro-nihilism and toward what I call an "enactive" understanding of the self.[10]

I-MAKING

In Indian philosophy, the term for the feeling that "I am," especially the sense that I have of being an individual through time who has a unique first-person perspective and who is a thinker of thoughts and a doer of deeds, literally means "I-making" (*ahaṃkāra*). I am going to borrow this term in order to propose a way of thinking about the self as emergent from and dependent on a variety of "I-making" processes. To put the idea another way, when I say that the self is not a thing but a process, what I mean is that the self is a process of "I-ing," a process that enacts an "I" and in which the "I" is no different from the I-ing process itself, rather like the way dancing is a process that enacts a dance and in which the dance is no different from the dancing.[11]

To implement this proposal about how to understand the self, I will describe a variety of ways that I-making happens at multiple levels—biological, psychological, and social. My theoretical tool will be the concept of a "self-specifying system." A self-specifying system is a collection of processes that mutually specify each other so that they constitute the system as a self-perpetuating whole in relation to the environment.[12] For example, a living cell is a collection of chemical processes that mutually produce each other so that they constitute the cell as a self-perpetuating whole in relation to its environment. Stated another way, the chemical processes that make up a cell enact or bring forth a relational self/nonself distinction, whereby the cell has a unique identity or "self" in relation to the environment or what is "not-self."

I will start with self-specifying systems in the domain of biological life and build up to an account of the self as we know it from our own experience. First, it will be helpful to summarize the logical structure of this enactive approach to the self:

THE ENACTIVE APPROACH TO THE SELF

Concept: I-making—the sense of being an "I" who endures through time and who is a thinker of thoughts and a doer of deeds.

Proposal: The self is a process of I-ing—an ongoing process that enacts an "I" and in which the "I" is no different from this process itself.

Theoretical Tool: Self-specifying system—a collection of processes that mutually specify each other and thereby constitute the system as a self-perpetuating whole in relation to the wider environment.

Implementation: Description of the self-specifying systems that constitute I-making at multiple levels (biological, psychological, and social).

SELF-MAKING CELLS

Imagine you're looking at a new biochemical specimen through a high-speed digital confocal microscope. You're able to observe a variety of interacting processes. Upon closer inspection, you discover that some of these processes mutually condition or enable each other, so that they form an interlocking network that is self-organizing and self-perpetuating. Furthermore, some of these chemical reactions produce a boundary that encloses and provides a home for the other reactions. The boundary is permeable but selective; it lets in certain molecules from the environment but not others, and it sends out other molecules into the environment. At this point, you realize that what you're looking at must be some kind of cell, a living individual that actively relates to its environment.

In complex systems theory, the term used to describe this kind of biochemical self-production is "autopoiesis."[13] Neurobiologists Humberto Maturana and his student, Francisco Varela, introduced the concept in the 1970s.[14] "Autopoiesis" literally means self-making or self-producing (from the Greek *autos*, meaning "self," and *poiein*, meaning "to make" or "to produce"). An autopoietic system is a biochemical system that produces its own molecular components, including a boundary that defines what's inside versus what's outside the system. Not all biochemical systems are autopoietic. A virus, for example, is a bounded entity with a protein coat, but the molecular constituents of the virus (nucleic acids) aren't generated inside the virus, only outside it in a host cell. A virus has no metabolism, so it doesn't produce and maintain itself in the autopoietic sense. In contrast, all living cells are autopoietic. Thus autopoiesis is a key property that distinguishes cellular life from other kinds of biochemical systems.

The living cell, by virtue of being autopoietic, provides the minimal and fundamental case of a self-specifying system. A cell stands out of a molecular soup by specifying the boundary that sets it apart from what it is not. At the same time, this boundary specification happens through internal chemical transformations that the boundary itself makes possible. The boundary and the internal transformations specify each other, and in this way the cell emerges as a figure out of a chemical background. Should something interrupt this self-specifying process, the cellular components will gradually diffuse back into a molecular soup and will no longer form a distinct whole.[15]

It's worthwhile to compare the living cell to other self-organizing systems that aren't self-specifying. A tornado arises from a huge aggregation of air and water molecules getting sucked into the global pattern of a vortex, while this pattern influences how the individual molecules locally behave. Analogously, we might think of the self as an emergent global pattern of the complex dynamical behavior of aggregated mind-body states.[16] Or take a candle flame. In this case, instead of the molecules shifting their global patterns of aggregation, as in the tornado, they are continuously transformed by the combustion process, yet the flame persists as a flow of air that brings in fresh oxygen and gets rid of waste. Analogously, we might think of the self as an emergent macrostate of mind-body processes that undergo a continuous transformation.[17] In Indian philosophy, the naturalist or materialist school, called Cārvāka, as well as the Buddhist Vātsīputrīya school, held such "tornado" or "flame" views of the self, as Jonardon Ganeri, a contemporary English philosopher, describes in his monumental work of cross-cultural philosophy, *The Self*.[18] Ganeri draws on the flame view in developing his own account of the self. Yet neither the tornado nor the flame has any internal chemical reaction network that synthesizes its own molecular components. Nor does either system control how matter and energy flow through it and around it in the way a cell does, thanks to its metabolism. Unlike the tornado or the flame, the cell creates its own molecular being as an individual whole. This kind of self-specification makes the cell strongly emergent:[19] its parts reciprocally produce each other while the cell as a whole provides the necessary setting for these chemical reactions. Since this kind of self-specification is absent from the tornado and the flame, these systems aren't full-fledged individuals but simply macrostates of underlying microprocesses. For these reasons, the living cell works better as an analogy for the emergent and dependently arisen self than either the tornado or the flame does. What I want to do now is to build on this analogy.

SENSE-MAKING IN PRECARIOUS CONDITIONS

Wherever there's a self, there's a world on which that self depends and to which it must relate by finding and creating meaning. Similarly, wherever there's a living being, there's an environment on which that being depends and to which it must relate by finding and creating sense. In short, living is sense-making in precarious conditions.[20]

Take bacteria, the evolutionarily oldest and structurally simplest living beings. Many bacteria swim by means of rotating flagella embedded in their membranes. These bacteria can detect around fifty distinct chemicals, including sugars and amino acids that attract the cells so they swim toward them, as well as acids and heavy metals that repel the cells so they swim away. The bacteria swim by coordinating the rotation of their flagella so they form a propeller; when the flagella rotation is uncoordinated, the cells tumble about randomly. As they move, they're able to register differences over time in the levels of attractants and repellents—for example, in the rate of change in the concentration of sucrose or aspartate, which they can feed on. The cells maintain their direction as long as they detect an increase in the nutrient level over time. If the nutrient level decreases, then the cells go into their random tumbling mode until they hit on an orientation where they again detect an increase, at which point they go off in that direction. By repeating these behaviors—swimming in the same direction as long as conditions are improving or not getting any worse, and tumbling when conditions start deteriorating—bacteria can travel long distances toward favorable locales.

This type of behavior, known as chemotaxis, illustrates how living is a process of sense-making. Bacteria have specialized receptor molecules embedded in their membranes that enable them to sense certain things as beneficial and others as harmful. Yet this kind of significance—beneficial versus harmful—doesn't exist ready-made in the physical structure of chemicals such as sucrose (an attractant) or alcohol (a repellent). Rather, it comes into being only in relation to the bacteria, given their structure and lifestyle. The bacterial lifestyle endows some molecules with significance as food and other molecules with significance as repellents. In this way, the bacteria literally make sense of the physical environment, and their sense-making is directly embodied in their behavior—swim in the same direction, swim away, or tumble about.

Yet, as this example shows, living isn't just sense-making; it's sense-making in precarious conditions. Imagine you're very small, so you're continually buffeted by water molecules and bumped off course, while the watery contents inside you are in constant motion. Such is the external and internal milieu of bacteria, the microworld of molecular diffusion and Brownian motion (the random motion of particles suspended in a fluid). How do you manage to hold together? You depend on strong and weak chemical bonds, but you also hold together because your innards are entangled in a self-specifying way—every one of your constituent processes both enables and is enabled by one or more of your other constituent processes. Your condition is precarious because these processes are always in danger of running down, and none of them can sustain itself for long in the absence of the system they form together. In other words, remove these processes from the support group they form together and they waste away. Any living process is precarious in this sense: break open a cell and its metabolic constituents diffuse back into a molecular soup; take an ant out of a colony and it eventually dies; remove a person from a relationship and she or he may cease to flourish.

DEPENDENT ARISING

This vision of living beings as self-specifying and sense-making beings works well as an illustration of the Buddhist concept of "dependent arising," especially as Tibetan expositors of Madhyamaka elaborate it. According to their understanding, dependent arising happens at three levels.[21]

The first level of dependence is causal. A phenomenon depends on causes and conditions for its existence—not only those causes that bring it into existence but also those that bring about its ceasing to exist. In the case of the cell, many kinds of causes and conditions—environmental, genetic, metabolic, and so forth—contribute to its coming into existence, its continuing to exist for a time, and its ceasing to exist.

The second level of dependence is whole/part dependence, or what philosophers call "mereological dependence" (mereology is the theory of part-part and part-whole relations). A phenomenon depends on its parts for its existence. Thus the cell depends on its membrane, organelles, molecular constituents, and so on. In the case of a self-specifying system, the whole system as a network depends on the mutually enabling relations among its constituent processes.

In Madhyamaka, this second level of dependent arising is usually described as one-way—the whole depends on its parts, but not vice versa. Complex systems theory, however, requires that we include the dependence of the parts on the whole.[22] In a self-specifying system, the whole depends on the parts and the parts depend on the whole. For example, in the cell, the specific metabolic pathways and the molecules they synthesize depend on the functioning of the entire cell for their existence, for these pathways and molecules can't persist for long outside of the special environment the cell maintains within its membrane. Thus part and whole co-arise and mutually specify each other.

The third level of dependent arising is conceptual dependence. This level is the subtlest. According to the subschool of Madhyamaka called Prāsaṅgika Madhyamaka, the identity of something as a single whole depends on how we conceptualize it and refer to it with a term. We can also add that its identity depends on a scale of observation. Thus the cell has no intrinsic identity independent of a conceptual scheme and scale of observation that individuate it as a unit. As biologist Neil Theise points out in a 2005 essay in *Nature*, although we've thought of the cell as the basic building block of the body ever since the microscope revealed the cell membrane, from another conceptual and observational perspective, the body is a fluid continuum:

> On one level, cells are indivisible things; on another they dissolve into a frenzied, self-organizing dance of smaller components. The substance of the body becomes self-organized fluid-borne molecules, which know nothing of such delineating concepts as "intracellular" and "extracellular." . . . The cell as a definable unit exists only on a particular level of scale. Higher up, the cell has no observational validity. Lower down, the cell as an entity vanishes, having no independent existence. The cell as a thing depends on perspective and scale: "now you see it, now you don't," as a magician might say.[23]

We can also use the concept of a self-specifying system in order to illustrate the third level of dependent arising as conceptual dependence. Here the identity of the interlocking network as one self-specifying system depends on how we cognitively frame things, that is, on our decision to focus on those conditioning relations that mutually specify each other in a recursive way. Depending on our interests or explanatory purposes, we could decide to frame things differently by focusing on other conditioning relations. What we mark off as a system depends on our cognitive frame of reference and the concepts we have available.

Such conceptual dependence doesn't mean that nothing exists apart from our words and concepts, or that we make up the world with our minds. On the contrary, in order to designate something as a cell or as a fluid, or more generally as a system of whatever kind, there must be some basis for that designation in what we observe. The subtle point, however, is that what shows up for us as a system of whatever kind depends not just on a basis of designation but also on how we conceptualize that basis and use words to talk about it.

For this reason, the full statement of dependent arising as conceptual dependence, according to Prāsaṅgika Madhyamaka, is that whatever is dependently arisen depends for its existence on a basis of designation, a designating cognition, and a term used to designate it.[24]

This formulation of dependent arising gives us a crucial conceptual ingredient we need to build from the idea of a self-specifying system to a full-fledged I-making system—one that has a sense of itself as an I who endures through time as a thinker of thoughts and a doer of deeds. That crucial ingredient is what I will call a "self-designating system"—one that conceptually designates itself as a self and where changing body-mind states serve as the basis of designation. But we have a way to go before we get to that part of the story. First we need to look at how self-specifying processes underlie perception, action, and feeling, as well as how this cognitive science framework relates to the Buddhist framework of the five aggregates.

SENSORIMOTOR SELFHOOD

What do the brain and nervous system add to the basic self-making and sense-making capacities of life?

If you're an animal on the move, then you need to hold together so your cells don't go their separate ways as you swim, run, or fly around. One way you keep things together is by having specialized cells with long extensions that interconnect distant parts of your body. These are the nerve cells or neurons, whose long fibers (axons) and branching projections (dendrites) interconnect nerve cells to each other and to the body's sensory surfaces (skin, sense organs, nerve endings) and effectors (muscles, glands, and organs). Neurons are also electrically excitable cells that communicate across the body through electrochemical signals that travel much faster than molecules can diffuse. For these reasons, wherever quick movement over large distances is essential to the lifestyle of a multicellular organism, there is the corresponding development of a nervous system.

Any nervous system operates according to a basic "neurologic," a pattern that continues and elaborates the "biologic" of autopoiesis. The basic logic of the nervous system is to link sensory activity and motor activity in an ongoing and self-specifying way.[25] Think of the relationship between what you see and how your eyes and head move. Suppose you're watching the path of a bird in flight. While the sensory stimulation your eyes receive brings about your eye and head movements, your eye and head movements bring about the sensory stimulation your eyes receive. In this way, your visual tracking of the bird comprises a sensorimotor cycle, whereby what you see depends directly on how you move and how you move depends directly on what you see. The linkage the nervous system establishes between your body's sensory systems and motor systems makes possible this ongoing sensorimotor cycle.

We're now in position to see exactly how this sensorimotor cycle is self-specifying. Consider that when you watch the bird in flight, you need to be able to tell the difference between sensory changes arising from your own eye movements and sensory changes arising from the bird's motion across the sky. In other words, you need to be able to distinguish perceptually between what happens as a result of something you do and what happens as a result of something going on in your surroundings. Stated in general terms, any animal needs to be able to tell the difference between the sensory stimulation it receives as a consequence of its own motor actions (self) and the sensory stimulation it receives as a consequence of changes going on in its environment (not-self). The nervous system distinguishes between them by systematically relating the motor commands (efferent signals) for the production of an action (an eye, head, or body movement) to the sensory stimuli (afferent signals) arising from the execution of that action (the flow of sensory feedback from motor action). This process is known as sensorimotor integration. According to various models of sensorimotor integration going back to the pioneering work of German physiologist Erich von Holst (1908–62), the central nervous system contains a specialized comparator system that receives a copy of the motor command and compares it with the incoming sensory signals arising from the motor action. For example, when you move your fingers across the keyboard, your cortex sends signals for the motor movements to the motor neurons in the brainstem and spinal cord while also sending a copy of these signals to the cerebellum. Meanwhile, the cerebellum also receives sensory signals from the muscles, joints, and tendons that convey information about the finger movements you've executed. By comparing this sensory information about finger positions and movements to the information about the motor command, the cerebellum enables the brain to differentiate between sensory changes caused by motor action and sensory changes caused by the environment, so that the brain can also keep track of how well the action is being done. Stated in general terms, such a comparator system enables the nervous system to integrate efferent (motor) and afferent (sensory) signals in a way that systematically distinguishes between "reafference"—sensory signals arising as a result of the animal's own motor actions (self)—and "exafference"—sensory signals arising as a result of environmental events (not-self). This constant integration of efferent signals allowing the performance of an action with reafferent signals specific to this action makes the sensorimotor cycle a self-specifying process.[26]

The crucial point to appreciate here is that the sensorimotor cycle is self-specifying not in the sense that it points to a self that is separate from perception and action, but rather in the sense that it enacts a self that is no different from the process of perceiving and acting. Consider that the reafference you receive is specific to you because it's intrinsically related to your own bodily action (there is no such thing as reafference that is not specific to the perceiving subject and acting agent). Thus, by relating the motor signals for action (efference) to the sensory signals arising from action (reafference), the nervous system enacts a self/not-self distinction in perception and action. On one side lies a unique sensorimotor perspective that constitutes the subject of perception and the agent of action. On the other side lies the environment as the meaningful locale of perception and action. In this way, sensorimotor I-making and sensorimotor sense-making arise together and are inseparable; they're dependently co-arisen.

FEELING YOUR WAY

Where and how does feeling enter this story about the biology of I-making? Does the single bacterial cell feel anything as it makes its way around its world? Or does feeling require having a nervous system with a brain? Is sentience—the ability to feel—a basic capacity of all life, or does it arise with the evolution of the nervous system?

The short answer is that we don't really know how to answer these questions. At the present time, we have no deep or fundamental understanding of how feeling could arise from a biological process or be realized in such a process. This lack of understanding is an aspect of what philosophers call the "explanatory gap" between our scientific theories of biological nature and our experiential knowledge of consciousness (see chapter 3).

Sentience has seemed to some biologists to be coeval with life itself. This is the view that the late pioneering microbiologist Lynn Margulis and her son, Dorion Sagan, took in their book *What Is Life?* In their words:

> Not just animals are conscious, but every organic being, every auto-poietic cell is conscious. In the simplest sense, consciousness is an awareness of the outside world. And this world need not be the world outside one's mammalian fur. It may also be the world outside one's cell membrane. Certainly some level of awareness, of responsiveness owing to that awareness, is implied in all autopoietic systems.[27]

However, many neuroscientists prefer to think of feelings as arising with brains that constantly register the organism's changing body states and keep those states within a range that facilitates self-maintenance.[28] According to this viewpoint, whether a feeling arises from an external situation—such as fear in a dangerous situation or happiness in a pleasant one—or from an internal state of the body—such as hunger or fatigue—it reflects a state of the body in which the condition of the internal organs plays a key role. From this perspective, sentience is an evolved capacity of certain forms of animal life—ones in which the brain has specialized systems for regulating the body's changing internal states.[29]

I have a great deal of sympathy for the idea that all cellular life possesses some degree of sentience, an idea I will come back to shortly. Nevertheless, the point I want to call attention to now is that the link between internal body states and the brain processes that register them is a self-specifying one and hence contributes to the sense of self. In this case, the self-specifying processes support an interoceptive sense of self—a feeling from within of one's body as that which inhabits the sensorimotor perspective from which one perceives and acts.[30]

For example, the back-and-forth cycle between the motor regulation of breathing (an efferent process) and the feeling of breathing (a reafferent sensory process) is self-specifying and contributes to our basic bodily sense of self. It's therefore no surprise that so many types of meditation begin with focused attention to the sensation of the breath as a way of centering our awareness on our embodied being in the present moment.

More generally, such interoceptive, efferent-reafferent loops are crucial for homeostasis—the body's maintenance of its internal milieu in a condition conducive to life preservation and advantageous behavior. To this end, the brain continually relates reafferent (sensory) information about the body's internal state—its chemical composition, temperature, nutrient levels, degree of visceral muscle contraction, and so on—to the corresponding internal efferent (motor) processes that keep the body's interior state within a viable range. This self-specifying interoceptive system enacts a bodily sense of self by maintaining the body's integrity (self) in relation to the environment (not-self) and by sustaining the feeling of the body from within.

THE FIVE AGGREGATES REVISITED

The biological perspective I've been presenting can be used to cast new light on the ancient Buddhist concept of the five aggregates (form, feeling, perception/cognition, inclination, and conscious-ness). From this perspective, the five aggregates can be seen as five sorts of basic psychophysical activities that make up a sentient being or an individual person.[31]

Form, the first aggregate, includes not simply the material con-stitution of the body but also the body as living and experientially lived. Specifically, form includes the body's sensory systems and the sensible qualities to which they're sensitive.[32] At a functional level of description, form consists in one of the exteroceptive or interocep-tive senses making "contact" with a physical stimulus that it thereby registers. Following Jonardon Ganeri, we can think of the first aggre-gate as the psychophysical activity of "registering" some discrim-inable sensory quality.[33]

According to one contemporary neuroscience view, the exterocep-tive or interoceptive registering of an outer or inner sensory stimulus immediately triggers a host of bodily changes and reactions, which in turn elicit feelings that are either pleasant, unpleasant, or neutral.[34] For example, seeing a snake elicits changes in heart rate, breathing, and hormonal secretion, as well as the action tendency of avoidance along with the unpleasant feeling of fear.

Similarly, the second aggregate of feeling consists in the immedi-ate experience of any event of sensory registration as being pleasant, unpleasant, or neither pleasant nor unpleasant. From the Buddhist perspective, however, feeling isn't restricted to sensory events and the bodily changes and reactions to which they give rise, for any mental event—such as a thought or a memory—has its associated feeling of being pleasant, unpleasant, or neutral. Thus feelings can be thought of as immediate and low-level affective "appraisings."[35] Whatever we register we also appraise as agreeable, disagreeable, or indifferent.

Closely intertwined with feelings or affective appraisings are incli-nations to react in particular ways to what feels pleasant, unpleasant, or neutral. For example, when I was a little kid, I was terrified of the water, so when my parents would take me to a swimming pool I'd refuse to go into it; even seeing the Charles River from the back seat of our car as we drove from Watertown to Cambridge, Massachusetts, made me feel uneasy. Such habitual reactions fall under the heading of the fourth aggregate of inclinations. Inclinations can be described as "readyings" of the mind and body that strongly shape how we perceive, think, and act.[36]

At the same time, inclinations depend on how we perceptually and cognitively identify whatever we register and feel. Such identification is the province of the third aggregate. In order to have experienced the swimming pool as frightening and to be avoided, I had to have identified it as a pool of water. The third aggregate consists in identifying something according to some perceptual category or cognitive schema. Such a category or schema enables one to recognize that thing or things of that kind on future occasions. In Ganeri's terms, this activity amounts to a kind of "stereotyping."[37]

Finally, all the psychophysical activities included in these four aggregates—form, feeling, cognition, and inclination—both condition and are conditioned by an awareness of the presence of something via an exteroceptive or interoceptive sense modality or a mental mode of awareness such as introspection or memory. Such forms of awareness fall under the heading of the fifth aggregate of consciousness. Since "consciousness" in this context means awareness of the presence of something, it is implicitly selective in the minimal sense of being attentionally oriented toward one thing rather than another. For this reason, Ganeri describes the fifth aggregate as "attending" (even though, strictly speaking, attention is an inclinational factor that belongs to the fourth aggregate).[38]

One of the key insights of the five-aggregates framework is that these five types of psychophysical processes—form (registering), feeling (appraising), cognition (stereotyping), inclination (readying), and consciousness (orienting or attending)—lie beneath the level of ordinary first-person reports of experience, such as "I see the river" or "I'm afraid of the water," and together make up experience understood at that everyday level. This insight makes it possible to build bridges from the five-aggregates framework to cognitive science, which also analyzes complex cognitive and emotional phenomena into their more elementary constituents.[39] To quote from Ganeri's version of this kind of Buddhism-science bridge building:

Proto-cognitive and proto-affective processes, themselves realized in underlying neural systems, combine to constitute states of conscious intentional experience, experience which presents the world as one in which *attention* falls on objects which are perceptually *registered* as falling under schematic *stereotypes*, as organized in a hodological *appraisal* space of affordance and obstacle, in ways that shape the diachronic flow by *readying* for future experience.[40]

Ganeri writes of "underlying neural systems," but given our biological conception of living as a process of sense-making in precarious conditions, we can ask whether any of these processes—registering, appraising, stereotyping, readying, and attending—or their precursors are found in simpler organisms without neural systems.[41]

Consider again the evolutionarily oldest and structurally simplest living beings, the bacteria. The first aggregate of form already appears in these simple organisms. In this case, "form" means molecular shape, and "registering" means what happens when molecules of different shapes "contact" the appropriate chemoreceptors in the bacterial membrane and bind to them. In this way, these single-cell organisms can register around fifty different chemicals, and move toward or away from them.

We can also see evolutionary precursors to "appraising," "stereotyping," and "readying." Bacteria identify certain kinds of molecules as attractants and others kinds as repellents. Attractants elicit approach— the bacteria swim toward the area of greatest molecular concentration— whereas repellents elicit avoidance—the bacteria swim away. This kind of behavior—bacterial chemotaxis—depends on a rudimentary form of memory.[42] The bacteria keep track of the rate of change in the concentration of molecules as they swim, so that they can move either toward or away from the areas of greatest concentration. In short, bacteria discriminate and remember physically distinct elements of the environment (proto-stereotyping); they respond to these elements as either favorable or unfavorable (proto-appraising); and they incline either toward or away from these elements (proto-readying).

We might wonder, however, whether "appraising" in bacteria involves any kind of feeling, or more generally, whether bacteria are sentient. Do attractants feel pleasant and repellents feel unpleasant? Although bacteria certainly look energized as they swim toward an attractant and frantic as they swim away from a repellent, do the cells feel vigorous in the one case and distressed in the other? More generally, do bacteria have any awareness of the presence of what they sense?

These questions came up at the first Mind and Life dialogue with the Dalai Lama in 1987. The Dalai Lama asked Francisco Varela whether he would consider a one-celled being, such as an amoeba or a bacterium, a sentient being. It seems fair to say that science hasn't advanced beyond the answer that Varela gave:

VARELA: The behavior of the bacterium or amoeba is one of avoiding some things and seeking others, much like the behavior of clearly sentient beings like cats and humans. Hence I have no basis for saying that the behavior is not of the same kind, although I would say there is no *consciousness* of pain or pleasure. The amoeba intrinsically manifests a differentiation between what it likes and what it doesn't like. In that sense, there is sentience. Why do I say that a cat feels pleasure and pain and seeks satisfaction and is a sentient being? There is no way that I can know what the experience of a cat is.

DALAI LAMA: Yes, that's right.

VARELA: Exactly the same argument applies to the amoeba or bacterium. I cannot know what the experience of a bacterium is, but if I observe its behavior, it is of the same kind. This is why, as a scientist, I can say that the behavior of the bacterium is cognitive behavior in that it makes discriminations through this form of sensory-motor correlations I have described [bacterial chemotaxis]. The mechanism is the same as, say, in cats. I know psychologists shudder when I say this, but I'm talking as a neuroscientist. People who have studied bacterial behavior as a way to understand the simplest forms of behavior . . . have no hesitation in referring to it by terms such as behavior, perception, and instinct. I must say, I find their approach convincing. Now, it is certainly true that from there to what we normally call cognition, which is some form of awareness, there's a large leap. But the question remains open as to whether this is not a continuum.[43]

When Varela said, "there is no *consciousness* of pain or pleasure," he was probably using the term "consciousness" to mean not merely the qualitative feeling of pain or pleasure but also the ability to cognize that feeling, that is, to be mentally aware of it as a state of oneself. According to contemporary Western philosophy of mind, whereas the mere feeling of pleasure or pain would count as a case of "phenomenal consciousness," cognitively grasping that feeling as a state of oneself would count as a case of "access consciousness." This terminology had not yet been introduced into the neuroscience of consciousness at the time Varela made these remarks. Similarly, in the Buddhist framework of the five aggregates, there's a difference between the mere feeling of a sensory stimulus as pleasant or unpleasant (the second aggregate of feeling), and the ability to be aware of the feeling as a mental phenomenon (the fifth aggregate of consciousness, specifically inner mental consciousness), including being able to identify or recognize the kind of feeling it is (the third aggregate of recognitional cognition). Whether bacteria can feel—whether they're sentient or phenomenally conscious—remains a difficult question, but if they are sentient, it certainly seems that they have no mental awareness of their condition or any kind of cognitive access to it (they have no physical apparatus to support this kind of cognitive awareness, nor do they exhibit any behavior that manifests this kind of awareness).

CONSCIOUSNESS IN LIFE

At the end of his answer to the Dalai Lama, Varela raised the question of whether there is a continuum from bacterial perception to human awareness and cognition. If we suppose that bacteria are sentient—or that all cellular life is sentient—can we trace a path from the sentience of life to animal consciousness and the human sense of self?

Free-moving single-cell organisms, such as bacteria, paramecia, and amoebae, are excitable cells, that is, cells that create electrical currents when they're stimulated and that form miniature bioelectrical fields. These cells have membranes that are selectively permeable to charged ions, which create an electrical current as they flow across the membrane. In multicellular animal organisms, there are three kinds of excitable cells—sensory cells, muscle cells, and nerve cells. In the case of neurons, these cells are selectively permeable to sodium, potassium, calcium, and chloride, so they form differential electrical fields and interact with the fields of other neurons through the action potential—the burst of electrical activity that travels along the axon when sodium ions cross the membrane and the charge concentration between the ion solutions outside and inside the cell exceeds a certain threshold. When many neurons are interconnected to form neuronal networks or brains, the sum or superposition of their electrical fields generates macroscopic neuroelectrical fields at a higher level of system complexity.[44] As we've seen in earlier chapters, these large-scale neuroelectrical fields, measured using EEG, shift in distinct ways across the states of waking, dreaming, and deep and dreamless sleep, as well as in various kinds of meditation.

From this bioelectrical perspective, evolution occurs not just in structures, organs, and bodies but also in dynamic electrical fields.[45] Whereas life in general comprises the emergence of self-organizing bioelectrical fields, animal evolution comprises the emergence of self-organizing neuroelectrical fields. As neuroscientist Roman Bauer observes, "Life in general can be considered as an outcome of the self-organization of molecules to cells and of cells to organs and organisms, and in the same way mind and consciousness can be considered as a manifestation of the self-organization of elementary bio-electrical fields to neuro-electrical macrofields in brains."[46]

Consider again the action potential—the burst of electrical activity that travels down the axon. Although usually described as a mechanism of cell-to-cell signaling or communication, it's also a concrete biochemical event whereby the neuron opens to the outside world and lets a huge number of positive ions—one hundred million per membrane channel per second—flow into the cell and out again.

In this way, the neuron undergoes a massive change in its internal state while directly sensing the charge state of its immediate outside environment. Neuroscientist Norman Cook calls this kind of sensing "neuron-level sentience."[47] Individually, each neuron senses only its local electrochemical state; collectively, however, neurons synchronize their action potentials, both locally and across large distances, and this temporal synchronization of an enormous number of action potentials produces the coherent and large-scale electrodynamical states of the brain that correlate with various modes of consciousness. Hence, as Cook proposes, neural synchrony—the temporal synchronization of numerous action potentials—can be seen not only as a mechanism for cognition but also as the way the sentience of individual autopoietic neurons self-organizes into the brain-level phenomenon of consciousness.

According to this way of thinking, sentience depends fundamentally on electrochemical processes of excitable living cells while consciousness depends fundamentally on neuroelectrical processes of the brain. Consciousness isn't an abstract informational property, such as Giulio Tononi's "integrated information" (see chapter 8), it's a concrete bioelectrical phenomenon. It follows—as philosopher John Searle has long argued—that consciousness can't be instantiated in an artificial system simply by giving the system the right kind of computer program, for consciousness depends fundamentally on specific kinds of electrochemical processes, that is, on a specific kind of biological hardware.[48] This view predicts that only artificial systems having the right kind of electrochemical constitution would be able to be conscious.

Recall now the Dalai Lama's statement that the physical basis for consciousness is a subtle energy whose presence can be felt in the body. This energy—called *prāṇa* in Sanskrit and *lüng* in Tibetan—is said to carry all excitation and movement, including at the level of cells. Although the Dalai Lama suggested that the scientific concept of matter may need to be modified in order to appreciate this energy, I'm inclined to think that this energy is already known to science as the electromagnetic fields produced by living cells, especially the neuroelectrical fields produced by the brain and the bioelectrical fields produced by the heart. What has barely begun to be investigated, however, is how meditative practices sensitize one to subtle experiential aspects of these bioelectrical fields, as well as how these practices enable one to alter bioelectromagnetic processes in the brain, the heart, and the rest of the body.[49]

SOCIAL SELF-MAKING

So far we've seen how the nervous system is self-specifying—how it enacts a unique sensorimotor perspective that functions as the subject of perception and the agent of action—and how dynamic neuroelectrical fields may be the crucial substrate for consciousness. Together, these biological processes are enough to bring about a unique embodied perspective on the world, in which you feel your body from within as it defines the center of the surrounding space in which you perceive and act. Yet having this kind of basic experiential perspective isn't enough to give you the feeling of being a self who is a thinker of thoughts and a doer of deeds. To have such a sense of self, you also must be able to attend to your experiences and conceive of yourself as a subject of experience.

These abilities are what I had in mind when I said earlier that the crucial ingredient that takes us from a self-specifying system to a full-fledged I-making system is that of being a "self-designating" system. A self-designating system is one that can designate itself as a self. This means it can attend to its changing experiential states and conceive of itself as the subject of those states.

Consider what happens when you recognize your face in a mirror. Your perception takes place from the particular subjective perspective your body defines, but you don't see just a visual image located in front of you in the surrounding space. In addition, you see an image that you recognize as an image of you—that you attribute to yourself or mentally designate as an image of you. At the same time, you're aware of being the one who is looking at the image. You need to be aware of yourself in this way—as the subject who sees the mirror image—in order to be aware of the image as an image of you, the one seeing. To recall our earlier terminology, your self-experience includes both perceiving yourself as something in the world—the self-as-object—and being aware of yourself as an experiential subject—the self-as-subject.

We know from ethology that only certain animals are able to recognize themselves in a mirror. The way to test for this ability is to put a visible colored dot on the animal's body while it's asleep. The mark needs to be placed on an out-of-view area that the animal can see only with the aid of a mirror. If the animal inspects or touches the area when it wakes up and sees its image in the mirror, then we can infer that the animal considers the mirror image to be an image of itself. Hundreds of animal species have been tested in this way, but only a few besides humans pass this mirror test for self-recognition. To date these species include the four great apes (bonobos, chimpanzees, gorillas, and orangutans), bottlenose dolphins, Asian elephants, and the Eurasian magpie.[50]

Mirror self-recognition is associated with a number of important cognitive abilities. Around the time that human children first pass the mirror test—at about eighteen months of age—they begin to show the ability to look at the world from another person's perspective, including the ability to understand themselves as an object of someone else's attention.[51] The nonhuman animal species that pass the mirror test are all highly social animals, and some of them—the great apes—have been shown to have high levels of empathy.[52] Social cognition, perspective taking, and empathy are closely related to the ability to think of oneself as a self, as an individual subject of experience distinct from other such individuals.[53]

Indeed, being able to think of oneself as a self seems inseparable from being able mentally to grasp an outside view of oneself, that is, from the vantage point of the other. Studies from developmental psychology indicate that these two mental abilities arise together and build on the capacity for shared or joint attention. Joint attention emerges around nine months of age and comprises the threefold structure of a child, an adult, and something to which they know they are both giving attention. It includes activities such as gaze following (reliably following where one or the other person is looking), acting together with shared objects (such as toys), and imitative learning (the child's acting on or using things the way adults do). Developmental psychologist Michael Tomasello describes how coming to have an outside view of oneself emerges in shared attentional situations:

> As infants begin to follow into and direct the attention of others to outside entities at nine to twelve months of age, it happens on occasion that the other person whose attention an infant is monitoring focuses on the infant herself. The infant then monitors that person's attention to her in a way that was not possible previously, that is, previous to the nine-month social-cognitive revolution. From this point on the infant's face-to-face interactions with others—which appear on the surface to be continuous with her face-to-face interactions from early infancy—are radically transformed. She now knows she is interacting with an intentional agent who perceives her and intends things toward her. . . . After coming to this understanding, the infant can monitor the adult's intentional relation to the world including herself. . . . By something like this same process infants at this age also become able to monitor adults' emotional attitudes toward them as well—a kind of social referencing of others' attitudes to the self. This new understanding of how others *feel* about me opens up the possibility for the development of shyness, self-consciousness, and a sense of self-esteem. . . . Evidence for this is the fact that within a few months after the social-cognitive revolution, at the first birthday, infants begin showing the first signs of shyness and coyness in front of other persons and mirrors.[54]

Let's relate these ideas back to our concepts of I-making and self-specifying processes. Of concern now is the kind of I-making that happens at a cognitive and emotional level, that is, the creation of a cognitive and affective sense of being an individual "I" among other such individuals. Here the self-specifying processes are joint attentional activities that specify each attentional agent as the focus of their shared attention: I attend to you attending to me, and vice versa. What these mutually specifying processes enact or bring forth isn't you or me as separate entities, but rather the dyad we form together.

It seems likely that being able to participate in such dyads is required for being able mentally to grasp an outside view of oneself and hence for being able to think of oneself as a self. Certainly this seems to be the case as a matter of cognitive development. To quote Tomasello again:

> As the child begins to monitor adults' attention to outside entities, that outside entity sometimes turns out to be the child herself—and so she begins to monitor adults' attention to her and thus to see herself from the outside, as it were. She also comprehends the role of the adult from this same outside vantage point, and so, overall, it is as if she were viewing the whole scene from above, with herself as just one player in it. This is as opposed to the way other primate species and six-month-old human infants view the social interaction from an "inside" perspective, in which the other participants appear in one format (third-person exteroception) and "I" appears in another different format (first-person proprioception . . .).[55]

As we've seen in previous chapters, occupying an outside vantage point where one views oneself from above happens in various ways in memory, dreams, and out-of-body experiences. We're now in position to appreciate that this kind of self-experience emerges from I-making processes that are fundamentally social and intersubjective.

SELF-PROJECTION

Take a minute to think about something you did yesterday or that happened to you. Now think about something you plan to do later today or tomorrow. In both cases you need to shift your perspective from the immediate present to an alternative perspective, either to your remembered personal past or to your imagined future. Psychologists call this kind of perspective switch "self-projection."[56]

In self-projection, you mentally project yourself into an alternative situation, and whatever you think about you mentally represent in relation to yourself. To be precise, whatever you think about you mentally represent in relation to your mental representation of yourself. For example, when you remember something by imagining yourself as seen from the outside in the past scene, you mentally represent yourself from a mentally represented third-person perspective. In this case, your memory includes a mental image of yourself-as-object. And when you remember something by seeing it again through your own eyes, you mentally represent the past scene as seen from a mentally simulated first-person perspective. In this case, your memory includes a mental simulation of yourself-as-subject in the past. These two perspectives—the first-person or field perspective and the third-person or observer perspective—also structure how we project ourselves into the future. We can envision a future scene with ourselves included in it as seen from the outside or as seen through our own eyes. This is one reason self-projection into the future, or prospection, is sometimes called "memory of the future."[57]

Memory and prospection are the crucial mental capacities enabling you to think of yourself as an "I" who endures through time as a thinker of thoughts and a doer of deeds. Memory and prospection create a personal and historical sense of self because they enable you to think of yourself as having a unique story line through time. In this kind of self-projection—also known as mental time travel—every memory or expectation you encounter normally presents itself as yours, as belonging to you, where you feel as if you're one and the same self who endures through time as the subject of these experiences. Thus self-projection exemplifies the sense of self that consists in the feeling of being a distinct individual with a unique personal identity and a protracted existence in time. Neuroscientist Antonio Damasio calls this the "autobiographical self"; phenomenologists call it the "narrative self."[58]

Self-projection, and hence the autobiographical or narrative sense of self, depends on a network of brain areas that are especially developed in human beings.[59] This network includes frontal lobe areas, which traditionally have been associated with planning, and medial temporal-parietal areas, traditionally associated with memory. The self-projection network overlaps closely with the brain's so-called "default network," a network of regions that are active when outer demands on your attention are low, such as when you're resting in a brain scanner with no particular task to do, but that decrease their activity when you need to perform an outer-directed, attention-demanding task.[60] The connection between self-projection and the default network is that during resting or passive situations, when the default network is active, spontaneous thoughts are at their peak, and these often take the form of musing about past happenings, making plans for the future, daydreaming about yourself in fantasized situations, and so on—in other words, self-projection thoughts. Conversely, during attention-demanding tasks, both spontaneous thoughts and default-network activity decrease.

In addition, brain regions that lie along the midline of the prefrontal cortex and belong to both the self-projection and default networks have been shown to be selectively active in situations where you're required to decide whether something describes you or belongs to you, such as whether a body part you see belongs to your body, whether a word is your name, or whether an adjective such as "anxious" or "hopeful" describes your personality.[61] Psychologists call this kind of cognition "self-related processing," because you have to evaluate or judge something in relation to how you perceive or think about yourself.

During meditation, self-related thoughts and emotions arise spontaneously, especially when your mind is restless or drowsy. You run through memories, fantasies, and plans—reliving yesterday's conversation, daydreaming about the other things you could have said, and planning what you'll say next time. Eventually you come to notice that your mind has been wandering in this way. At that moment, you have the opportunity to disengage from identifying with the contents of these thoughts—specifically, from identifying with the mentally imagined "I" who is the central character—and to shift your attention to the thoughts simply as thoughts, mental events that come and go within or against the background of a larger field of awareness. You can begin to notice and investigate how the arising of thoughts is keyed to pleasant and unpleasant feelings in the body, as well as changes in the breath, and how such changes condition the arising of self-related thoughts. With repeated experience of this dynamic—getting lost in self-related thinking or self-projection and waking up to where you are, what your mind is doing, and how your body feels as this is happening—the frequency of spontaneously arising thoughts seems to lessen and you seem to notice them more quickly when they do arise. You also feel the difference between identifying with the content of a self-related thought—with the "I" as you mentally represent it—and identifying that a thought is occurring while experiencing the larger field of awareness in which it arises.

Given that self-projection and self-related cognition depend on specific neural networks, including parts of the default network, it stands to reason that changes to self-experience through meditation should be linked to changes in these neural networks. This connection between meditation and self-related networks in the brain is the last topic we need to examine before coming back to our title question about whether the self is an illusion.

FROM MIND WANDERING TO MINDFULNESS

According to a recent study published in *Science* by Harvard University psychologists Matthew Killingsworth and Daniel Gilbert, almost half our waking thoughts have little relation to what we're currently doing.[62] Although in general it's clearly useful to be able to think about things that aren't present here and now, and although mind wandering in particular can facilitate creative problem solving, it is also linked to negative emotions and unhappiness.[63] As psychologist Jonathan Smallwood and his colleagues have shown, negative moods lead the mind to wander.[64] As Killingsworth and Gilbert discovered, people are less happy when their minds are wandering than when they're focusing on what they're doing.[65] Furthermore,

although people are more likely to mind wander to pleasant topics than to unpleasant or neutral ones, people are no happier when
thinking about pleasant topics than when they focus on the task at
hand, and they're less happy when they mind wander to neutral topics than when they focus on their current activity. As Killingsworth
and Gilbert conclude, "a human mind is a wandering mind, and a
wandering mind is an unhappy mind. The ability to think about what
is not happening is a cognitive achievement that comes at an emotional cost."[66]

Mind wandering is known to be associated with the brain's default
network—the areas that show increased activity during task-free,
resting periods and whose activity decreases during outer-directed,
attention-demanding tasks.[67] Given this association, it's not surprising that focused attention and open awareness forms of meditation,
which involve stabilizing awareness while developing meta-awareness of ongoing mental activities, affect the brain's default network.

In one study, Wendy Hasenkamp and Lawrence Barsalou of Emory
University used focused attention meditation to model the dynamics
between mind wandering and attention—specifically, how individuals become aware of mind wandering and shift their attention so they
can remain focused on a task—as well as to determine the neural
correlates of these mental processes.[68] According to their model, the
dynamics of mind wandering and attention has a cyclic structure.
When you try to sustain attention on an object, such as the breath,
you inevitably experience mind wandering. At some time during
the mind wandering, you become aware that your attention isn't on
the object, at which point you disengage from the current train of
thought and shift your attention back to the object. You stay focused
again for some period of time, which then leads to another phase of
mind wandering, and so on.

Hasenkamp and Barsalou asked fourteen meditation practitioners
experienced in a variety of styles of Buddhist meditation to perform twenty minutes of focused-attention breath meditation while
undergoing fMRI scanning. They were instructed to press a button
whenever they realized their mind had wandered, and then to return
their attention to the breath. The button presses served as temporal

markers for abstracting four periods from the cyclic dynamic of mind wandering and focused attention—mind wandering, awareness of mind wandering, shifting attention back to the breath, and maintaining the focus of attention on the breath. The researchers hypothesized that default network regions would be active during mind wandering periods, whereas distinct regions belonging to the task-directed attentional network would be selectively active during the awareness versus shifting versus focusing periods.

The fMRI brain scans supported this prediction. During mind wandering, brain regions belonging to the default network were active. Another set of regions—the dorsolateral prefrontal cortex and lateral inferior parietal cortex—was active during the periods of shifting attention to the breath and maintaining attention on the breath. These regions belong to the so-called "executive network," which helps to reorient and direct attention while you maintain a goal. A third set of regions—the anterior insula and dorsal anterior cingulate cortex—was active during the period of the awareness of mind wandering before the period of shifting attention back to the breath. These regions belong to the so-called "salience network," which supports the present-moment feeling of the whole body from within as well as the detection of salient events. Finally, individuals with more meditation experience showed lower levels of neural activity in the executive network regions during the period of shifting attention than did individuals with less meditation experience. Given that the performance of well-learned activities, including meditation, requires less neural activity than does the performance of unfamiliar or less well-practiced activities, this finding suggests that experienced meditators require less neural activity from executive regions in order to disengage from mind wandering.[69]

In another study, Judson Brewer and his colleagues at Yale University compared brain activity in experienced meditators with brain activity in novice meditators as they practiced three different types of Theravada Buddhist meditation—concentration (focused attention), loving-kindness, and choiceless awareness (open awareness).[70] Compared to the novices, the experienced meditators reported less mind wandering during the meditation periods.

In addition, the main nodes of the default network—the posterior cingulate cortex and the medial prefrontal cortex—were less active in the experienced meditators across all three types of meditation. Finally, brain regions known to be involved in the monitoring and control of mental processes—notably, the dorsal anterior cingulate cortex—were more active in the experienced meditators than in the novice meditators when default network regions were active. This finding supports the hypothesis that during meditation practice, experienced meditators coactivate different brain regions when their default network is activated, compared to nonmeditators. The finding also suggests that the mental processes the default network supports—mind wandering and self-related processing—may be more accessible to monitoring and control in experienced meditators than in novice meditators. As Brewer and his colleagues speculate, this kind of accessibility of mind wandering to cognitive control may eventually be able to bring about a new default mode of brain activity, not only during meditation but also during passive resting states, marked by less uncontrolled mind wandering and less negative mood states.

One other study is especially relevant to our concern with the self. Norman Farb and Adam Anderson of the University of Toronto investigated the relationship between mindfulness practice and the neural systems underlying two different modes of self-experience—present-moment awareness of the body versus the narrative or autobiographical sense of self.[71] Farb and Anderson were particularly interested in the difference between nonjudgmental awareness of one's embodied being in the here and now and evaluating something in relation to one's mental concept of oneself.

The problem is that these two kinds of self-experience are typically mixed together from moment to moment. How you feel in your body in the present moment triggers some self-related thought, which leads to mind wandering, which in turn leads to changes in how you feel in your body. Given this rapid fluctuation and mutual influence between present-moment body awareness and narrative thinking about the self, we need some way to disentangle them in order to investigate their neural basis.

Farb and Anderson relied on individuals who had taken an eight-week course in Mindfulness-Based Stress Reduction (MBSR). MBSR incorporates elements from focused attention and open awareness meditation, as well as meditative body practices such as hatha yoga.[72] Farb and Anderson compared individuals who had taken an MBSR course with those who hadn't taken the course while the individuals adopted either a nonjudgmental "experiential focus" or an evaluative "narrative focus."

To appreciate the difference between these two types of focus, imagine someone praises you for something you've just done, saying how kind and generous you are. Or imagine they criticize you, accusing you of being cowardly and selfish. Faced with words like these—"kind," "generous," "selfish," "cowardly"—we habitually react by focusing on what the word means to us and either welcoming or rejecting that meaning as a description of ourselves. Later we may daydream or ruminate about what was said, so that our mind wanders from what we're doing in the present moment and gets caught up in thoughts about a certain positive or negative image of ourselves. In this way, we focus on our imagined self in the past or future, mentally traveling backward to recall what we did and what was said, imagining other ways things could have gone, or projecting forward to anticipate or fantasize about what we'll do and what may be said about us afterward. In other words, we mentally spin a story or narrative about our self, in which we attribute certain qualities to our self while withholding others.

Compare this kind of "narrative focus" with the "experiential focus" of being in the present moment without these kinds of mental judgments. Here we dispassionately observe our thoughts, emotions, and body feelings as they happen from moment to moment, trying not to get lost in them and all the ways we can mentally elaborate them, but also without trying to suppress or actively inhibit them. When we notice our minds wandering, we calmly return our attention to our present experience without berating ourselves for losing focus. In this way, our attention becomes progressively anchored to our bodily being in the here and now rather than immersed in the narrative self we mentally elaborate.

Farb and Anderson used fMRI to map the brain activities associated with the experiential focus versus the narrative focus. Participants in the experiment adopted one or the other kind of focus while reading words describing positive or negative personality traits (such as "cheerful," "resentful," and "fearless"). In the narrative focus, they tried to decide whether the word described their personalities, a task that triggers narrative thinking about the self and rumination. In the experiential focus, they nonjudgmentally noticed their reactions to the words and calmly returned their attention to the present moment whenever they realized they had become distracted.

The individuals with no MBSR training showed little change in brain activity from task to task. Their brains showed activity mostly in the medial prefrontal cortex, a region that belongs to the default network and supports evaluation of things in relation to one's mental representation of oneself (self-related processing). The individuals with MBSR training, however, showed significant changes in brain activity when they shifted from the narrative focus to the experiential focus: significantly reduced activity in the medial prefrontal cortex during the experiential focus and increased activity in a network of regions known to support internal awareness of the body and present-centered awareness (the right lateral prefrontal cortex, insula, and secondary somatosensory cortex).

These findings suggest that it's easier to disengage from narrative forms of self-identification when we have the kind of training in present-centered awareness that mindfulness practices provide. Although we need narrative thinking to understand ourselves as individuals with personal histories and plans for the future, and as members of traditions and communities, we can easily get stuck in worrisome rumination about our past and future selves, or become attached to some mental representation of ourselves. Individuals with mindfulness training seem better able to adopt an experiential focus and avoid getting stuck in the narrative focus. In other words, they seem able to move flexibly between narrative thinking about themselves and present-centered, embodied awareness, and imaging their brains accordingly brings to light the distinct neural systems supporting these two kinds of self-experience.

IS THE SELF AN ILLUSION?

I've been sketching a way of thinking about the self as enacted by self-specifying and self-designating processes. According to this way of thinking, the self isn't an independent thing or entity; it's a process. To end this chapter, I want to explain how this view is different from saying that there really is no self or that the appearance of there being a self is nothing but an illusion.

One of the most sophisticated no-self views comes from the Yogācāra school of Indian Buddhism (see chapter 2).[73] Its basic idea is that an individual stream of consciousness mentally represents itself as belonging to a self, but in reality no self exists, so the mental impression of there being a self is an illusion. Here "self" means an independent "I" or ego that's wholly present at each moment and owns the experiences that make up the stream of consciousness. The illusion of self comes about through a particular form of what I've called a self-designating process. One part of the mental stream designates another part of the stream as a self. But since no part of the stream or the stream as a whole is a self, the designation is mistaken. In this way, the result of the self-designating process is an error—the delusion that there's a self to which the mental stream belongs.

Let me explain in more detail how Yogācāra thinks this error is made. In the Yogācāra model of the workings of consciousness, an individual mental stream that's capable of conceiving of itself as a subject of experience does so by drawing on a subliminal repository of psychological information about itself while attending to its own mental states and preattentively experiencing itself as the conscious subject of this attentive activity.[74] For example, suppose the thought, "I feel anxious," arises in the mental stream. The thought comes about because there's an implicit and involuntary propensity to experience certain situations as anxiety producing and certain body sensations as anxiety, attention is drawn to the elicited feeling of anxiety, and there's a preattentive awareness of attending in this way. Yogācāra calls the subliminal repository of psychological propensities the "store consciousness" (ālayavijñāna), the attention to mental states "mental consciousness" (manovijñāna), and the preattentive awareness "mind" (*manas*). I'll call them the mental repository, the inner mental awareness, and the preattentive mind. Whereas the inner mental awareness takes mental states—thoughts, emotions, and so on—as objects of attention, the preattentive mind provides the feeling of being a conscious subject. Given the presence of this feeling, whenever the inner mental awareness attends to a mental state, that state is experienced not as floating freely but as belonging to the mental stream. So, in our example, when mental attention is given to the feeling of anxiety, the anxiety appears not as belonging to no one but as belonging to the same mental stream as does the mental attention to it. In other words, from the inside perspective of the mental stream, the anxiety appears as "mine." In this way, the preattentive mind functions as a base or support for the mental stream's ability to attend to its own states (via the inner mental awareness) and to be aware of them precisely as its own—to feel that they are "mine."

At the same time, the mental repository—the store consciousness—functions as a long-term support for the preattentive mind. The mental repository contains all the habits, tendencies, and propensities of an individual being. Traditionally, it's described as containing "seeds" or latent dispositions that eventually "sprout" or manifest in one's mental life, given the right conditions. In modern terms, it can be described as a kind of data bank or repository of information belonging to an individual stream of consciousness, a "first-person mental file."[75] It ensures not only that there's mental continuity across the gaps or breaks in ordinary consciousness but also that there's mental continuity across the gaps or breaks resulting from certain deep meditative states known as "cessations." In these states, any sense of being a conscious subject is said to disappear, and the body is sometimes said to enter a state of suspended animation.[76] Indeed, the concept of the store consciousness was probably first introduced in order to account for the mental continuity across the gaps that cessations produce in the stream of consciousness.[77] Although traditionally the store consciousness is said to carry on from one lifetime to the next, from a cognitive science perspective it makes sense to think of the concept of the store consciousness as pointing toward what today we know as the huge amount of cognitive and affective functioning going on in our bodies and brains beneath the surface and across the gaps of consciousness.[78]

We're now ready to see how the self-designating error enters the story. The preattentive mind mistakenly designates the mental repository as a self by mistakenly attributing to the mental repository the role of being an "I" or ego that's wholly present at each moment and that owns the mental stream.[79] In reality, however, the mental repository is a subliminal data bank, not an ego, and it's a constantly changing process, not a substantial thing. Hence the impression that there's a self is a mental fabrication, and what the fabrication represents doesn't exist. Since the preattentive mind is responsible for this error, the preattentive mind is also called the "afflicted mind" (*kliṣṭa-manas*). In its afflictive role, it doesn't function merely as a mode of preattentive self-awareness; it functions also as the delusion that a substantially real self underwrites this mode of awareness.

Consider again the thought, "I feel anxious." This thought has a present-tense "I-Me-Mine" form and seems to refer to an inner self that has the feeling and accounts for the fact that the feeling feels like "mine." "I-Me-Mine" thoughts can also take a self-projection form, such as "I am going to be so happy when I go on vacation," "I can see myself at that meeting and I know it's not going to go well," "I remember being a shy and anxious kid," and so on. Such thoughts seem to refer to one and the same inner self that existed in the past, exists now, and will exist in the future. According to Yogācāra, however, although the pronoun "I" in such thoughts seems to refer to or to designate a self, it doesn't refer at all, for there's no self to be its referent—no independent ego that was wholly present in the past, is wholly present now, and will be wholly present in the future. Such a self simply doesn't exist; what exists is only the mental representation of such a self superimposed on changing mental and physical states. It follows that "I-Me-Mine" thoughts are never literally true, for there's no self of which they could be true. Nevertheless, you habitually and involuntarily take yourself to be referring to your self when you think such thoughts. In this way, you're caught in the grip of a deep and fundamental error.

Although the enactive account of the self that I'm proposing is close in one way to the Yogācāra account, it also differs from it in another important way. Although I agree with Yogācāra that our sense of self or "I-Me-Mine" is mentally constructed, I don't think it follows that there is no self or that the appearance of the self is nothing but an illusion. Although some illusions are constructions, not all constructions are illusions. The self is a case in point. To say that the sense of self is a mental construction—or rather that it's a process under constant mental and bodily construction—doesn't logically imply that there is no self or that the sense of self presents an illusion.[80] As Jonardon Ganeri points out, the Yogācāra claim that the sense of self is an error doesn't logically follow from the Yogācāra model of the stream of consciousness as dependent on a mental repository and as including both a preattentive awareness of itself and a mental capacity to bring attention to bear on its own states.[81] Instead, we can take this model as contributing to an analysis of how the self—understood as a process and not a thing—comes to be constructed.

Part of the issue here is whether, as some Buddhist and Western philosophers claim, thinking of a stream of consciousness as "mine" is an error, or in other words, whether experiencing the stream of consciousness from within as being "mine" is a delusion. I want to explain now why I think there's a basic and natural sense of the "mineness" of experience that isn't a delusion.[82]

Ordinarily, when you're aware of a thought, emotion, perception, or sensation, you feel it as your own. For example, sitting in meditation, I suddenly realize I've been daydreaming about a planned yoga vacation in the Bahamas. I mentally note, "fantasizing," and return my attention to the breath. Yet even though I note the fantasizing in an impersonal way—thinking, "fantasizing occurring," instead of "I'm fantasizing"—there's a basic way in which the fantasizing feels mine. I don't mean in the way that the content of the fantasy feels mine when I inhabit the fantasy and identify myself with its main character. Nor do I mean in the way that it would feel were I to go on to identify with the fantasizing by thinking, "I always daydream when I try to meditate" or "I'm a great daydreamer but not a very good meditator." Self-related processing proliferates in these thoughts, and meditation practice involves learning to notice how and when this happens, and how to disengage from it. What I mean by saying that the fantasizing feels mine is that it shows up as an event in my field of awareness and nowhere else. So too does the witnessing and the mental noting of both the fantasizing and the subsequent self-evaluation. All these mental events happen in *this* field of awareness *here*—the one that feels mine.

This sense of mineness isn't a function of where attention happens to be focused, for it's more basic than selective attention. In particular, it can't be based on introspectively attending to mental states or experiences and identifying them as mine. In order to identify something as mine, I need to recognize some characteristic property the thing has, know that the property pertains to me, and know that I'm the one identifying this property. But how do I know these things? In particular, how do I know that the act of identifying the mental state is my act of identification? If we say I know this because I can in turn introspectively identify this act, then we're headed off on an infinite regress, because now I need to know that this second act of meta-identification is mine too.[83] For example, suppose there's some internal cognitive process that tags experiences with the self-referential label "mine" so that when I attend to an experience it feels like mine by virtue of my "reading" the label. This process will work only if I know that I'm the one reading the label. But if we say that the way I know that the reading of the label is my reading is by my reading another label that tells me the first reading is mine, then we're facing a vicious infinite regress, because now I need to know that this second reading of this second label is also my reading, and so on. Similarly, for a mental state to seem mine when I attend to it, I need to be aware that I'm the one attending in this way and must already experience this awareness as mine.

The upshot is that it can't be right to say that what makes a mental state or experience appear as mine is that I attend to it and identify it as mine on the basis of some characteristic property or label. Rather, there must be a more basic, preattentive, and nonidentifying way that I experience the mental stream as mine.[84] The preattentive mind in the Yogācāra account of consciousness plays precisely this role. For this reason, the preattentive mind can be described as a preattentive mode of self-awareness.[85]

Yogācāra also claims, however, that this preattentive self-awareness is afflicted insofar as it generates "I-Me-Mine" thoughts that mistakenly refer to a self. The preattentive mode of self-awareness is the awareness the mental stream has of itself, not the awareness of a self. This claim that "I-Me-Mine" thoughts are always necessarily mistaken takes the further step of assuming that the right way to interpret these thoughts is as essentially making reference to a substantial self. Neuro-nihilism today makes the same assumption that we intuitively but mistakenly experience ourselves as having or being a self-substance, and it assumes that the only way "I-Me-Mine" thoughts could be true is if there were a self-substance to serve as their referent.[86] But we needn't accept this assumption for two reasons.

First, the minimal notion of self that's crucial for "I-Me-Mine" thinking is that of a subject of experience and an agent of action, not that of a substantially existent ego. Thinking of myself this way— as a subject and agent—enables me to think of some experiences and actions as mine and not yours, and of some experiences and actions as yours and not mine. This provides a perfectly legitimate and valuable notion of self and doesn't require thinking of you or me as substantially existent entities. It also allows for the fact that we experience ourselves as changing beings who are neither the same as a stream of consciousness impersonally or anonymously conceived, nor different from the stream of consciousness in the way a substantial ego would have to be.

Of course, if there are experiences in which any preattentive sense of "mineness" completely disappears—as happens perhaps in certain meditative states—then these experiences can't properly be characterized as experientially "mine" or "yours," except perhaps retrospectively or prospectively. (One might be able to know retrospectively that the stream of consciousness that presently feels "mine" was for a time in such a state—maybe by remembering the outer edges of entering and exiting the state—or be able to know prospectively that the stream of consciousness that presently feels "mine" is going to enter such a state.) Such truly selfless states wouldn't be able to support even the minimal form of "I-Me-Mine" thinking just described (in which one thinks of oneself as a subject of experience but not as a substantial ego).

I see no reason to deny the possibility of such states, but I do see reason to deny that they show that there is no self or that "I-Me-Mine" thoughts are intrinsically mistaken. What they do show is that there can be experiences in which no sense of self is present. But the absence of a sense of self in some experiences doesn't logically imply that there is no self or that "I-Me-Mine" thoughts are intrinsically incorrect. On the contrary, if the self is a construction, then we should expect that it could be dismantled, even while some of its constituent processes—such as bare sentience or phenomenal consciousness—remain present. To put it another way, given that the self is a process and not a substantial thing, it may be possible to shut down this process under certain conditions and then start it up again.

Second, the pronoun "I" doesn't get its meaning by functioning in an ordinary referential way, that is, by indicating or denoting an object. Instead, the word functions in a performative way, that is, by performing an act or carrying out an activity. In thinking or saying "I," one carries out a self-individuating act—the act of making oneself stand out as the thinker or speaker. One also performs a self-appropriating act—the act of appropriating or making something one's own. To think or say "I am happy" or "I am nervous" is to lay claim to a feeling of being happy or being nervous that preattentively presents itself as mine and to which my attention has been turned.[87] "I-Me-Mine" statements are what philosophers and linguists call "performative utterances"—they perform an act or create a state of affairs by the fact of their being uttered under the appropriate circumstances (as when you say, "I promise I'll be there tomorrow night," and thereby perform the act of promising).

This way of thinking about the meaning of "I" comes from philosopher Jonardon Ganeri, who uses it to explain the views of the sixth-century C.E. Buddhist philosopher Candrakīrti. In Ganeri's words:

The utterance of "I" serves an appropriative function, to claim possession of, to take something as one's own. The appropriation in question is to be thought of as an *activity* of laying claim to, not the making of an *assertion* of ownership. . . . When I say "I am in pain," I do not *assert* ownership of a particular painful experience; rather, I *lay claim* to the experience within a stream. This is a performativist account of the language of self, in which "I" statements are performative utterances, and not assertions, and the function of the term "I" is *not* to refer.[88]

If the function of the term "I" isn't to refer, then searching for a referent for the word, especially in the form of a thing or entity or substance, is misguided. It's not the case that the job of the word "I" is to refer to a self and that the word fails because there is no self. Rather, the function of the term "I" is to enact a self. To think or say "I" is to engage in a self-individuating and self-appropriating form of I-making. One individuates oneself as a subject of experience and agent of action by laying claim to thoughts, emotions, and feelings—as well as commitments and social practices—and thereby enacts a self that is no different from the self-appropriating activity itself. Again, the self isn't an object or thing; it's a process—the process of "I-ing" or ongoing self-appropriating activity.[89] Ganeri's image is an enactive one—the self is a "whirlpool of self-appropriating action."[90]

Candrakīrti belonged to the philosophical school of Madhyamaka founded by Nāgārjuna, and later Tibetan philosophers regarded him as one of the principal exponents of the subschool Prāsaṅgika Madhyamaka. His view of the self was different from the Yogācāra view;[91] it is especially relevant to us today because it provides an important corrective to neuro-nihilism.

Candrakīrti presents his view in the form of a commentary on Nāgārjuna's chapter on the self—the text quoted at the beginning of this chapter—and he repeats Nāgārjuna's basic argument: if the self independently exists, then it's either really different from the mind-body aggregates or really the same as them, but since it's neither, it doesn't independently exist.

On the one hand, if the self were really different from the five aggregates—form (registering), feeling (appraising), perception/

cognition (stereotyping), inclination (readying), and consciousness (orienting or attending)—then it could be identified and described without reference to them, but it cannot be so identified and described. The self can be conceived of only in relation to the aggregates and as dependent on the aggregates.

On the other hand, if the self were really the same as the aggregates, then the self and the aggregates would be conceived to have all the same attributes. But the self is conceived of and is experienced as one thing, whereas the aggregates are plural.

Since the self can't be conceived of in either of these two ways—as the same as or different from the aggregates—it can't be a real entity, that is, an independently existent thing.

But here's the crucial point—Candrakīrti doesn't conclude that there is no self. This would be to succumb to the nihilist extreme, which says that since the self has no independent existence, it has no existence at all. Instead, Candrakīrti concludes that the self is dependently arisen. In other words, the self exists dependent on causes and conditions, including especially how we mentally construct it and name it in language.

Recall the Prāsaṅgika Madhyamaka idea that whatever is dependently arisen depends for its existence on a basis of designation, a designating cognition, and a term used to designate it. In the case of the self, the five aggregates are the basis of designation, the thought that projects "self" onto the aggregates is the designating cognition, and the pronoun "I" is the term used to designate it. Notice that this designation isn't an ordinary referential one; it's performative—it's how the mind-body aggregates self-individuate as "I" and self-appropriate as "me" and "mine."

Since the self arises as a mental projection onto the mind-body aggregates, it's not different from the mind-body aggregates in the sense of existing independently of them; it's dependent on them as its basis. Yet the self isn't the same as the mind-body aggregates, for it exists only in relation to the cognition that projects it, and what that cognition projects is the idea of a whole or unitary self, not an impersonal composite of mind-body processes. The self is like an image in a mirror. The image depends for its existence on the mirror—the mirror is the basis for the image—but the image isn't one and the same thing as the mirror, nor is it made of the same stuff as the mirror, for as an image it exists only in relation to an observer. Notice that the mirror image, though observer-dependent, isn't a subjective illusion. So too the self, though mind-dependent, isn't a subjective illusion.

Nevertheless, the way that the self appears does involve an illusion, even if it's not the case that there is no self or that the appearance of the self is nothing but an illusion. The illusion—or delusion—is taking the self to have an independent existence, like taking the mirror image to be really in the mirror. Notice that the image as such isn't an illusion; it's the taking of the image to exist in the mirror that's the illusion. Similarly, it's not the appearance of the self as such that's the illusion; it's taking the self to exist independently that's the illusion.[92]

Notice too that contrary to neuro-nihilism, the illusion isn't that the self appears to be a self-substance. That view of the self is theoretical and doesn't accurately describe our experience. The conception of the self as a substance isn't a cognitive illusion; it's a false belief that derives from philosophy (Descartes in the West and the Nyāya school in India), not from everyday experience. Neuro-nihilism mistakenly diagnoses our self-experience as being committed to a certain philosophical conception of the self and thereby overintellectualizes our experience.

Candrakīrti, however, says that the fundamental illusion is that we take the self to exist by virtue of itself or by virtue of its own being, when in reality its existence is dependent. The illusion is cognitive and existential.

Another important point is that undoing this illusion—through highly developed meditative concentration combined with acute analytical insight—doesn't mean destroying the appearance of the self as independent; it means not being taken in by the appearance and believing that the self is independent. This ignorant and deep-seated belief, not the appearance of the self as such, habitually deludes us into thinking, feeling, and acting as if the self were independent.[93]

In my view, the appearance of the self as independent is entirely natural. Whenever we perceive a mind-body stream, either from within or from without, the idea of self—a subject of experience and an agent of action—naturally arises. And when this happens, the self must appear as something distinct from the stream, because that's precisely what a self is—the dependently arisen and constructed appearance of an independent subject of experience and action.

What I take from this perspective—and here I state my own view and make no claim that any other Indian yogic philosopher would agree—is that "enlightenment" or "liberation"—at least in any sense I would want to affirm—doesn't consist in dismantling our constructed sense of self, as may happen in certain meditative states. Rather, it consists in wisdom that includes not being taken in by the appearance of the self as having independent existence while that appearance is nonetheless still there and performing its important I-making function. Nor does "enlightenment" or "liberation" consist in somehow abandoning all I-making or "I-ing"—all self-individuating and self-appropriating activity—though it does include knowing how to inhabit that activity without being taken in by the appearance of there being an independent self that's performing the activity and controlling what happens. We could say that the wisdom includes a kind of awakening—a waking up to the dream of independent existence without having to wake up from the dreaming.

CODA

When buddhas don't appear
And their followers are gone,
The wisdom of awakening
Bursts forth by itself.[94]

NOTES

PROLOGUE: THE DALAI LAMA'S CONJECTURE

1. The complete conference, including my keynote speech, can be seen on four video DVDs: *Mind and Life XI: Investigating the Mind: Exchanges Between Buddhism and the Biobehavioral Sciences on How the Mind Works*, available from the Mind and Life Institute (www.mindandlife.org). A complete transcription of the conference is also available as the book, *The Dalai Lama at MIT*, edited by Anne Harrington and Arthur Zajonc (Cambridge, MA: Harvard University Press, 2006).

2. See Evan Thompson, "Neurophenomenology and Francisco Varela," in *The Dalai Lama at MIT*, 19–24.

3. See Eugenio Rodriguez et al., "Perception's Shadow: Long-Distance Synchronization of Human Brain Activity," *Nature* 397 (1999): 430–433, and Francisco J. Varela et al., "The Brainweb: Phase Synchronization and Large-Scale Integration," *Nature Reviews Neuroscience* 2 (2001): 229–239.

4. See Francisco J. Varela, "Neurophenomenology: A Methodological Remedy for the Hard Problem," *Journal of Consciousness Studies* 3 (1996): 330–350. See also Antoine Lutz and Evan Thompson, "Neurophenomenology: Integrating Subjective Experience and Brain Dynamics in the Neuroscience of Consciousness," *Journal of Consciousness Studies* 10 (2003): 31–52.

5. See Antoine Lutz et al., "Guiding the Study of Brain Dynamics by Using First-Person Data: Synchrony Patterns Correlate with Ongoing Conscious States During a Simple Visual Task," *Proceedings of the National Academy of Sciences USA* 99 (2002): 1586–1591.

6. See Francisco Varela, Evan Thompson, and Eleanor Rosch, *The Embodied Mind: Cognitive Science and Human Experience* (Cambridge, MA: MIT Press, 1991; expanded edition, 2015), and Franciso J. Varela and Jonathan Shear, eds., *The View from Within: First-Person Approaches to the Study of Consciousness* (Thorverton: Imprint Academic, 1991).
7. Evan Thompson and Francisco J. Varela, "Radical Embodiment: Neural Dynamics and Consciousness," *Trends in Cognitive Sciences* 5 (2001): 418–425.
8. See *The Dalai Lama at MIT*, 95–96.
9. Dalai Lama, *The Universe in a Single Atom: The Convergence of Science and Spirituality* (New York: Morgan Road Books, 2005), 125.
10. See Franz Reichle's film, *Montegrande: What Is Life?*, available on DVD from www.montegrande.ch/eng/home.php and also the Mind and Life Institute (www.mindandlife.org).

INTRODUCTION

1. In this book, I use the terms "yogic traditions" and "yogic philosophies" in a broad sense that includes Buddhism. For justification of this usage, see Stephen Phillips, *Yoga, Karma, and Rebirth: A Brief History and Philosophy* (New York: Columbia University Press, 2009), 4–5.
2. Patrick Olivelle, *Upaniṣads* (Oxford: Oxford University Press, 1996); Valerie J. Roebuck, *The Upaniṣads* (London: Penguin, 2003).
3. See Antoine Lutz et al., "Attention Regulation and Monitoring in Meditation," *Trends in Cognitive Sciences* 12 (2008): 163–169.
4. See Antoine Lutz et al., "Meditation and the Neuroscience of Consciousness: An Introduction," in Philip David Zelazo et al., eds., *The Cambridge Handbook of Consciousness* (Cambridge: Cambridge University Press, 2007), 499–553. See also James H. Austin, *Zen and the Brain: Toward an Understanding of Meditation and Consciousness* (Cambridge, MA: MIT Press, 1999), and James H. Austin, *Selfless Insight: Zen and the Meditative Transformations of Consciousness* (Cambridge, MA: MIT Press, 2009).
5. See Pierre Hadot, *Philosophy as a Way of Life: Spiritual Exercises from Socrates to Foucault,* ed. and with an introduction by Arnold Davidson (Malden, MA: Blackwell Publishing, 1995). See especially "Part II: Spiritual Exercises."

1. SEEING: WHAT IS CONSCIOUSNESS?

1. I quote from the translation by Valerie J. Roebuck, *The Upaniṣads* (London: Penguin, 2003), but I have also consulted the translation by Patrick Olivelle, *Upaniṣads* (Oxford: Oxford University Press, 1996). The dialogue between

Yājñavalkya and King Janaka occurs in the *Bṛhadāraṇyaka Upaniṣad* ("The Great Forest Teaching"), 66–76 of the Roebuck translation, 58–68 of the Olivelle translation. My presentation of this dialogue is indebted to Ben-Ami Scharfstein, *A Comparative History of World Philosophy: From the Upanishads to Kant* (Albany, NY: State University of New York Press, 1998), 62–65, as well as two books by Bina Gupta: *Cit: Consciousness* (New Delhi: Oxford University Press, 2003), chapter 2, and *The Disinterested Witness: A Fragment of Advaita Vedānta Phenomenology* (Evanston, IL: Northwestern University Press, 1998), 18–27.

2. Roebuck, *The Upaniṣads*, 67.
3. Ibid., 68–69.
4. Ibid., 69.
5. Ibid., 69.10.
6. David J. Chalmers, "Facing Up to the Problem of Consciousness," *Journal of Consciousness Studies* 2 (1995): 200–219, at 203.
7. See Roebuck, *The Upaniṣads*, 345–348, and Olivelle, *Upaniṣads*, 288–290.
8. Roebuck, *The Upaniṣads*, 347.
9. See Andrew Fort, *The Self and Its States: A States of Consciousness Doctrine in Advaita Vedānta* (Delhi: Motilal Banarsidass, 1990).
10. Ibid.
11. See Fort, *The Self and Its States*, 39. For Śaṅkara's commentary on the *Māṇḍūkya Upaniṣads*, see Swami Gambhirananda, trans., *Eight Upaniṣads, Volume Two* (Kolkata, India: Advaita Ashrama, 1958), 167–405.
12. Roebuck, *The Upaniṣads*, 72.
13. Ibid.
14. Ibid., 73.
15. Ibid., 76.
16. See my *Colour Vision: A Study in Cognitive Science and the Philosophy of Perception* (London: Routledge, 1995).
17. See Arvind Sharma, *Sleep as a State of Consciousness in Advaita Vedānta* (Albany, NY: State University of New York Press, 2004); Gupta, *The Disinterested Witness*, 79–80, 84–90; Fort, *The Self and Its States*. See also Eliot Deutsch, *Advaita Vedānta: A Philosophical Reconstruction* (Honolulu: University Press of Hawaii, 1969), 61–65.
18. See Sharma, *Sleep as a State of Consciousness in Advaita Vedānta*, 45–48 and chapter 4; Fort, *The Self and Its States*.
19. See Mark Siderits et al., eds., *Self, No Self? Perspectives from Analytical, Phenomenological, and Indian Traditions* (Clarendon: Oxford University Press, 2010). See also Chakravarthi Ram-Prasad, *Indian Philosophy and the Consequences of Knowledge: Themes in Ethics, Metaphysics, and Soteriology* (Hampshire, England and Burlington, VT: Ashgate, 2007), chapter 2.

20. See Matthew D. Mackenzie, "The Illumination of Consciousness: Approaches to Self-Awareness in the Indian and Western Traditions," *Philosophy East and West* 57: 40–62.

21. William Irwin Thompson, *Coming Into Being: Artifacts and Texts in the Evolution of Consciousness* (New York: St. Martin's Press, 1996, 1998), 236.

2. WAKING: HOW DO WE PERCEIVE?

1. See *Bṛhadāraṇyaka Upaniṣad* I.4.7. Valerie J. Roebuck, *The Upaniṣads* (London: Penguin, 2003), 20; Patrick Olivelle, *Upaniṣads* (Oxford: Oxford University Press, 1996), 14.

2. See Richard Gombrich, *What the Buddha Thought* (London and Oakville, CT: Equinox, 2009).

3. Bhikkhu Ñāṇamoli and Bhikkhu Bodhi, *The Middle Length Discourses of the Buddha: A Translation of the Majjhima Nikāya* (Sommerville, MA: Wisdom, 1995), 350–351.

4. Bhikkhu Bodhi, *In the Buddha's Words: An Anthology of Discourses from the Pāli Canon* (Somerville, MA: Wisdom, 2005), 67–68.

5. Anālayo, "*Nāma-rūpa*," in A. Sharma, ed., *Encyclopedia of Indian Religions* (Berlin: Springer Science+Business Media, 2015).

6. Ibid.

7. Bhikkhu Bodhi, *The Connected Discourses of the Buddha: A New Translation of the Saṃyutta Nikāya* (Somerville, MA: Wisdom, 2002), 608–609.

8. Anālayo, "*Nāma-rūpa*."

9. See Christof Koch, *The Quest for Consciousness: A Neurobiological Approach* (Greenwood Village, CO: Roberts & Company, 2007).

10. See Nikos K. Logothetis, "Single Units and Conscious Vision," *Philosophical Transactions of the Royal Society of London B* 353 (1998): 1801–1818.

11. See Randolph Blake and Nikos K. Logothetis, "Visual Competition," *Nature Reviews Neuroscience* 3 (2002): 13–21.

12. Nikos K. Logothetis, "Vision: A Window on Consciousness," *Scientific American* 281 (1999): 68–75, at 74.

13. Diego Cosmelli et al., "Waves of Consciousness: Ongoing Cortical Patterns During Binocular Rivalry," *Neuroimage* 23 (2004): 128–140. See also Diego Cosmelli and Evan Thompson, "Mountains and Valleys: Binocular Rivalry and the Flow of Experience," *Consciousness and Cognition* 16 (2007): 623–641.

14. Eugenio Rodriguez et al., "Perception's Shadow: Long-Distance Synchronization of Human Brain Activity," *Nature* 397 (1999): 430–433.

15. Lucia Melloni et al., "Synchronization of Neural Activity Across Cortical Areas Correlates with Conscious Perception," *Journal of Neuroscience* 27 (2007): 2858–2865.

16. Satu Palva et al., "Early Neural Correlates of Conscious Somatosensory Perception," *Journal of Neuroscience* 25 (2005): 5248–5258; Raphaël Gaillard et al., "Converging Intracranial Markers of Conscious Access," *PloS Biology* 7 (3) (2009): e1000061. doi:10.1371/journal.pbio.1000061.

17. Sam M. Doesburg et al., "Rhythms of Consciousness: Binocular Rivalry Reveals Large-Scale Oscillatory Network Dynamics Mediating Visual Perception," *PLoS ONE* 4 (7) (2009): e6142. doi:10.1371/journal.pone.0006142.

18. William James, *The Principles of Psychology* (Cambridge, MA: Harvard University Press, 1982), 233.

19. Louis de la Vallée Poussin, "Notes sur le moment ou ksnana des bouddhistes," in H. S. Prasad, ed., *Essays on Time in Buddhism* (Delhi: Sri Satguru, 1991), trans. Georges Dreyfus, as quoted in Georges Dreyfus and Evan Thompson, "Asian Perspectives: Indian Theories of Mind," in Philip David Zelazo et al., eds., *The Cambridge Handbook of Consciousness* (New York and Cambridge: Cambridge University Press, 2007), 89–114, at 95.

20. For introductory treatments, see Dreyfus and Thompson, "Asian Perspectives: Indian Theories of Mind," and Rupert Gethin, *The Foundations of Buddhism* (Oxford and New York: Oxford University Press, 1998). For a philosophical treatment, see Mark Siderits, *Buddhism as Philosophy* (Indianapolis, IN: Hackett, 2007). For Vasubandhu's classic text, see *Abhidharmakośabhāṣyam*, by *Louis de la Vallée Poussin. Vols. 1–4*, trans. Leo M. Pruden (Berkeley, CA: Asian Humanities Press, 1991). For the Pāli Theravada Ahbidhamma, see Bhikkhu Bodhi, ed., *A Comprehensive Manual of Abhidhamma* (Onalaska, WA: Buddhist Publication Society, 1993, 1999).

21. See *Abhidharmakośabhāṣyam* I: 16, 74.

22. Geshe Rabten, *The Mind and Its Functions*, trans. Stephen Batchelor (Mt. Pelerin, Switzerland: Tharpa Choeling, 1981), 52.

23. Bhikkhu Bodhi, *A Comprehensive Manual of Abhidharma*, 78.

24. Bhikkhu Bodhi, *A Comprehensive Manual of Abhidhamma*, 81.

25. See J. Kevin O'Regan and Alva Noë, "A Sensorimotor Account of Vision and Visual Consciousness," *Behavioral and Brain Sciences* 24 (2001): 939–1031.

26. See Edward Conze, *Buddhist Thought in India* (Ann Arbor: University of Michigan Press, 1962), 134–137, 282; Gethin, *The Foundations of Buddhism*, 221–222; de la Vallée Poussin, "Notes sur le moment ou ksnana des bouddhistes."

27. *Abhidharmakośabhāṣyam* III: 85, II: 475.

28. Bhikkhu Bodhi, *A Comprehensive Manual of Abhidhamma*, 156.

29. Francisco J. Varela et al., "Perceptual Framing and Cortical Alpha Rhythm," *Neuropsychologia* 19 (1981): 675–686.

30. Francisco J. Varela, Evan Thompson, and Eleanor Rosch, *The Embodied Mind: Cognitive Science and Human Experience* (Cambridge, MA: MIT Press, 1991; expanded ed., 2014), 75.

31. Michel Gho and Francisco J. Varela, "A Quantitative Assessment of the Dependency of the Visual Temporal Frame Upon the Cortical Alpha Rhythm," *Journal of Physiology-Paris* 83 (1988–1989): 95–101. See also Ruffin Van Rullen and Christof Koch, "Is Perception Discrete or Continuous?" *Trends in Cognitive Sciences* 7 (2003): 207–213, and Ruffin Van Rullen et al., "Ongoing EEG Phase as a Trial-by-Trial Predictor of Perceptual and Attentional Variability," *Frontiers in Psychology* 2 (2011): 1–9.

32. See Van Rullen and Koch, "Is Perception Discrete or Continuous?", and Van Rullen, "Ongoing EEG Phase as a Trial-by-Trial Predictor of Perceptual and Attentional Variability."

33. Niko A. Busch et al., "The Phase of Ongoing EEG Oscillations Predicts Visual Perception," *Journal of Neuroscience* 29 (2009): 7869–7876; Kyle E. Mathewson et al., "To See or Not to See: Prestimulus α Phase Predicts Visual Awareness," *Journal of Neuroscience* 29 (2009): 2725–2732.

34. See Jin Fan et al., "The Relation of Brain Oscillations to Attentional Networks," *Journal of Neuroscience* 27 (2007): 6197–6206.

35. See Ruffin Van Rullen et al., "The Blinking Spotlight of Attention," *Proceedings of the National Academy of Sciences* U.S.A. 104 (2007): 19204–19209; Niko A. Busch and Ruffin Van Rullen, "Spontaneous EEG Oscillations Reveal Periodic Sampling of Visual Attention," *Proceedings of the National Academy of Sciences* 107 (2010): 16048–16053; Timothy J. Buschman and Earl K. Miller, "Shifting the Spotlight of Attention: Evidence for Discrete Computations in Cognition," *Frontiers in Human Neuroscience* 4 (2010): 1–9.

36. Busch and Van Rullen, "Spontaneous EEG Oscillations Reveal Periodic Sampling of Visual Attention."

37. See Antoine Lutz et al., "Guiding the Study of Brain Dynamics by Using First-Person Data: Synchrony Patterns Correlate with Ongoing Conscious States During a Simple Visual Task," *Proceedings of the National Academy of Sciences USA* 99 (2002): 1586–1591.

38. Remigiusz Szczepanowski and Luiz Pessoa, "Fear Perception: Can Objective and Subjective Awareness Measures Be Dissociated?" *Journal of Vision* 7 (2007): 1–17.

39. See Luiz Pessoa et al., "Target Visibility and Visual Awareness Modulate Amygdala Responses to Fearful Faces," *Cerebral Cortex* 16 (2006): 366–375; Szczepanowski and Pessoa, "Fear Perception: Can Objective and Subjective Awareness Measures Be Dissociated?"

40. Szczepanowski and Pessoa, "Fear Perception: Can Objective and Subjective Awareness Measures Be Dissociated?"

41. James, *Principles of Psychology*, 380–381.

42. Ibid., 949n31.

43. Heleen Slagter et al., "Mental Training Affects Distribution of Limited Brain Resources," *PLoS Biology* 5 (6) (2007): e138. doi:10.1371/journal.pbio.0050138.

44. See Antoine Lutz et al., "Attention Regulation and Monitoring in Meditation," *Trends in Cognitive Sciences* 12 (2008): 163–169.

45. Heleen Slagter et al., "Theta Phase Synchrony and Conscious Target Perception: Impact of Intensive Mental Training," *Journal of Cognitive Neuroscience* 21 (2009): 1536–1549.

46. Antoine Lutz et al., "Mental Training Enhances Attentional Stability: Neural and Behavioral Evidence," *Journal of Neuroscience* 29 (2009): 13418–13427.

47. In addition to the studies cited in notes 43, 45, and 46, notable studies include the following: Julie A. Brefczynski-Lewis et al., "Neural Correlates of Attentional Expertise in Long-Term Meditation Practitioners," *Proceedings of the National Academy of Sciences U.S.A.* 104 (2007): 11483–11488; Amishi P. Jha et al., "Mindfulness Training Modifies Subsystems of Attention," *Cognitive, Affective, and Behavioral Neuroscience* 7 (2007): 109–119; Katherine A. Maclean et al., "Intensive Meditation Training Improves Perceptual Discrimination and Sustained Attention," *Psychological Science* 21 (2010): 829–839; Yi-Yuan Tang et al., "Short-Term Meditation Training Improves Attention and Self-Regulation," *Proceedings of the National Academy of Sciences U.S.A.* 104 (2007): 17152–17156; Yi-Yuan Tang et al., "Central and Autonomic Nervous System Interaction Is Altered by Short-Term Meditation," *Proceedings of the National Academy of Sciences U.S.A.* 106 (2009): 8865–8870.

48. Olivia Carter et al., "Meditation Alters Perceptual Rivalry in Tibetan Buddhist Monks," *Current Biology* 15 (2005): R412–R413.

49. Ibid., R413.

50. Ibid.

51. Daniel C. Dennett, *Consciousness Explained* (Boston: Little, Brown, 1991), 356.

52. William S. Waldron, *The Buddhist Unconscious: The Ālayavijñāna in the Context of Indian Buddhist Thought* (London and New York: RoutledgeCurzon, 2003), 55–67.

53. Bhikkhu Bodhi, *A Comprehensive Manual of Abhidharma,* 123.

54. See Gethin, *The Foundations of Buddhism,* 215.

55. See Steven Colins, *Selfless Persons: Imagery and Thought in Theravāda Buddhism* (Cambridge: Cambridge University Press, 1982), 245, and Waldron, *The Buddhist Unconscious,* 82.

56. See Waldron, *The Buddhist Unconscious.*

57. The Vietnamese Zen teacher Thich Nhat Hanh uses the term "manifestation" instead of "impression," which is the usual English translation for the Sanskrit term *vijñapti*. The Yogācāra school is also known as the *Vijñaptimātra* or "Impressions Only" school. See Thich Nhat Hanh, *Understanding Our Mind* (Berkeley, CA: Parallax Press, 2006). Thich Nhat Hanh's use of "manifestation" is influenced by the Chinese development of the Yogācāra and serves also to align the Yogācāra with certain themes in Edmund Husserl's

Phenomenology (see 97). For more on the relationship between Yogācāra and Phenomenology, see Dan Lusthaus, *Buddhist Phenomenology: A Philosophical Investigation of Yogācāra Buddhism and the Ch'eng Wei-shih lun* (London and New York: RoutledgeCurzon, 2002).

58. See Georges Dreyfus, "Self and Subjectivity: A Middle Way Approach," in Mark Siderits et al., eds., *Self, No Self? Perspectives from Analytical, Phenomenological, and Indian Traditions* (Clarendon: Oxford University Press, 2010), 114–156.

59. See Tim Bayne, "Conscious States and Conscious Creatures: Explanation in the Scientific Study of Consciousness," *Philosophical Perspectives* 21 (2007): 1–22.

60. John Searle, "Consciousness," *Annual Review of Neuroscience* 23 (2000): 557–578.

61. Ibid., 572.

62. Ibid.

3. BEING: WHAT IS PURE AWARENESS?

1. Anne Harrington and Arthur Zajonc, eds., *The Dalai Lama at MIT* (Cambridge, MA: Harvard University Press, 2006), 95–96.

2. Dalai Lama, *The Universe in a Single Atom: The Convergence of Science and Spirituality* (New York: Morgan Road Books, 2005).

3. Ibid., 29–31.

4. See Jeremy W. Hayward and Francisco J. Varela, eds., *Gentle Bridges: Conversations with the Dalai Lama on the Sciences of Mind* (Boston: Shambhala Press, 1992).

5. See Daniel Goleman, *Destructive Emotions: A Scientific Dialogue with the Dalai Lama* (New York: Bantam, 2003).

6. Ibid., 305–333. See also Eugenio Rodriguez et al., "Perception's Shadow: Long-Distance Synchronization of Human Brain Activity," *Nature* 397 (1999): 430–433.

7. Antoine Lutz et al., "Long-Term Meditators Self-Induce High Amplitude Gamma Synchrony During Mental Practice," *Proceedings of the National Academy of Sciences* 101 (2004): 16369–16373.

8. See Francisco J. Varela et al., "The Brainweb: Phase Synchronization and Large-Scale Integration," *Nature Reviews Neuroscience* 2 (2001): 229–239.

9. See the reviews in Antoine Lutz et al., "Meditation and the Neuroscience of Consciousness: An Introduction," in Philip David Zelazo et al., eds., *The Cambridge Handbook of Consciousness* (Cambridge: Cambridge University Press, 2007) , 499–553, and Juergen Fell et al., "From Alpha to Gamma:

Electrophysiological Correlates of Meditation-Related States of Consciousness," *Medical Hypotheses* 75 (2010): 218–224. For gamma activity in Vipassanā meditation, see B. Rael Cahn et al., "Occipital Gamma Activation During Vipassana Meditation," *Cognitive Processing* 11 (2010): 39–56.

10. Antoine Lutz et al., "Changes in the Tonic High-Amplitude Gamma Oscillations During Meditation Correlate with Long-Term Practitioners' Verbal Reports," Association for the Scientific Study of Consciousness Annual Meeting, Poster Presentation, 2006.

11. Dalai Lama, *The Universe in a Single Atom*, 90.

12. Dalai Lama, "On the Luminosity of Being," *New Scientist* (May 24, 2003): 42.

13. See Georges Dreyfus, *The Sound of Two Hands Clapping: The Education of a Tibetan Buddhist Monk* (Berkeley: University of California Press, 2003). See also Georges Dreyfus, *Recognizing Reality: Dharmakīrti's Philosophy and Its Tibetan Interpretations* (Albany: State University of New York Press, 1997), and John D. Dunne, *Foundations of Dharmakīrti's Philosophy* (Somerville, MA: Wisdom, 2004).

14. See Dan Arnold, "Dharmakīrti's Dualism: Critical Reflections on a Buddhist Proof of Rebirth," *Philosophy Compass* 3 (2008): 1079–1096, and Richard Hayes, "Dharmakīrti on *Punarbhava*," in Egaku Mayeda, ed., *Studies in Original Buddhism and Mahayana Buddhism. Volume 1* (Kyoto: Nagata Bunshodo, 1993), 111–129.

15. See Dalai Lama, *The Universe in a Single Atom*, 147–148. For discussion of the Dalai Lama's version of this argument, see Owen Flanagan, *The Bodhisattva's Brain: Buddhism Naturalized* (Cambridge, MA: MIT Press, 2011), 85–86.

16. See Dalai Lama, *The Universe in a Single Atom*, 131–132. See also Dalai Lama, *Sleeping, Dreaming, and Dying: An Exploration of Consciousness with the Dalai Lama* (Boston: Wisdom, 1996), 119–120, and Hayward and Varela, eds., *Gentle Bridges,* 153–154.

17. Dalai Lama, *The Universe in a Single Atom*, 110.

18. Lati Rinpoche and Jeffrey Hopkins, *Death, Intermediate State and Rebirth in Tibetan Buddhism* (Ithaca, NY: Snow Lion, 1981), 42.

19. See Hayward and Varela, eds., *Gentle Bridges*, 159–162, and Dalai Lama, *Sleeping, Dreaming, and Dying*, 118–126, 164–171.

20. Dalai Lama, *Sleeping, Dreaming, and Dying*, 229.

21. Hayward and Varela, eds., *Gentle Bridges*, 161.

22. Dalai Lama, *Sleeping, Dreaming, and Dying*, 127–130.

23. For the "constructivist" position that interpretation shapes experience, see Robert H. Sharf, "Buddhist Modernism and the Rhetoric of Meditative Experience," *Numen* 42 (1995): 228–283, and Robert H. Sharf, "The Rhetoric of Experience and the Study of Religion," *Journal of Consciousness Studies* 7 (2000): 267–287. For the view that contemplative experience taps into

universal aspects of consciousness, see B. Alan Wallace, *The Taboo of Subjectivity: Toward a New Science of Consciousness* (New York: Oxford University Press, 2000).

24. For examinations of Buddhism's encounter with modernity, see Donald Lopez, Jr., *Buddhism and Science: A Guide for the Perplexed* (Chicago: University of Chicago Press, 2008), and David L. McMahan, *The Making of Buddhist Modernism* (New York: Oxford University Press, 2008).

25. See Lopez, *Buddhism and Science*, and David L. McMahan, *The Making of Buddhist Modernism*, especially chapter 4. See also José Ignacio Cabezón, "Buddhism and Science: On the Nature of the Dialogue," in B. Alan Wallace, ed., *Buddhism and Science: Breaking New Ground* (New York: Columbia University Press, 2003), 35–68. The idea that Buddhism offers a unique "mind science" is a strong theme in B. Alan Wallace's writings. See his *Contemplative Science: Where Buddhism and Neuroscience Converge* (New York: Columbia University Press, 2007).

26. Dalai Lama, *Dzogchen: The Heart Essence of the Great Perfection*, trans. Thupten Jinpa and Richard Barron (Chökyi Nyima) (Ithaca, NY: Snow Lion, 2000), 168.

27. Dalai Lama, *The Universe in a Single Atom*, 2–3.

28. Thupten Jinpa, "Science as an Ally or a Rival Philosophy? Tibetan Buddhist Thinkers' Enagement with Modern Science," in B. Alan Wallace, ed., *Buddhism and Science*, 71–85, at 77–78.

29. Dalai Lama, *Sleeping, Dreaming, and Dying*, 94.

30. David J. Chalmers, "On the Search for a Neural Correlate of Consciousnes," in Stuart R. Hameroff et al., eds., *Toward a Science of Consciousness II* (Cambridge, MA: MIT Press, 1998), 219–229, at 220.

31. See Adrian M. Owen and Martin R. Coleman, "Functional Neuroimaging of the Vegetative State," *Nature Reviews Neuroscience* 9 (2008): 235–243, and Damian Cruse et al., "Bedside Detection of Awareness in the Vegetative State: A Cohort Study," *The Lancet* 378 (2011): 2088–2094.

32. Ned Block and Cynthia MacDonald, "Consciousness and Cognitive Access," *Proceedings of the Aristotelian Society* CVIII, Part 3 (2008): 289–316. See also Ned Block, "Consciousness, Accessibility, and the Mesh Between Psychology and Neuroscience," *Behavioral and Brain Sciences* 30 (2007): 481–548.

33. See Evan Thompson, *Mind in Life: Biology, Phenomenology, and the Sciences of Mind* (Cambridge, MA: Harvard University Press, 2007), chapter 13.

34. Here and in what follows I am indebted to Michel Bitbol, "Is Consciousness Primary?" *NeuroQuantology* 6 (2008): 53–71, and to Michel Bitbol and Pier-Luigi Luisi, "Science and the Self-Referentiality of Consciousness," *Journal of Cosmology* 14 (2011): 207–223. See also Piet Hut and Roger Shepard, "Turning the 'Hard Problem' Upside Down and Sidways," *Journal of Consciousness Studies* 3 (1996): 313–329.

35. See Michel Bitbol, "Is Consciousness Primary?" This conception of science comes from Edmund Husserl, *The Crisis of European Sciences and Transcendental Phenomenology*, trans. David Carr (Evanston, IL: Northwestern University Press, 1970).

36. The term "neurophysicalism" comes from Owen Flanagan, *The Bodhisattva's Brain*. Other terms for this position are "reductive materialism," "mind-brain identity theory," and "psychoneural identity theory."

37. This idea has been a central theme of my work going back to my first book, coauthored with Francisco Varela and Eleanor Rosch, *The Embodied Mind: Cognitive Science and Human Experience* (Cambridge, MA: MIT Press, 1991; expanded edition, 2014). See also *Mind in Life*. For more recent discussions, see Diego Cosmelli and Evan Thompson, "Envatment Versus Embodiment: Reflections on the Bodily Basis of Consciousness," in John Stewart et al., eds., *Enaction: Towards a New Paradigm for Cognitive Science* (Cambridge, MA: MIT Press, 2010), 361–386, and Evan Thompson and Diego Cosmelli, "Brain in a Vat or Body in a World? Brainbound Versus Enactive Views of Experience," *Philosophical Topics* 39 (2011): 163–180.

38. Bhikkhu Bodhi, *The Connected Discourses of the Buddha: A New Translation of the Saṃyutta Nikāya* (Somerville, MA: Wisdom, 2002), 608–609.

39. For discussion of emergentism, see my book *Mind in Life*, chapter 3 and appendix B.

40. See Colin McGinn, *The Mysterious Flame: Conscious Minds in a Material World* (New York: Basic Books, 2000).

41. Galen Strawson, "Realistic Monism: Why Physicalism Entails Panpsychism," *Journal of Consciousness Studies* 13 (2006): 3–31, revised and reprinted in Galen Strawson, *Real Materialism and Other Essays* (Oxford: Clarendon Press, 2008), 53–74.

42. See Hut and Shepard, "Turning the 'Hard Problem' Upside Down and Sideways."

43. Francisco J. Varela, "Neurophenomenology: A Methodological Remedy for the Hard Problem," *Journal of Consciousness Studies* 3 (1996): 330–349.

4. DREAMING: WHO AM I?

1. Jean-Paul Sartre, *The Imaginary: A Phenomenological Psychology of the Imagination*, trans. Jonathan Webber (London: Routledge, 2004), 37–49.

2. The conference, "Mind and Life XII: Neuroplasticity: The Neuronal Substrates of Learning and Transformation," took place October 18–22, 2004. It is the subject of Sharon Begley's book, *Train Your Mind, Change Your Brain* (New York: Ballantine, 2007).

3. For more on neuroplasticity, see Norman Doidge, *The Brain That Changes Itself: Stories of Personal Triumph from the Frontiers of Brain Science* (New York: Penguin, 2007).

4. For more on downward causation, see Nancey Murphy et al., eds., *Downward Causation and the Neurobiology of Free Will* (Berlin: Springer Verlag, 2009).

5. See Shanti Ganesh et al., "How the Human Brain Goes Virtual: Distinct Cortical Regions of the Person-Processing Network Are Involved in Self-Identification with Virtual Agents," *Cerebral Cortex* 22 (2012): 1577–1585. See also Jayne Gackenbach and Matthew Rosie, "Presence in Video Game Play and Nighttime Dreams: An Empirical Inquiry," *International Journal of Dream Research* 4 (2011): 98–109.

6. Alfred Maury, "Des hallucinations hypnagogiques ou des erreurs des sens dans l'etat intermédiaire entre la veille et le sommeil" [Hypnagogic hallucinations or sensory errors in the intermediate state between wakefulness and sleep], *Annales Medico-Psychologiques du système nerveux* 11 (1848): 26–40.

7. Frederick H.M. Myers, *Human Personality and Its Survival of Bodily Death* (London: Longmans, Green, 1903).

8. Robert Frost, *The Poetry of Robert Frost: The Collected Poems, Complete and Unabridged* (New York: Henry Holt, 1969), 68.

9. Andrew Marvell, "The Garden": In M. H. Abrams et al., eds., *The Norton Anthology of English Literature, Volume I*, 6th ed. (New York and London: Norton, 1993), 1428–1429.

10. For an excellent discussion of the hypnagogic state, see Jeff Warren, *Head Trip: Adventures on the Wheel of Consciousness* (Toronto: Random House Canada, 2007), Chapter 1.

11. Marcel Proust, *The Way by Swann's*, trans. Lydia Davis (London: Penguin, 2003), 7.

12. Ibid.

13. Johannes Müller's description of the hypnagogic state appears in his book *Ueber die phantastischen Gesichtserscheinungen. Eine physiologische Untersuchung* [On the fantastic phenomena of vision], as quoted by Jiri Wackermann et al., "Brain Electrical Activity and Subjective Experience During Altered States of Consciousness: Ganzfeld and Hypnagogic States," *International Journal of Psychophysiology* 46 (2002): 123–146, at 123–124.

14. Robert Stickgold et al., "Replaying the Game: Hypnagogic Images in Normals and Amnesics," *Science* 290 (2000): 350–353.

15. See also Phileppe Stenstrom et al., "Mentation During Sleep Onset Theta Bursts in a Trained Participant: A Role for NREM Stage 1 Sleep in Memory Processing?" *International Journal of Dream Research* 5 (2012): 37–46.

16. For Nielsen's self-observation method, see Tore A. Nielsen, "Describing and Modeling Hypnagogic Imagery Using a Systematic Self-Observation

Procedure," *Dreaming* 5 (1995): 75–94. For his experiments measuring electrical brain activity associated with hypnagogic imagery, see Anne Germain and Tore A. Nielsen, "EEG Power Associated with Early Sleep Onset Images Differing in Sensory Content," *Sleep Research Online* 4 (2001): 83–90.

17. See Charles Tart, ed., *Altered States of Consciousness* (New York: Anchor Books, Doubleday, 1972), 75.

18. For descriptions and instructions, see Rubin R. Naiman, *Healing Night: The Science and Spirit of Sleeping, Dreaming, and Awakening* (Minneapolis: Syren Book Company, 2006).

19. These examples are quoted in Andreas Mavromatis, *Hypnagogia: The Unique State of Consciousness Between Wakefulness and Sleep* (London and New York: Routledge and Kegan Paul, 1987), 97. For Herbert's original paper, see "Report on a Method of Eliciting and Observing Certain Symbolic Hallucination Phenomena," reprinted in D. Rapaport, ed., *Organization and Pathology of Thought* (New York: Columbia University Press, 1951). For discussion, see Daniel L. Schacter, "The Hypnagogic State: A Critical Review of the Literature," *Psychological Bulletin* 83 (1976): 452–481.

20. See David W. Foulkes, "Dream Reports from Different Stages of Sleep," *Journal of Abnormal and Social Psychology* 65 (1962): 14–25, and David W. Foulkes and Gerald Vogel, "Mental Activity at Sleep Onset," *Journal of Abnormal Psychology* 70 (1965): 231–243.

21. For accessible overviews of sleep science, see J. Allan Hobson, *Dreaming: An Introduction to the Science of Sleep* (Oxford: Oxford University Press, 2002) and Jim Horne, *Sleepfaring: A Journey Through the Science of Sleep* (Oxford: Oxford University Press, 2006).

22. For an accessible overview of this controversy, see Andrea Rock, *The Mind at Night: The New Science of How and Why We Dream* (New York: Basic Books, 2004), chapter 3. For the relevant scientific review papers, see J. Allan Hobson et al., "Dreaming and the Brain: Toward a Cognitive Neuroscience of Conscious States," *Behavioral and Brain Sciences* 23 (2000); 793–842; Mark Solms, "Dreaming and REM Sleep Are Controlled by Different Brain Mechanisms," *Behavioral and Brain Sciences* 23 (2000): 843–850; and Tore A. Nielsen, "A Review of Mentation in REM and NREM Sleep: 'Covert' REM Sleep as a Possible Reconciliation of Two Opposing Models," *Behavioral and Brain Sciences* 23 (2000): 851–866. See also the discussion in J. F. Pagel, *The Limits of Dream: A Scientific Exploration of the Mind-Brain Interface* (Oxford: Elsevier/Academic Press, 2008).

23. See Yuval Nir and Guilio Tononi, "Dreaming and the Brain: From Phenomenology to Neurophysiology," *Trends in Cognitive Sciences* 14 (2009): 88–100, at 95.

24. See Mavromatis, *Hypnagogia,* 79.

25. See Tadao Hori et al., "Topographical EEG Changes and the Hypnagogic Experience," in *Anonymous Sleep Onset: Normal and Abnormal Processes* (Washington, D.C.: American Psychological Association, 1994), 237–253, and Hideki Tanaka et al., "Topographical Characteristics and Principal Component Structure of the Hypnagogic EEG," *Sleep* 20 (1997): 523–534.

26. Germain and Nielsen, "EEG Power Associated with Early Sleep Onset Images Differing in Sensory Content."

27. See Mavromatis, *Hypnagogia*, chapter 5, and James Austin, *Zen and the Brain* (Cambridge, MA: MIT Press, 1999), 379–386.

28. See Antoine Lutz et al., "Meditation and the Neuroscience of Consciousness: An Introduction," in Philip David Zelazo et al., eds., *The Cambridge Handbook of Consciousness* (Cambridge: Cambridge University Press, 2007), 499–553.

29. Austin, *Zen and the Brain*, 383.

30. Vladimir Nabokov, *Speak, Memory: An Autobiography Revisted* (New York: Vintage International, 1989), 33–34.

31. Austin, *Zen and the Brain*, 43–47.

32. See Jennifer M. Windt, "The Immersive Spatiotemporal Hallucination Model of Dreaming," *Phenomenology and the Cognitive Sciences* 9 (2010): 295–316.

33. Tomas Tranströmer, "Dream Seminar," in *Selected Poems 1954–1986* (Hopewell, NJ: Ecco Press, 1987), 171–172.

34. As quoted in Mavromatis, *Hypnagogia*, 193.

35. See Gerald Vogel et al., "Ego Functions and Dreaming During Sleep Onset," in Tart, ed., *Altered States of Consciousness*, 77–94. For critical discussion, see Schacter, "The Hypnagogic State: A Critical Review of the Literature," 476–477.

36. Mavromatis, *Hypnagogia*, 12, 68–71, 168–172, 267–270.

37. Ibid., 12.

38. See Joseph Goldstein, *The Experience of Insight* (Boston: Shambhala, 1976), 105–106.

39. Austin, *Zen and the Brain*, 373–376, and Philip Kapleau Roshi, *The Three Pillars of Zen* (New York: Anchor Books, Doubleday, 1989), 41–44.

40. Sartre, *The Imaginary*, 43.

41. I owe this phrasing to Robert Sokolowski, *An Introduction to Phenomenology* (New York: Cambridge University Press, 2000), 67.

42. Sartre, *The Imaginary*, 44.

43. In Abrams et al., eds., *The Norton Anthology of English Literature,* 1357.

44. See Windt, "The Immersive Spatiotemporal Hallucination Model of Dreaming."

45. See David Foulkes, *Children's Dreaming and the Development of Consciousness* (Cambridge, MA: Harvard University Press, 2002).

46. Sartre, *The Imaginary*, 165–166.

47. Homer, *The Iliad*, trans. Robert Fitzgerald (London: Everyman's Library, 1992), 521.

48. See Daniel Schacter, *Searching for Memory: The Brain, the Mind, and the Past* (New York: Basic Books, 1996), 18–22.

49. William Irwin Thompson, *Nightwatch and Dayshift* (Princeton, NJ: Wild River Review Books, 2014).

50. See Evan Thompson, *Mind in Life: Biology, Phenomenology, and the Sciences of Mind* (Cambridge, MA: Harvard University Press, 2007), chapter 13.

51. See Patrick McNamara et al., "'Theory of Mind' in REM and NREM Dreams," in Patrick McNamara and Deirdre Barrett, eds., *The New Science of Dreaming, Volume II: Content, Recall, and Personality Correlates* (Westport, CT: Praeger, 2007), 201–220.

52. See Georgia Nigro and Ulric Neisser, "Point of View in Personal Memories," *Cognitive Psychology* 15 (1983): 467–482; Heather K. McIsaac and Eric Eich, "Vantage Point in Episodic Memory," *Psychological Science* 9 (2002): 146–150.

53. See Eric Eich et al., "Neural Systems Mediating Field and Observer Memories," *Neuropsychologia* 47 (2009): 2239–2251.

54. See Jennifer Michelle Windt and Thomas Metzinger, "The Philosophy of Dreaming and Self-Consciousness: What Happens to the Experiential Subject During the Dream State?" in Patrick McNamara and Deirdre Barrett, eds., *The New Science of Dreaming, Volume III: Cultural and Theoretical Perspectives on Dreaming* (Westport, CT: Praeger, 2007), 193–247, at 205. Sartre makes a similar point in *The Imaginary*, 160.

55. In Jorge Luis Borges, *Everything and Nothing* (New York: New Directions, 1989), 79–81.

56. Dorothée Legrand, "Pre-reflective Self-as-Subject from Experiential and Empirical Perspectives," *Consciousness and Cognition* 16 (2007): 583–599.

57. Windt and Metzinger, "The Philosophy of Dreaming and Self-Consciousness."

58. See Thien Thanh Dang-Vu et al., "Neuroimaging of REM Sleep and Dreaming," in Patrick McNamara and Deirdre Barrett, eds., *The New Science of Dreaming, Volume I: The Biology of Dreaming* (Westport, CT: Praeger, 2007), 95–113; Pierre Maquet et al., "Human Cognition During REM Sleep and the Cortical Activity Profile Within Frontal and Parietal Cortices: A Reappraisal of Functional Neuroimaging Data," *Progress in Brain Research* 150 (2005): 219–227; and Sophie Schwartz and Pierre Maquet, "Sleep Imaging and the Neuropsychological Assessment of Dreams," *Trends in Cognitive Sciences* 6 (2002): 23–30.

59. See Amir Muzur et al., "The Prefrontal Cortex in Sleep," *Trends in Cognitive Sciences* 6 (2002): 475–481, and Edward F. Pace-Schott, "The Frontal Lobes and Dreaming," in Patrick McNamara and Deirdre Barrett, eds., *The*

New Science of Dreaming, Volume I: The Biology of Dreaming (Westport, CT: Praeger, 2007), 115–154.

5. WITNESSING: IS THIS A DREAM?

1. Here my view differs from that of lucid dream researcher Stephen LaBerge. In his book with Howard Rheingold, *Exploring the World of Lucid Dreaming* (New York: Ballantine, 1980), he writes (31): "The person, or dream ego, that we experience being in the dream is the same as our waking consciousness. It constantly influences the events of the dream through its expectations and biases, just as it does in waking life. The essential difference in the lucid dream is that the ego is aware that the experience is a dream." I agree that the person influences the events of the dream through its expectations and biases, but I think there is an important conceptual and phenomenological distinction to draw between the ego within the dream and the self who is the dreamer.

2. See Tadas Stumbrys et al., "Induction of Lucid Dreams: A Systematic Review of the Evidence," *Consciousness and Cognition* 21 (2012): 1456–1475.

3. For a translation of the original text of the first edition, see Sigmund Freud, *The Interpretation of Dreams*, trans. Joyce Crick (New York: Oxford University Press, 1999). For a translation that includes Freud's many subsequent additions, see Sigmund Freud, *The Interpretation of Dreams,* trans. and ed. James Strachey (New York: Basic Books, 1955).

4. See Freud, *The Interpretation of Dreams,* James Strachey translation, 571.

5. Frederik Van Eeden, "A Study of Dreams," reprinted in Charles Tart, ed., *Altered States of Consciousness* (New York: Anchor Books, 1969), 147–160; also available at the homepage of the Lucidity Institute: http://www.lucidity.com/vanEeden.html. For discussion of Freud's meeting and correspondence with Frederik van Eeden, see Bob Rooksby and Sybe Terwee, "Freud, Van Eeden and Lucid Dreaming," http://www.spiritwatch.ca/LL%209(2)%20web/Rooksby_Terwee%20paper.htm.

6. Freud, *The Interpretation of Dreams,* James Strachey translation, 353, 493–494, 570–571. The following quotations are from 353.

7. Marcel Proust, *The Way by Swann's,* trans. Lydia Davis (London: Penguin, 2003), 31.

8. Freud, *The Interpretation of Dreams*, James Strachey translation, 253.

9. Ibid., 570.

10. Friedrich Nietzsche, *The Birth of Tragedy and Other Writings,* ed. Raymond Geuss and Ronald Speirs (Cambridge: Cambridge University Press, 1999), 15–16.

11. Ibid.

12. Freud, *The Interpretation of Dreams*, James Strachey translation, 493–494.

13. Nietzsche, *The Birth of Tragedy*, 25.

14. See Freud, *The Interpretation of Dreams*, James Strachey translation, 571.

15. Marquis d'Hervey de Saint-Denys, *Dreams and How to Guide Them* (London: Duckworth, 1982).

16. Freud, *The Interpretation of Dreams*, James Strachey translation, 571.

17. Daniel C. Dennett, "The Onus Re Experiences: A Reply to Emmett," *Philosophical Studies* 35 (1979): 315–318; the quotation appears on 316. This article is a reply to Kathleen Emmett, "Oneiric Experiences," *Philosophical Studies* 34 (1978): 445–450, which is a critical discussion of Dennett's "Are Dreams Experiences?" reprinted in his *Brainstorms: Philosophical Essays on Mind and Psychology* (Cambridge, MA: MIT Press/A Bradford Book, 1981), 129–148.

18. The passage from *Dreams and How to Guide Them* (London: Duckworth, 1982) is quoted from Andreas Mavromatis, *Hypnagogia: The Unique State of Consciousness Between Wakefulness and Sleep* (London and New York: Routledge and Kegan Paul, 1987), 91–92.

19. LaBerge and Rheingold, *Exploring the World of Lucid Dreaming*, 99. See also Paul Tholey, "Techniques for Inducing and Manipulating Lucid Dreams," *Perceptual and Motor Skills* 57 (1983): 79–90.

20. See LaBerge and Rheingold, *Exploring the World of Lucid Dreaming*, 96–98.

21. Venerable Gyatrul Rinpoche, *Meditation, Transformation, and Dream Yoga*, trans. B. Alan Wallace and Sangye Khandro (Ithaca, NY: Snow Lion, 1993 and 2002), 109. This passage is Gyatrul Rinpoche's commentary on a seventeenth-century text; see 81.

22. Mavromatis, *Hypnagogia*, 106. See also the excellent discussion of the differences between lucid dreaming and dreaming you're dreaming in Janice E. Brooks and Jay A. Vogelsong, *The Conscious Exploration of Dreaming: Discovering How We Create and Control Our Dreams* (Bloomington, IN: The International Online Library, 2000), 25–35.

23. Oliver Fox, *Astral Projection* (New Hyde Park, NY: University Books, 1962), 32–33, as quoted by LaBerge and Rheingold, *Exploring the World of Lucid Dreaming*, 40–41.

24. "Dreamtigers," in Jorge Luis Borges, *Selected Poems*, ed. Alexander Coleman (New York: Penguin, 1999), 75.

25. Jennifer Michelle Windt and Thomas Metzinger, "The Philosophy of Dreaming and Self-Consciousness: What Happens to the Experiential Subject During the Dream State?" in Patrick McNamara and Deirdre Barrett, eds., *The New Science of Dreaming, Volume III: Cultural and Theoretical Perspectives on Dreaming* (Westport, CT: Praeger, 2007), 193–247, at 212–213.

26. William C. Dement and Nathaniel Kleitman, "The Relation of Eye Movements During Sleep to Dream Activity: An Objective Method for the Study of Dreaming," *Journal of Experimental Psychology* 53 (1957): 89–97.

27. Keith M.T. Hearne, *Lucid Dreams: An Electrophysiological and Psychological Study,* Ph.D. diss., University of Liverpool, 1978; Stephen LaBerge et al., "Lucid Dreaming Verified by Volitional Communication During REM Sleep," *Perceptual and Motor Skills* 52 (1981): 727–731; Stephen LaBerge et al., "Psychophysiological Correlates of the Initiation of Lucid Dreaming," *Sleep Research* 10 (1981): 149. For description of these studies and subsequent research, see Stephen LaBerge, "Lucid Dreaming," in Patrick McNamara and Deirdre Barrett, eds., *The New Science of Dreaming, Volume II: Content, Recall, and Personality Correlates of Dreams* (Westport, CT: Praeger, 2007), 307–328.

28. LaBerge, "Lucid Dreaming."

29. Ibid.

30. Daniel Erlacher and Michael Schredl, "Time Required for Motor Activities in Lucid Dreams," *Perceptual and Motor Skills* 99 (2004): 1239–1242.

31. Martin Dresler et al., "Dreamed Movement Elicits Activation in the Sensorimotor Cortex," *Current Biology* 21 (2011): 1–5.

32. LaBerge, "Lucid Dreaming."

33. This unpublished study by J. Strelen is described in Daniel Erlacher and Michael Schredl, "Do REM (Lucid) Dreamed and Executed Actions Share the Same Neural Substrate?" *International Journal of Dream Research* 1 (2008): 7–14.

34. Daniel Erlacher and Heather Chapin, "Lucid Dreaming: Neural Virtual Reality as a Mechanism for Performance Enhancement," *International Journal of Dream Research* 3 (2010): 7–10, at 9.

35. Ursula Voss et al., "Lucid Dreaming: A State of Consciousness with Features of Both Waking Consciousness and Non-Lucid Dreaming," *Sleep* 32 (2009): 1191–1200.

36. See J. Allan Hobson, *Consciousness* (New York: Scientific American Library, 1999), 22, and J. Allan Hobson et al., "Dreaming and the Brain: Toward a Cognitive Neuroscience of Conscious States," *Behavioral and Brain Sciences* 23 (2000); 793–842, at 834.

37. Martin Dresler et al., "Neural Correlates of Dream Lucidity Obtained from Contrasting Lucid Versus Non-lucid REM Sleep: A Combined EEG/fMRI Study," *Sleep* 35 (2012): 1017–1020.

38. Justin L. Vincent et al., "Evidence for a Frontoparietal Control System Revealed by Intrinsic Connectivity," *Journal of Neurophysiology* 100 (2008): 3328–3342.

39. Van Eeden, "A Study of Dreams," 149–150.

40. See Voss et al., "Lucid Dreaming." See also Allan Hobson, "The Neurobiology of Consciousness: Lucid Dreaming Wakes Up," *International Journal of Dream*

Research 2 (2009): 41–44, and J. Allan Hobson, "REM Sleep and Dreaming: Towards a Theory of Protoconsciousness," *Nature Review Neuroscience* 10 (2009): 803–813.

41. See Aristotle, *"De Somniis* (On Dreams)," in Richard McKeon, ed., *The Basic Works of Aristotle* (New York: Random House, 1941), 618–625, at 624.

42. Stephen LaBerge, "Signal-Verified Lucid Dreaming Proves That REM Sleep Can Support Reflective Consciousness," *International Journal of Dream Research* 3 (2010): 26–27.

43. Philip M. Bromberg, *Awakening the Dreamer: Clinical Journeys* (Mahwah, NJ: The Analytic Press, 2006), 2.

44. Ibid., 40–41.

45. LaBerge and Rheingold, *Exploring the World of Lucid Dreaming*, 1–2.

46. See Celia Green and Charles McCreery, *Lucid Dreaming: The Paradox of Consciousness During Sleep* (London: Routledge Press, 1994), 30–31.

47. See Venerable Gyatrul Rinpoche, *Meditation, Transformation, and Dream Yoga.* See also Chogyal Namkhai Norbu, *Dream Yoga and the Practice of the Natural Light* (Ithaca, NY: Snow Lion, 1992); Dzogchen Ponlop, *Mind Beyond Death* (Ithaca, NY: Snow Lion, 2006); Tenzin Wangyal Rinpoche, *The Tibetan Yogas of Dream and Sleep* (Ithaca, NY: Snow Lion, 1998).

48. See Stephen LaBerge, "Lucid Dreaming and the Yoga of the Dream State: A Psychophysiological Perspective," in B. Alan Wallace, ed., *Buddhism and Science: Breaking New Ground* (New York: Columbia University Press, 2003), 233–258.

6. IMAGINING: ARE WE REAL?

1. Venerable Gyatrul Rinpoche, *Meditation, Transformation, and Dream Yoga,* trans. B. Alan Wallace and Sangye Khandro (Ithaca, NY: Snow Lion, 1993 and 2002), 77.

2. See Stephen LaBerge, "Lucid Dreaming and the Yoga of the Dream State: A Psychophysiological Perspective," in B. Alan Wallace, ed., *Buddhism and Science: Breaking New Ground* (New York: Columbia University Press, 2003), 233–258.

3. Lynne Levitan and Stephen LaBerge, "Testing the Limits of Dream Control: The Light and Mirror Experiment," *Nightlight* 5 (2) (Summer 1993), http://www.lucidity.com/NL52.LightandMirror.html.

4. See Jan Westerhoff, *Twelve Examples of Illusion* (New York: Oxford University Press, 2010), 69–82.

5. See B. Alan Wallace and Brian Hodel, *Dreaming Yourself Awake: Lucid Dreaming and Tibetan Dream Yoga for Insight and Transformation* (Boston: Shambhala, 2012). See also B. Alan Wallace, *Genuine Happiness: Meditation as a Path to Fulfillment* (Hoboken, NJ: Wiley, 2005), 183–195.

6. For Wallace's view on the nature of the substrate consciousness, including the differences between the Yogācāra and Dzogchen understandings, see B. Alan Wallace, *Contemplative Science: Where Buddhism and Neuroscience Converge* (New York: Columbia University Press, 2007), 14–24.

7. See Evan Thompson, *Mind in Life: Biology, Phenomenology, and the Sciences* of Mind (Cambridge, MA: Harvard University Press, 2007), 417–441; Nancey Murphy et al., eds., *Downward Causation and the Neurobiology of Free Will* (Berlin: Springer, 2009).

8. H. Spitzer et al., "Increased Attention Enhances Both Behavioral and Neuronal Performance," *Science* 15 (1988): 338–340.

9. See R. Christopher deCharms et al., "Control Over Brain Activation and Pain Learned by Using Real-Time Functional MRI," *Proceedings of the National Academy of Sciences USA* 102 (2005): 18626–18631.

10. See Stephen LaBerge and Howard Rheingold, *Exploring the World of Lucid Dreaming* (New York: Ballantine, 1990), 146–147, and Paul Tholey, "Techniques for Inducing and Manipulating Lucid Dreams," *Perceptual and Motor Skills* 57 (1983): 79–90.

11. LaBerge and Rheingold, *Exploring the World of Lucid Dreaming*, 140–144.

12. See Daniel Erlacher and Michael Schredl, "Do REM (Lucid) Dreamed and Executed Actions Share the Same Neural Substrate?" *International Journal of Dream Research* 1 (2008): 7–14, and Daniel Erlacher and Michael Schredl, "Cardiovascular Responses to Dreamed Physical Exercise During REM Lucid Dreaming," *Dreaming* 18 (2008): 112–121.

13. Stephen LaBerge, "Lucid Dreaming," in Patrick McNamara and Deirdre Barrett, eds., *The New Science of Dreaming, Volume II: Content, Recall, and Personality Correlates of Dreams* (Westport, CT: Praeger, 2007), 307–328, at 323.

14. See J. Allan Hobson et al., "Dreaming and the Brain: Toward a Cognitive Neuroscience of Conscious States," *Behavioral and Brain Sciences* 23 (2000): 793–842. For the original formulation of the activation-synthesis model, see J. Allan Hobson and Robert McCarley, "The Brain as a Dream State Generator: An Activation-Synthesis Hypothesis of the Dream Process," *American Journal of Psychiatry* 134 (1977): 1335–1348. For the role of the prefrontal cortex, see Amir Muzur et al., "The Prefrontal Cortex in Sleep," *Trends in Cognitive Sciences* 6 (2002): 475–481.

15. See J. Allan Hobson, *13 Dreams Freud Never Had: The New Mind Science* (New York: Pi Press, 2005).

16. See also Owen Flanagan, *Dreaming Souls: Sleep, Dreams, and the Evolution of the Conscious Mind* (New York: Oxford University Press, 2000).

17. J. Allan Hobson, *Dreaming: An Introduction to the Science of Sleep* (Oxford: Oxford University Press, 2002), 64.

18. For an overview, see Andrea Rock, *The Mind at Night* (New York: Basic Books, 2004), 1–60. For the scientific debates, see the commentaries published together with the article by Hobson et al., "Dreaming and the Brain." See also Mark Solms and Oliver Turnbull, *The Brain and the Inner World: An Introduction to the Neuroscience of Subjective Experience* (New York: Other Press, 2002), chapter 6, and J. F. Pagel, *The Limits of Dream: A Scientific Exploration of the Mind/Brain Interface* (Oxford: Elsevier/Academic Press, 2008).

19. Chögyal Namkhai Norbu, *Dream Yoga and the Practice of the Natural Light* (Ithaca, NY: Snow Lion, 1992), 47.

20. See Thomas Metzinger, *The Ego Tunnel: The Science of the Mind and the Myth of the Self* (New York: Basic Books, 2009), 156–157.

21. Mark Solms, "Dreaming and REM Sleep Are Controlled by Different Brain Mechanisms," *Behavioral and Brain Sciences* 23 (2000): 843–850, and Solms and Turnbull, *The Brain and the Inner World*.

22. Solms, "Dreaming and REM Sleep Are Controlled by Different Brain Mechanisms," 849, note 1.

23. Hobson et al., "Dreaming and the Brain," 799.

24. I am using the term "imagination" in a wide sense that allows perception, imagination, hallucination, and dreaming to form a continuum. See Nigel J.T. Thomas, "The Multidimensional Spectrum of Imagination: Images, Dreams, Hallucinations, and Active, Imaginative Perception," http://www.imagery-imagination.com/Spectrum.pdf. The philosophical source for the more restricted imagination conception of dreaming is Jean-Paul Sartre, *The Imaginary: A Phenomenological Psychology of the Imagination,* trans. Jonathan Webber (London and New York: Routledge, 2004), 159–175. For recent philosophical discussions that develop the imagination conception, see Colin McGinn, *Mindsight: Image, Dream, Meaning* (Cambridge, MA: Harvard University Press, 2004); Ernest Sosa, "Dreams and Philosophy," *Proceedings and Addresses of the American Philosophical Association* 79 (2) (2005): 7–18; and Jonathan Ichikawa, "Dreaming and Imagination," *Mind and Language* 24 (2009): 103–121. My conception of dreaming as imagination draws from these works but also departs from them in various ways and includes aspects of Jennifer Windt's recent model of dreaming as "immersive spatiotemporal hallucination." See Jennifer M. Windt, "The Immersive Spatiotemporal Hallucination Model of Dreaming," *Phenomenology and the Cognitive Sciences* 9 (2010): 295–316.

25. Hobson, *Dreaming*, 107–108.

26. See McGinn, *Mindsight*, 26–29, 92.

27. Ibid., 92.

28. The evidence I cite comes from Solms and Turnbull, *The Brain and the Inner World*, 208–211, and Yuval Nir and Guilio Tononi, "Dreaming and the Brain:

From Phenomenology to Neurophysiology," *Trends in Cognitive Sciences* 14 (2009): 88–100, at 96–97.

29. See David Foulkes, *Children's Dreaming and the Development of Consciousness* (Cambridge, MA: Harvard University Press, 1999).

30. See Erlacher and Schredl, "Do REM (Lucid) Dreamed and Executed Actions Share the Same Neural Substrate?"

31. Martin Dresler et al., "Dreamed Movement Elicits Activation in the Sensorimotor Cortex," *Current Biology* 21 (2011): 1–5.

32. See Daniel L. Schacter et al., "Remembering the Past to Imagine the Future: The Prospective Brain," *Nature Reviews Neuroscience* 8 (2007): 657–661.

33. See Rodolfo R. Llinás and Urs Ribary, "Perception as an Oneiric-Like State Modulated by the Senses," in Christof Koch and Joel L. Davis, eds., *Large-Scale Neuronal Theories of the Brain* (Cambridge, MA: MIT Press, 1994), 111–124. See also Rodolfo Llinás and Urs Ribary, "Coherent 40-Hz Oscillation Characterizes Dream State in Humans," *Proceedings of the National Academy of Sciences* USA 90 (1993): 2078–2081.

34. See also Antti Revonsuo, *Inner Presence: Consciousness as a Biological Phenomenon* (Cambridge, MA: MIT Press, 2006) and Metzinger, *The Ego Tunnel*.

35. Rodolfo Llinás, *I of the Vortex: From Neurons to Self* (Cambridge, MA: MIT Press, 2002), 94.

36. Thomas Metzinger, *Being No One: The Self-Model Theory of Subjectivity* (Cambridge, MA: MIT Press/A Bradford Book, 2003), 52.

37. See his *Meditations on First Philosophy* in *The Philosophical Writings of Descartes, Volume 2*, trans. John Cottingham, Robert Stoothoff, and Dugald Murdoch (Cambridge: Cambridge University Press, 1985), 13.

38. See Flanagan, *Dreaming Souls*, 163–174, and Sartre, *The Imaginary*, 159–162.

39. Jennifer Windt and Thomas Metzinger, "The Philosophy of Dreaming and Self-Consciousness: What Happens to the Experiential Subject During the Dream State?" in Patrick McNamara and Deirdre Barrett, eds., *The New Science of Dreaming, Volume III: Cultural and Theoretical Perspectives on Dreaming* (Westport, CT: Praeger, 2007), 193–247, at 232–237.

40. Frederik van Eeden, "On the Study of Dreams," in Charles Tart, ed., *Altered States of Consciousness* (New York: Anchor, 1969), 147–160, at 155.

41. Windt and Metzinger, "The Philosophy of Dreaming and Self-Consciousness," 235.

42. Wendy Doniger O'Flaherty, *Dreams, Illusion and Other Realities* (Chicago: University of Chicago Press, 1984), 37–52, 175–205.

43. Sukasah Syadan's poem "Dream" can be found at his blog "Tjipoetat Quill," http://tjipoetatquill.blogspot.com/search?q=dream.

44. For selections of the original texts by Śaṅkara and his predecessor Gauḍapāda, see Eliot Deutsch and Rohit Dalvi, eds., *The Essential Vedānta: A New Source*

Book of Advaita Vedānta (Bloomington, IN: World Wisdom, 2004); for a philosophical treatment, see Eliot Deutsch, *Advaita Vedānta: A Philosophical Reconstruction* (Honolulu: University Press of Hawaii, 1969). See also Andrew Fort, "Dreaming in Advaita Vedānta," *Philosophy East and West* 35 (1985): 377–386.

45. See Chakravarthi Ram Prasad, "Dreams and Reality: The Śankarite Critique of Vijñānavāda," *Philosophy East and West* 43 (1993): 405–455.

46. This translation of the "Butterfly Dream" parable is Hans-Georg Moeller's from his excellent book, *Daoism Explained: From the Dream of the Butterfly to the Fishnet Allegory* (Chicago and LaSalle, IL: Open Court, 2004), 48. My reading of this parable closely follows Moeller's. For a complete translation of Zhuang Zi, along with selections from numerous Chinese commentators, see *Zhuangzi: The Essential Writings*, trans. Brook Ziporyn (Indianapolis, IN: Hackett, 2009).

47. Moeller, *Daoism Explained*, 48.

48. *Zhuangzi*, 163.

49. Ibid.

50. Some translations avoid the question by having Zhou wake up and not know whether *he* has been dreaming *he* was a butterfly, or whether *he's* now a butterfly dreaming *he's* Zhou. These translations imply the waking Zhou remembers the dream, but this reading goes against Zhou's and the butterfly's mutual ignorance, which is central to the parable, at least on Guo Xiang's traditional and highly influential interpretation. Here I follow Moeller's reading in *Daoism Explained*.

51. See *Zhuangzi*, 162. See also Moeller, *Daoism Explained*.

7. FLOATING: WHERE AM I?

1. See Olaf Blanke and Christine Mohr, "Out-of-Body Experience, Heautoscopy, and Autoscopic Hallucination of Neurological Origin: Implications for Neurocognitive Mechanisms of Corporeal Awareness and Self-Consciousness," *Brain Research Reviews* 50 (2005): 184–199.

2. See Michael N. Marsh, *Out-of-Body and Near-Death Experiences: Brain-State Phenomena or Glimpses of Immortality?* (Oxford: Oxford University Press, 2010), chapters 6 and 7.

3. Susan J. Blackmore, *Beyond the Body: An Investigation of Out-of-the-Body Experiences* (Chicago: Academy Chicago, 1982, 1992), 5.

4. Ibid., 1–5.

5. For the importance of body state and posture, see Blackmore, *Beyond the Body*, 61–63, and for methods of inducing out-of-body experiences, see chapter 10.

6. Olaf Blanke and Shahar Arzy, "The Out-of-Body Experience: Disturbed Self-Processing at the Temporo-Parietal Junction," *The Neuroscientist* 11 (2005): 16–24, at 18.

7. Ibid., 22.

8. Thomas Metzinger, "Out-of-Body Experiences as the Origin of the Concept of a 'Soul,'" *Mind & Matter* 3 (2005): 57–84, at 68.

9. Here I follow Dorothée Legrand, "Myself with No Body? Body, Bodily-Consciousness, and Self-Consciousness," in Daniel Schmicking and Shaun Gallagher, eds., *Handbook of Phenomenology and Cognitive Science* (New York, Heidelberg, London: Springer, 2010), 181–200.

10. Metzinger, "Out-of-Body Experiences as the Origin of the Concept of a 'Soul,'" 70.

11. See Olaf Blanke, "Multisensory Brain Mechanisms of Bodily Self-Consciousness," *Nature Reviews Neuroscience* 13 (2012): 556–571.

12. Olaf Blanke et al., "Stimulating Illusory Own-Body Perceptions," *Nature* 419 (2002): 269–270.

13. Olaf Blanke et al., "Out-of-Body Experience and Autoscopy of Neurological Origin," *Brain* 127 (2004): 243–258, at 247.

14. Olaf Blanke et al., "Linking Out-of-Body Experience and Self-Processing to Mental Own-Body Imagery at the Temporoparietal Junction," *Journal of Neuroscience* 25 (2005): 550–557.

15. Shahar Arzy et al., "Neural Basis of Embodiment: Distinct Contributions of Temporoparietal Junction and Extrastriate Body Area," *Journal of Neuroscience* 26 (2006): 8074–8081.

16. Blanke and Arzy, "The Out-of-Body Experience."

17. Bigna Lenggenhager et al., "Video Ergo Sum: Manipulating Bodily Self-Consciousness," *Science* 317 (2007): 1096–1099; H. Henrik Ehrsson, "The Experimental Induction of Out-of-Body Experiences," *Science* 317 (2007): 1048.

18. Bigna Lenggenhager et al., "Spatial Aspects of Bodily Self-Consciousness," *Consciousness and Cognition* 18 (2009): 110–117.

19. Silvio Ionta et al., "Multisensory Mechanisms in Temporo-Parietal Cortex Support Self-Location and First-Person Perspective," *Neuron* 70 (2011): 363–374.

20. See Valeria I. Petkova and H. Henrik Ehrsson, "If I Were You: Perceptual Illusion of Body Swapping," *PLoS ONE* 3 (12) (2008): e3832. doi:10.1371/journal.pone.0003832; and Mel Slater et al., "First Person Experience of Body Transfer in Virtual Reality," *PLoS ONE* 5(5) (2010): e10564. doi:10.1371/journal.pone.0010564

21. The classic text is Sylvan Muldoon and Hereward Carrington, *The Projection of the Astral Body* (New York: Samuel Weiser, 1969). This work was originally published in 1929. See also Robert Monroe, *Journeys Out of the Body* (New

York: Doubleday, 1971). For discussion of both these works, see Blackmore, *Beyond the Body*.

22. Pim van Lommel, *Consciousness Beyond Life: The Science of Near-Death Experience* (New York: Harper One, 2010), 121–122.

23. Thomas Metzinger, "Why Are Out-of-Body Experiences Interesting for Philosophers? The Theoretical Relevance of OBE Research," *Cortex* 45 (2009): 256–258, at 257.

24. See Blackmore, *Beyond the Body*, chapter 8.

25. Thomas Metzinger, *The Ego Tunnel: The Science of the Mind and the Myth of the Soul* (New York: Basic Books, 2009), 82–89.

26. See Blackmore, *Beyond the Body*.

27. Metzinger, "Out-of-Body Experiences as the Origin of the Concept of a 'Soul,'" 78.

28. Gregory Bateson, *Steps to an Ecology of Mind* (New York: Ballantine, 1972).

29. For an autobiographical description of this research with references to the original publications, see Charles Tart, *The End of Materialism: How Evidence of the Paranormal Is Bringing Science and Spirit Together* (Oakland, CA: New Harbinger Publications, 2009), chapter 12. Blackmore discusses this research in *Beyond the Body*, chapter 18.

30. See Blackmore, *Beyond the Body*, 41–42.

31. See Michael Sabom, *Light and Death* (Grand Rapids, MI: Zondervan, 1998), 37–51.

32. See Marsh, *Out-of-Body and Near-Death Experiences*, 19–27.

33. See Tart, *The End of Materialism*, 215–217, and Blackmore, *Beyond the Body*, 41–42.

34. Blackmore, *Beyond the Body*, chapters 16 and 22.

35. Metzinger, *The Ego Tunnel*, 87–89.

36. See Elizabeth Lloyd Mayer, *Extraordinary Knowing: Science, Skepticism, and the Inexplicable Powers of the Human Mind* (New York: Bantam, 2007).

37. Metzinger, "Out-of-Body Experiences as the Origin of the Concept of a 'Soul,'" 78.

38. Tart, *The End of Materialism*, 199.

39. Daniel Smilek et al., "When '3' is a Jerk and 'E' is a King: Personifying Inanimate Objects in Synesthesia," *Journal of Cognitive Neuroscience* 19 (2007): 981–992.

40. Frederik Van Eeden, "A Study of Dreams," reprinted in Charles Tart, ed., *Altered States of Consciousness* (New York: Anchor Books, 1969), 147–160, at 153. Also available at the homepage of the Lucidity Institute: http://www.lucidity.com/vanEeden.html.

41. Tart, *The End of Materialism*, chapter 12.

42. Lynne Levitan et al., "Out-of-Body Experiences, Dreams, and REM Sleep," *Sleep and Hypnosis* 1 (1999): 186–196.

43. Metzinger, "Out-of-Body Experiences as the Origin of the Concept of a 'Soul,'" 68.

8. SLEEPING: ARE WE CONSCIOUS IN DEEP SLEEP?

1. Edmund Husserl, *On the Phenomenology of the Consciousness of Internal Time (1893–1917)*, trans. John Brough (Dordrecht: Kluwer Academic Publishers, 1991). For an accessible introduction to Husserl's phenomenology of time consciousness, see Shaun Gallagher and Dan Zahavi, *The Phenomenological Mind*, 2nd ed. (London and New York: Routledge, 2012), chapter 4.
2. Marcel Proust, *The Way by Swann's*, trans. Lydia Davis (London: Penguin, 2003), 9.
3. Tomas Tranströmer, "The Name," in *Selected Poems 1954–1986* (Hopewell, NJ: Ecco Press, 1987), 93.
4. Ibid.
5. Daniel Heller-Roazen, *The Inner Touch: Archaeology of a Sensation* (New York: Zone Books, 2007), 76.
6. Ibid.
7. Jane Hirshfield, "Moment," in *Given Sugar, Given Salt* (New York: Harper Perennial, 2002), 59.
8. Ibid.
9. See Ramesh Kumar Sharma, "Dreamless Sleep and Some Related Philosophical Issues," *Philosophy East and West* 51 (2001): 210–231.
10. See Daniel Raveh, "*Ayam aham asmīti*: Self-Consciousness and Identity in the Eighth Chapter of the *Chāndogya Upaniṣad* vs Śankara's *Bhāṣya*," *Journal of Indian Philosophy* 36 (2008): 319–333.
11. For a translation of the *Yoga Sūtras* with Vyāsa's commentary, see Sāṃkhya-yogāchāra Swāmi Hariharānanda Āraṇya, *Yoga Philosophy of Patañjali*, rendered into English by P. N. Mukerji (Albany: State University of New York Press, 1983). Other useful translations can be found in Pandit Usharbudh Arya, *Yoga-Sūtras of Patañjali with the Exposition of Vyāsa. Volume I: Samādhi-pāda.* (Honesdale, PA: The Himalayan International Institute, 1989); Edwin F. Bryant, *The Yoga Sūtras of Patañjali* (New York: North Point Press, 2009); Christopher Key Chapple, *Yoga and the Luminous: Patañjali's Spiritual Path to Freedom* (Albany: State University of New York Press, 2008); B.K.S. Iyengar, *Light on the Yoga Sūtras of Patañjali* (London: Thorsons, 1996); and Stephen Phillips, *Yoga, Karma, and Rebirth: A Brief History and Philosophy* (New York: Columbia University Press, 2009).
12. See Arya, *Yoga-Sūtras of Patañjali with the Exposition of Vyāsa*, 178–184; Bryant, *The Yoga Sūtras of Patañjali*, 41–43; and Iyengar, *Light on the Yoga Sūtras of Patañjali*, 59–60.

13. Sāṃkhya-yogāchāra Swāmi Hariharānanda Āraṇya, *Yoga Philosophy of Patañjali*, 30. See also Arya, *Yoga-Sūtras of Patañjali with the Exposition of Vyāsa*, 178.

14. Bryant, *The Yoga Sūtras of Patañjali*, 42.

15. Rāmānuja, *The Vedānta-Sūtras with the Commentary by Rāmānuja. Sacred Books of the East, Volume 48*, trans. George Thibaut, 1904. Available online at http://www.sacred-texts.com/hin/sbe48/index.htm. The quotation comes from the section, "The conscious subject persists in deep sleep." For discussion of the differences between the Advaita and Viśiṣṭādvaita conceptions of deep sleep, see Michael Comans, "The Self in Deep Sleep According to Advaita and Viśiṣṭādvaita," *Journal of Indian Philosophy* 18 (1990): 1–28.

16. My account of the Nyāya position and the Advaita Vedānta rebuttal relies heavily on two books by Bina Gupta, *Perceiving in Advaita Vedānta: An Epistemological Analysis and Interpretation* (Calcutta: Motilal Banarsidass, 1995), 56–66, 99 note 51, and *The Disinterested Witness: A Fragment of Advaita Vedānta Phenomenology* (Evanston, IL: Northwestern University Press, 1998), 84–86. My account simplifies a number of the complexities on both sides of the dispute.

17. See Ian Kesarcordi-Watson, "An Ancient Indian Argument for What I Am," *Journal of Indian Philosophy* 9 (1981): 259–272.

18. Ibid.

19. See Arvind Sharma, *Sleep as a State of Consciousness in Advaita Vedānta* (Albany: State University of New York Press, 2004), 44.

20. See Surendranath Dasgupta, *A History of Indian Philosophy. Volume I* (Cambridge: Cambridge University Press, 1922), 460–461.

21. For an overview of older Western philosophical treatments of dreamless sleep, see James Hill, "The Philosophy of Sleep: The Views of Descartes, Locke, and Leibniz," *Richmond Journal of Philosophy* 6 (2004): 1–7.

22. For two recent exceptions, see Corey Anton, "Dreamless Sleep and the Whole of Human Life: An Ontological Exposition," *Human Studies* 29 (2006): 181–202, and Nicolas de Warren, "The Inner Night: Towards a Phenomenology of (Dreamless) Sleep," in Dieter Lohmar, ed., *On Time: New Contributions to the Husserlian Problem of Time-Consciousness* (Dordrecht: Springer Verlag, 2010).

23. Giulio Tononi and Christof Koch, "The Neural Correlates of Consciousness: An Update," *Annals of the New York Academy of Sciences* 1124 (2008): 239–261, at 242.

24. Giulio Tononi, "Consciousness as Integrated Information: A Provisional Manifesto," *Biological Bulletin* 215 (2008): 216–242, at 216.

25. John R. Searle, "Consciousness," *Annual Review of Neuroscience* 23 (2000): 557–578, at 559.

26. Dzogchen Ponlop, *Mind Beyond Death* (Ithaca, NY: Snow Lion, 2006), 86; Dalai Lama, *Sleeping, Dreaming, and Dying: An Exploration of Consciousness with the Dalai Lama* (Boston: Wisdom, 1996), 40.

27. Tononi and Koch, "The Neural Correlates of Consciousness," 243. See also Tore A. Nielsen, "A Review of Mentation in REM an NREM Sleep: 'Covert' REM Sleep as a Possible Reconciliation of Two Opposing Models," *Behavioral and Brain Sciences* 23 (2000): 851–866.

28. Michael T. Alkire et al., "Consciousness and Anesthesia," *Science* 322 (2008): 876–880, at 876.

29. Marcello Massimini et al., "Breakdown of Cortical Effective Connectivity During Sleep," *Science* 309 (2005): 2228–2232. See also Giulio Tononi and Marcello Massimini, "Why Does Consciousness Fade in Early Sleep?" *Annals of the New York Academy of Sciences* 1129 (2008): 330–334.

30. Tononi, "Consciousness as Integrated Information."

31. Ibid., 232.

32. Ned Block, "Comparing the Major Theories of Consciousness," in Michael Gazzaniga, ed., *The Cognitive Neurosciences IV* (Cambridge, MA: MIT Press, 2009), 1111–1122.

33. See Susanne Diekelmann and Jan Born, "The Memory Function of Sleep," *Nature Reviews Neuroscience* 11 (2010): 114–126, and Matthew P. Walker, "The Role of Sleep in Cognition and Emotion," *Annals of the New York Academy of Sciences* 1156 (2009): 168–197.

34. See Björn Rasch et al., "Odor Cues During Slow-Wave Sleep Prompt Declarative Memory Consolidation," *Science* 315 (2007): 1426–1429. For Proust's description, see Proust, *The Way by Swann's*, 47–50.

35. Matthew A. Wilson and Bruce L. McNaughton, "Reactivation of Hippocampal Ensemble Memories During Sleep," *Science* 265 (1994): 676–679; Daoyun Jun and Matthew A. Wilson, "Coordinated Memory Replay in the Visual Cortex and Hippocampus During Sleep," *Nature Neuroscience* 10 (2007): 100–107.

36. Philippe Peigneux et al., "Are Spatial Memories Strengthened in the Human Hippocampus During Slow-Wave Sleep?" *Neuron* 44 (2004): 535–545.

37. See Gauḍapāda's commentary on the *Māṇḍūkya Upaniṣad* and Śaṅkara's commentary on Gauḍapāda's in Swami Gambhirananda, trans., *Eight Upanisads, Volume Two* (Kolkata, India: Advaita Ashrama, 1958), 209. For discussion of "seed sleep," see Gupta, *The Disinterested Witness*, 29–30; Sharma, *Sleep as a State of Consciousness in Advaita Vedānta*, 75, 91; and Andrew Fort, *The Self and Its States: A States of Consciousness Doctrine in Advaita Vedānta* (Delhi: Motilal Banarsidass, 1990), chapter 5.

38. See Walker, "The Role of Sleep in Cognition and Emotion," 170–175.

39. See Diekelmann and Born, "The Memory Function of Sleep," 118, 123–124.

40. György Buzsáki, *Rhythms of the Brain* (Oxford: Oxford University Press, 2006), 173–205.

41. Ibid., 208.

42. See Dalena van der Kloet et al., "Fragmented Sleep, Fragmented Mind: The Role of Sleep in Dissociative Symptoms," *Perspectives on Psychological Science* 7 (2012): 159–175.

43. Owen Flanagan, *Dreaming Souls: Sleep, Dreams, and the Evolution of the Conscious Mind* (New York: Oxford University Press, 2000), chapter 2.

44. J. Allan Hobson, *Consciousness* (New York: Scientific American Library, 1999), 142–143.

45. See Yuval Nir and Guilio Tononi, "Dreaming and the Brain: From Phenomenology to Neurophysiology," *Trends in Cognitive Sciences* 14 (2009): 88–100, at 95.

46. See Swami Satyananda Saraswati, *Yoga Nidra* (Munger, Bihar, India: Yoga Publications Trust, 6th ed., 1998). For a historical examination of the term *yoga nidrā*, see André Couture, "The Problem of the Meaning of Yoganidrā's Name," *Journal of Indian Philosophy* 27 (1999): 35–47.

47. See Padmasambhava, *Natural Liberation: Padmasambhava's Teachings on the Six Bardos*, trans. B. Alan Wallace (Boston: Wisdom, 1998). For contemporary Tibetan presentations of sleep yoga, see Chogyal Namkhai Norbu, *Dream Yoga and the Practice of the Natural Light* (Ithaca, NY: Snow Lion, 1992), 51–71; Dzogchen Ponlop, *Mind Beyond Death* (Ithaca, NY: Snow Lion, 2006), 65, 86–87; Tenzin Wangyal Rinpoche, *The Tibetan Yogas of Dream and Sleep* (Ithaca, NY: Snow Lion, 1998), 143–184.

48. Tenzin Wangyal Rinpoche, *The Tibetan Yogas of Dream and Sleep*, 146.

49. Dzogchen Ponlop, *Mind Beyond Death*, 86.

50. Ibid., 86–87.

51. See Ursula Voss et al., "Lucid Dreaming: A State of Consciousness with Features of Both Waking Consciousness and Non-lucid Dreaming," *Sleep* 32 (2009): 1191–1200, and my discussion of this study in chapter 4.

52. Alain Destexhe et al., "Are Corticothalamic 'Up' States Fragments of Wakefulness?" *Trends in Neurosciences* 30 (2007): 334–342.

53. Michel Le Van Quyen et al., "Large-Scale Microelectrode Recordings of High-Frequency Gamma Oscillations in Human Cortex During Sleep," *Journal of Neuroscience* 30 (2010): 7770–7782.

54. See James M. Krueger et al., "Sleep as a Fundamental Property of Neuronal Assemblies," *Nature Reviews Neuroscience* 9 (2008): 910–919.

55. Fabio Ferrarelli et al., "Experienced Mindfulness Meditators Exhibit Higher Parietal-Occipital EEG Gamma Activity During NREM Sleep," *PLoS ONE* 8 (8): e73417. doi:10.1371/journal.pone.0073417.

56. Ursula Voss et al., "Lucid Dreaming: A State of Consciousness with Features of Both Waking Consciousness and Non-lucid Dreaming," *Sleep* 32 (2009): 1191–1200.

57. L. I. Mason et al., "Electrophysiological Correlates of Higher States of Consciousness During Sleep in Long-Term Practitioners of the Transcendental Meditation Program," *Sleep* 20 (1997): 102–110.

58. See Ravindra P. Nagendra et al., "Meditation and Its Regulatory Role on Sleep," *Frontiers in Neurology* 3 (2012) Article 54: 1–3.
59. Gregory A. Tooley et al., "Acute Increases in Night-Time Plasma Melatonin Levels Following a Period of Meditation," *Biological Psychology* 53 (2000): 69–78.
60. Sathiamma Sulekha et al., "Evaluation of Sleep Architecture in Practitioners of Sadarshan Kriya Yoga and Vipassana Meditation," *Sleep and Biological Rhythms* 4 (2006): 207–214; Ravindra Pattanashetty et al., "Practitioners of Vipassana Meditation Exhibit Enhanced Slow Wave Sleep and REM Sleep States Across Different Age Groups," *Sleep and Biological Rhythms* 8 (2010): 34–41.

9. DYING: WHAT HAPPENS WHEN WE DIE?

1. Joan Halifax, *Being with Dying: Cultivating Compassion and Fearlessness in the Presence of Death* (Boston: Shambhala, 2008), xvi–xvii.
2. From the Indian epic *Mahābhārata*, as quoted in Halifax, *Being with Dying*, 6.
3. A description of the program can be found at http://www.upaya.org/bwd/index.php. See also Cynda Hylton Rushton et al., "Impact of a Contemplative End-of-Life Training Program: Being with Dying," *Palliative and Supportive Care* 7 (2009): 405–414.
4. One of our fellow participants, physician Gordon Giddings, wrote a book about his experience of the Being with Dying training. See Gordon Giddings, *Dying in the Land of Enchantment: A Doctor's Journey* (Big Pine, CA: Lost Borders Press, 2012).
5. Halifax, *Being with Dying*, 55–59.
6. Ibid., 172–177.
7. For a discussion of the traditional Tibetan practice and its cosmology by two modern Tibetan Buddhist teachers, see Dzogchen Ponlop, *Mind Beyond Death* (Ithaca, NY: Snow Lion, 2006), chapter 5 and appendix V, and Sogyal Rinpoche, *The Tibetan Book of Living and Dying* (San Francisco: HarperSanFrancisco, 1993), chapter 15.
8. Dzogchen Ponlop, *Mind Beyond Death*, 48.
9. See Dzogchen Ponlop, *Mind Beyond Death*, and Sogyal Rinpoche, *The Tibetan Book of Living and Dying*. For the classic Tibetan text on death, see Graham Coleman and Thupten Jinpa, eds., *The Tibetan Book of the Dead* (London: Penguin, 2007).
10. See Dalai Lama, *Sleeping, Dreaming, and Dying: An Exploration of Consciousness with the Dalai Lama* (Boston: Wisdom, 1996), 163. See also Dzogchen Ponlop, *Mind Beyond Death*, 139, and Sogyal Rinpoche, *The Tibetan Book of Living and Dying*, 254.

11. Dzogchen Ponlop, *Mind Beyond Death*, 139.

12. Sogyal Rinpoche, *The Tibetan Book of Living and Dying*, 108.

13. In Greek mythology, sleep (Hypnos) and death (Thanatos) are twin brothers born of the goddess of night (Nyx). In the Old Testament, David requests, "Consider and hear me, O Lord my God: lighten mine eyes, lest I sleep the sleep of death" (Psalms 13:3, King James Version). The New Testament presents this exchange between Jesus and his disciples: "These things He said, and after that He said to them, 'Our friend Lazarus sleeps, but I go that I may wake him up.' Then his disciples said, 'Lord, if he sleeps he will get well.' However, Jesus spoke of his death, but they thought He was speaking about taking rest in sleep. Then Jesus said to them plainly, 'Lazarus is dead'" (John 11:11–14, New King James Version). For the link between sleep and death in *The Epic of Gilgamesh*, see N. K. Sandars, trans., *The Epic of Gilgamesh* (London: Penguin, 1972), 107, 114–115.

14. Dalai Lama, *Sleeping, Dreaming, and Dying*, 169–170.

15. See Ian Stevenson, *Children Who Remember Previous Lives: A Question of Reincarnation*, rev. ed. (Jefferson, NC and London: McFarland and Company, 2000).

16. See Gananath Obeyesekere, *Imagining Karma: Ethical Transformation in Amerindian, Buddhist, and Greek Rebirth* (Berkeley: University of California Press, 2002). See also Owen Flanagan, *The Really Hard Problem: Meaning in a Material World* (Cambridge, MA: MIT Press, 2007), 94–99, and Owen Flanagan, *The Bodhisattva's Brain: Buddhism Naturalized* (Cambridge, MA: MIT Press, 2011), 222–225.

17. See Ian Stevenson, *European Cases of the Reincarnation Type* (Jefferson, NC and London: McFarland and Company, 2003).

18. Ian Stevenson, *Twenty Cases Suggestive of Reincarnation*, 2nd ed. (Charlottesville, VA and London: University of Virginia Press, 1974). See also Martin Willson, *Rebirth and the Western Buddhist* (London: Wisdom, 1987).

19. See Leonard Angel, "Reincarnation All Over Again: Evidence for Reincarnation Rests on Backward Reasoning," *The Skeptic* [USA] 9 (2002): 87–90.

20. See Jonathan Edelmann and William Bernet, "Setting Criteria for Ideal Reincarnation Research," *Journal of Consciousness Studies* 14 (2007): 92–101.

21. Stephen Batchelor, *Buddhism Without Beliefs: A Contemporary Guide to Awakening* (New York: Riverhead, 1997), 37–38.

22. Dzogchen Ponlop, *Mind Beyond Death*, 174.

23. Ibid., 173–197.

24. Here I am influenced by Robert H Sharf, "Buddhist Modernism and the Rhetoric of Meditative Experience," *Numen* 42 (1995): 228–283: "In etic terms, Buddhist meditation might best be seen as the ritualization of experience: it doesn't engender a specific experiential state so much as it enacts it. In

this sense Buddhist *mārga* treatises are not so much maps of inner psychic space as they are scripts for the performance of an eminently public religious drama" (269).

25. Ibid., 16.
26. The report can be viewed at http://tvnz.co.nz/sunday-news/coming-up-june-19–4231582. It can also be seen on Youtube: http://www.youtube.com/watch?v=xRAfGkqw_cU&feature=youtu.be and http://www.youtube.com/watch?v=6ndLv8VkUjo.
27. See Dalai Lama, *Sleeping, Dreaming, and Dying*, 163–164.
28. Sogyal Rinpoche, *The Tibetan Book of Living and Dying*, 266.
29. Ibid., 266–267.
30. See "Lama Putse's Passing," http://www.rangjung.com/authors/Lama_Putse's_passing.htm.
31. See "Chogye Trichen Rinpoche's Passing," http://blazing-splendor.blogspot.ca/2007/02/chogye-trichen-rinpoches-passing.html.
32. See "The Passing of Tenga Rinpoche," http://www.benchen.org/en/tenga-rinpoche/news/231-the-passing-of.html, and "Kyabje Tenga Rinpoche's Tukdam has ended," http://www.benchen.org/en/tenga-rinpoche/parinirvana/239-simply-amazing.html.
33. See "Former Ganden Tripa Stays on 'Thukdam' for 18 Days," http://www.pha-yul.com/news/article.aspx?id=22935.
34. See Arpad A. Vass et al., "Beyond the Grave: Understanding Human Decomposition," *Microbiology Today* 28 (2001): 190–192, and Arpad A. Vass, "Dust to Dust: The Brief, Eventful Afterlife of a Human Corpse," *Scientific American* (September 2010): 56–58.
35. See Sushila Blackman, *Graceful Exits: How Great Beings Die. Death Stories of Hindu, Tibetan Buddhist, and Zen Masters* (Boston: Shambhala, 1997).
36. Bruce Greyson, "Near-Death Experiences," in V. S. Ramachandran, ed., *The Encyclopedia of Human Behavior*, Second Edition, vol. 2 (Academic Press, 2012), 669–676.
37. Raymond A. Moody, Jr., *Life After Life: The Investigation of a Phenomenon—Survival of Bodily Death* (San Francisco: Harper, 1975, 2001).
38. Kenneth Ring, *Life at Death: A Scientific Investigation of Near-Death Experience* (New York: Conward, McCann & Geoghegan, 1980).
39. Bruce Greyson, "The Near-Death Experience Scale: Construction, Reliability, and Validity," *Journal of Nervous and Mental Disease* 171 (1983): 369–375.
40. Antoine Lutz and Evan Thompson, "Neurophenomenology: Integrating Subjective Experience and Brain Dynamics in the Neuroscience of Consciousness," *Journal of Consciousness Studies* 10 (2003): 31–52.
41. Pim van Lommel et al., "Near-Death Experience in Survivors of Cardiac Arrest: A Prospective Study in the Netherlands," *The Lancet* 358 (2001): 2039–2045.

42. Sam Parnia and Peter Fenwick, "Near Death Experiences in Cardiac Arrest: Visions of a Dying Brain or Visions of a New Science of Consciousness," *Resuscitation* 52 (2002): 5–11.

43. Van Lommel et al., "Near-Death Experience in Survivors of Cardiac Arrest," and Bruce Greyson, "Incidence and Correlates of Near-Death Experiences in a Cardiac Care Unit," *General Hospital Psychiatry* 25 (2003): 269–276.

44. See Pim van Lommel, *Consciousness Beyond Life: The Science of Near-Death Experience* (New York: Harper One, 2010), and Greyson, "Near-Death Experiences."

45. Bruce Greyson et al., "'There Is Nothing Paranormal About Near-Death Experiences' Revisited: Comment on Mobbs and Watt," *Trends in Cognitive Sciences* 16 (2012): 446. Greyson et al. are responding to Dean Mobbs and Caroline Watt, "There Is Nothing Paranormal About Near-Death Experiences: How Neuroscience Can Explain Seeing Bright Lights, Meeting the Dead, or Being Convinced You are One of Them," *Trends in Cognitive Sciences* 15 (2011): 447–449.

46. See Parnia and Fenwick, "Near Death Experiences in Cardiac Arrest."

47. See Christopher C. French, "Near-Death Experiences in Cardiac Arrest," *Progress in Brain Research* 150 (2005): 351–367, and Christopher C. French, "Near-Death Experiences and the Brain," in Craig D. Murray, ed., *Psychological Scientific Perspectives on Out of Body and Near Death Experiences* (Hauppauge, NY: Nova Science Publishers, 2009), 187–204.

48. See Olaf Blanke and Sebastian Dieguez, "Leaving Body and Life Behind: Out-of-Body and Near-Death Experience," in Steven Laureys and Giulio Tononi, eds., *The Neurology of Consciousness* (London: Academic Publishers, 2009), 303–325, and Audrey Vanhaudenhuyse, Marie Thonnard, and Steven Laureys, "Towards a Neuro-Scientific Explanation of Near-Death Experiences?" in Jean-Louis Vincent, ed., *Yearbook of Intensive Care and Emergency Medicine* (Berlin: Springer-Verlag, 2009), 961–968.

49. Holly L. Clute and Warren J. Levy, "Electroencephalographic Changes During Brief Cardiac Arrest in Humans," *Anesthesiology* 73 (1990): 821–825, as reported in French, "Near-Death Experiences and the Brain."

50. Mark Crislip, "Near Death Experiences and the Medical Literature," *The Skeptic* [USA] 14 (2008): 14–15.

51. G. M. Woerlee, "Setting the Record Straight. Commentary on an Article by Pim van Lommel," http://www.neardeath.woerlee.org/setting-the-record-straight.php.

52. Jimo Borjigin et al., "Surge of Neurophysiological Coherence and Connectivity in the Dying Brain," *Proceedings of the National Academy of Sciences USA* 110 (2013): 14432–14437.

53. Ibid., 14435.

54. See Jason J. Braithwaite, "Towards a Cognitive Neuroscience of the Dying Brain," *The Skeptic* [UK] 21 (2008): 8–15, and French, "Near-Death Experiences and the Brain."

55. Pierre Gloor, "Role of the Limbic System in Perception, Memory, and Affect: Lessons from Temporal Lobe Epilepsy," in Benjamin K. Doane and Kenneth E. Livingstone, eds., *The Limbic System: Functional Organization and Clinical Disorders* (New York: Raven Press), 159–169, as reported in French, "Near-Death Experiences and the Brain."

56. Eliane Koyabashi et al., "Widespread and Intense BOLD Changes During Brief Focal Electrographic Seizures," *Neurology* 66 (2006): 1049–1055, as reported in Braithwaite, "Towards a Cognitive Neuroscience of the Dying Brain" and French, "Near-Death Experiences and the Brain."

57. Daniel Kroeger et al., "Human Brain Activity Patterns Beyond the Isoelectric Line of Extreme Deep Coma," *PLoS ONE* 8(9): e75257. doi:10.1371/journal.pone.0075257.

58. See Michael Sabom, *Light and Death* (Grand Rapids, MI: Zondervan, 1998), 37–51. See also Pim van Lommel's discussion of this case in *Consciousness Beyond Life,* 169–176.

59. See G. M. Woerlee, "Pam Reynolds Near Death Experience," http://www.neardeath.woerlee.org/pam-reynolds-near-death-experience.php. See also G. M. Woerlee, "An Anaesthesiologist Examines the Pam Reynolds Story, Part I: Background Considerations," *The Skeptic* [UK] 18 (1) (2005): 14–17, and G. M. Woerlee, "An Anaesthesiologist Examines the Pam Reynolds Story. Part Two: The Experience," *The Skeptic* [UK] 18 (2) (2005): 16–20.

60. See Peter S. Sebel et al., "The Incidence of Awareness During Anesthesia: A Multicenter United States Study," *Anesthesia and Analgesia* 99 (2004): 833–839.

61. Ibid., 835.

62. See http://en.wikipedia.org/wiki/Pam_Reynolds_case, especially http://en.wikipedia.org/wiki/Pam_Reynolds_case#Timeline, where the timeline is presented.

63. See Woerlee, "Pam Reynolds Near-Death Experience." See also Michael N. Marsh's discussion of this case in his *Out-of-Body and Near-Death Experiences: Brain-State Phenomena or Glimpses of Immortality?* (Oxford: Oxford University Press, 2010), 19–27.

64. See Marsh, *Out-of-Body and Near-Death Experiences*, 19–27.

65. Another famous case is "the man with the dentures" reported in Pim van Lommel et al., "Near-Death Experience in Survivors of Cardiac Arrest: A Prospective Study in the Netherlands," 2041. For critical examination of this case, see G. M. Woerlee, "The Man with the Dentures," http://www.neardeath.woerlee.org/man-with-the-dentures.php

66. See the evidence collected in Janice Minder Holden, "Veridical Perception in Near-Death Experiences," in Janice Minder Holden et al., eds., *The Handbook of Near-Death Experiences: Thirty Years of Investigation* (Santa Barbara, CA: ABC-Clio, 2009), 185–211.

67. Ibid., 205–209.

68. For the dying brain model, see Susan Blackmore, *Dying to Live: Near-Death Experiences* (Buffalo, NY: Prometheus Books, 1993); G. M. Woerlee, *Mortal Minds: The Biology of Near-Death Experiences* (Amherst, NY: Prometheus Books, 2003); and Braithwaite, "Towards a Cognitive Neuroscience of the Dying Brain." For the reviving brain model, see Marsh, *Out-of-Body and Near-Death Experiences*. Marsh's book combines a neuropsychological account of near-death experiences with an evaluation of their spiritual veracity from a Christian theological perspective that believes in the general resurrection of the body. In Marsh's view, near-death experiences are brain-state phenomena, not glimpses of disembodied consciousness, let alone immortality. I find his neuropsychological case compelling, but I'm utterly unpersuaded by his theological position.

69. Braithwaite, "Towards a Cognitive Neuroscience of the Dying Brain."

70. James E. Whinnery, "Psychophysiologic Correlates of Unconsciousness and Near-Death Experiences," *Journal of Near-Death Studies* 15 (1997): 231–258, at 245, as quoted by French, "Near-Death Experiences and the Brain." See also Marsh, *Out-of-Body and Near-Death Experiences*, 76–79.

71. Thomas Lempert et al., "Syncope and Near-Death Experience," *The Lancet* 334 (1994): 829–830.

72. Van Lommel et al., "Near-Death Experience in Survivors of Cardiac Arrest," 2043.

73. See Braithwaite, "Towards a Cognitive Neuroscience of the Dying Brain," Crislip, "Near Death Experiences and the Medical Literature," French, "Near-Death Experiences and the Brain," and Blanke and Diguez, "Leaving Body and Life Behind," 315–317.

74. Willoughby B. Britton and Richard R. Bootzin, "Near-Death Experiences and the Temporal Lobe," *Psychological Science* 15 (2004): 254–258.

75. Blanke and Diguez, "Leaving Body and Life Behind," 320–321.

76. See also Borjigin et al., "Surge of Neurophysiological Coherence and Connectivity in the Dying Brain."

77. My thoughts in this section are indebted to an unpublished paper by Michel Bitbol, "Death from the First-Person Standpoint."

78. Martin Heidegger, *Being and Time*, trans. Joan Stambaugh (Albany: State University of New York Press, 1996), 223.

79. Todd May, *Death* (Stocksfield: Acumen Publishing, 2009), 9. See also J. J. Valberg, *Dream, Death, and the Self* (Princeton, NJ: Princeton University Press, 2007) for an extended philosophical discussion of this theme.

80. Leo Tolstoy, *The Death of Ivan Ilyich*, trans. Anthony Briggs (London: Penguin, 2006), 56–57.

81. See Valberg, *Dream, Death, and the Self*, 170.

82. Ibid., 61–62.

83. Robert Thurman, trans., *The Holy Teaching of Vimalakīrti: A Mahāyāna Scripture* (University Park and London: The Pennsylvania State University Press, 1976), 161.

84. Halifax, *Being with Dying*, xvii–xviii.

85. See Claire Petitmengin, "Describing One's Subjective Experience in the Second Person: An Interview Method for the Science of Consciousness," *Phenomenology and the Cognitive Sciences* 5 (2009): 229–269. See also the articles collected in Claire Petitmengin, ed., *Ten Years of Viewing from Within: The Legacy of F. J. Varela*, special issue of the *Journal of Consciousness* 16 (10–12) (October–December 2009).

86. See Claire Petitmengin et al., "Seizure Anticipation: Are Neurophenomenological Approaches Able to Detect Preictal Symptoms?" *Epilepsy and Behavior* 9 (2006): 298–306, and Claire Petitmengin et al., "Anticipating Seizure: Pre-Reflective Experience at the Center of Neuro-Phenomenology," *Consciousness and Cognition* 16 (2007): 746–764.

87. Yoel Hoffman, *Japanese Death Poems: Written by Zen Monks and Haiku Poets on the Verge of Death* (North Clarendon, VT: Charles E. Tuttle, 1986), 144. The story accompanying the poem is based also on Lucien Stryk et al., *Zen Poems of China and Japan* (New York: Grove Press, 1973), xxxiv.

10. KNOWING: IS THE SELF AN ILLUSION?

1. My paraphrase is inspired by Stephen Batchelor's poetic rendering of Nāgārjuna's text; see Stephen Batchelor, *Verses from the Center: A Buddhist Vision of the Sublime* (New York: Riverhead, 2000), 115–116. For a recent scholarly translation of Nāgārjuna's Sanskrit text, including philosophical commentary, see Mark Siderits and Shōryū Katsura, *Nāgārjuna's Middle Way: Mūlamadhyamakakārikā* (Somerville, MA: Wisdom, 2013). For a translation from the Tibetan, see Jay L. Garfield, *The Fundamental Wisdom of the Middle Way: Nāgārjuna's Mūlamadhyamakakārikā* (New York and Oxford: Oxford University Press, 1995); this book also includes Garfield's extensive philosophical commentary. For an older and still valuable translation from the Sanskrit, see Kenneth K. Inada, *Nāgārjuna: A Translation of His Mūlamadhyamakakārikā with an Introductory Essay* (Tokyo: The Hokuseido Press, 1970). For a philosophical study, see Jan Westerhoff, *Nāgārjuna's Madhyamaka: A Philosophical Introduction* (New York and Oxford: Oxford University Press, 2009).

2. For a detailed presentation of Nāgārjuna's reasoning summarized in the two previous paragraphs, see Westerhoff, *Nāgārjuna's Madhyamaka*, 154–158; Garfield, *The Fundamental Wisdom of the Middle Way*, 245–247; and Siderits and Katsura, *Nāgārjuna's Middle Way*, 195–196.

3. Anālayo, *Satipaṭṭāna: The Direct Path to Realization* (Cambridge, England: Windhorse Publications, 2003), 206.

4. Garfield, *The Fundamental Wisdom of the Middle Way*, 245.

5. Ibid., 246.

6. Thomas Metzinger, *The Ego Tunnel: The Science of the Mind and the Myth of the Self* (New York: Basic Books, 2009), 1.

7. Thomas Metzinger, *Being No One: The Self-Model Theory of Subjectivity* (Cambridge, MA: MIT Press/A Bradford Book, 2003), 1.

8. Batchelor, *Verses from the Center*, 69.

9. *Fundamental Stanzas on the Middle Way* 7:34, 17:31–32. See Garfield, *The Fundamental Wisdom of the Middle Way*, 176–177, 243–244, and Westerhoff, *Nāgārjuna's Madhyamaka*, 163–164.

10. Here I build on my earlier presentations of the enactive approach to the self. See Francisco Varela, Evan Thompson, and Eleanor Rosch, *The Embodied Mind: Cognitive Science and Human Experience* (Cambridge, MA: MIT Press, 1991; expanded edition 2015), and Evan Thompson, *Mind in Life: Biology, Phenomenology, and the Sciences of Mind* (Cambridge, MA: Harvard University Press, 2007). See also Matthew MacKenzie, "Enacting the Self: Buddhist and Enactivist Approaches to the Emergence of Self," in Mark Siderits et al., eds., *Self, No Self? Perspectives from Analytical, Phenomenological, and Indian Traditions* (Oxford: Oxford University Press, 2011), 239–273.

11. I take the idea of "I-ing" from Jonardon Ganeri's discussion of the Madhyamaka philosopher Candrakīrti (c. 600 C.E.). See Jonardon Ganeri, *The Concealed Art of the Soul: Theories of Self and Practices of Truth in Indian Ethics and Epistemology* (Oxford: Oxford University Press, 2007), 201.

12. See Thompson, *Mind in Life*, chapter 3.

13. For an overview of autopoiesis, see Thompson, *Mind in Life*, chapter 5.

14. Humberto R. Maturana and Francisco J. Varela, *Autopoiesis and Cognition: The Realization of the Living*, Boston Studies in the Philosophy of Science, vol. 42 (Dordrecht: D. Reidel, 1980).

15. See Francisco J. Varela, "The Creative Circle: Sketches on the Natural History of Circularity," in Paul Watzlavick, ed., *The Invented Reality: How Do We Know What We Believe We Know?* (New York: Norton, 1984), 309–323, and Thompson, *Mind in Life*, 46, 99.

16. See Jonardon Ganeri, *The Self: Naturalism, Consciousness, and the First-Person Stance* (Oxford: Oxford University Press, 2012), 43–48, 69–97, and Jonardon Ganeri, "Emergentisms: Ancient and Modern," *Mind* 120 (2011): 671–703.

17. Ganeri, "Emergentisms."
18. Ganeri, *The Self*.
19. See Thompson, *Mind in Life*, 64–65, 75, and appendix B.
20. See Evan Thompson, "Living Ways of Sense-Making," *Philosophy Today* SPEP Supplement 2011: 114–123, and Ezequiel Di Paolo, "Extended Life," *Topoi* 28 (2009): 9–21.
21. See Jeffrey Hopkins, *Meditations on Emptiness* (London: Wisdom, 1983), 167–168, and Westerhoff, *Nāgārjuna's Madhyamaka*, 27.
22. See Thompson, *Mind in Life*, 38, 65, 431.
23. Neil Theise, "Now You See It, Now You Don't," *Nature* 435 (2005): 1165.
24. Hopkins, *Meditations on Emptiness*, 168.
25. Humberto R. Maturana and Francisco J. Varela, *The Tree of Knowledge: The Biological Roots of Human Understanding* (Boston: Shambhala /New Science Library, 1987), 142–176.
26. See Dorothée Legrand and Perrine Ruby, "What Is Self-Specific? A Theoretical Investigation and Critical Review of Neuroimaging Results," *Psychological Review* 116 (2009): 252–282, and Kalina Christoff et al., "Specifying the Self for Cognitive Neuroscience," *Trends in Cognitive Sciences* 15 (2011): 104–112.
27. Lynn Margulis and Dorion Sagan, *What Is Life?* (New York: Simon and Schuster, 1995), 122.
28. See Antonio Damasio and Gil B. Carvalho, "The Nature of Feelings: Evolutionary and Neurobiological Origins," *Nature Reviews Neuroscience* 14 (2013): 143–152.
29. Ibid.
30. See A. D. Craig, "How Do You Feel? Interoception: The Sense of the Physiological Condition of the Body," *Nature Reviews Neuroscience* 3 (2002): 655–666, and A. D. Craig, "How Do You Feel—Now? The Anterior Insula and Human Awareness," *Nature Reviews Neuroscience* 10 (2009): 59–70.
31. See Jake H. Davis and Evan Thompson, "From the Five Aggregates to Phenomenal Consciousness: Towards a Cross-Cultural Cognitive Science," in Steven Emmanuel, ed., *A Companion to Buddhist Philosophy* (Hoboken, NJ: Wiley-Blackwell, 2013), 585–598. See also Ganeri, *The Self*, 127–138.
32. See Sue Hamilton, *Early Buddhism: A New Approach. The Eye of the Beholder* (Richmond, Surrey: Curzon Press, 2000), 29.
33. Ganeri, *The Self*, 127–138.
34. Damasio and Carvalho, "The Nature of Feelings."
35. Ganeri, *The Self*, 127–138.
36. The term "readying" for the fourth aggregate comes from Ganeri, *The Self*, 127–138.
37. Ibid.
38. Ibid.

39. See Varela, Thompson, and Rosch, *The Embodied Mind*, and Davis and Thompson, "From the Five Aggregates to Phenomenal Consciousness."

40. Ganeri, *The Self*, 130.

41. For a discussion relating the five aggregates to microbiology and the evolution of consciousness, see William Irwin Thompson, *Coming Into Being: Artifacts and Texts in the Evolution of Consciousness* (New York: St. Martin's Press, 1996), chapter 2.

42. Nikita Vladimirov and Victor Sourjik, "Chemotaxis: How Bacteria Use Memory," *Biological Chemistry* 390 (2009): 1097–1104.

43. Jeremy W. Hayward and Francisco J. Varela, eds., *Gentle Bridges: Conversations with the Dalai Lama on the Sciences of Mind* (Boston: Shambhala, 1992), 67–68.

44. See Roman Bauer, "In Search of a Neural Signature of Consciousness—Facts, Hypotheses, and Proposals," *Synthese* 141 (2004): 233–245.

45. Ibid., 237.

46. Ibid., 237–238.

47. N. D. Cook, "The Neuron Level Phenomena Underlying Cognition and Consciousness: Synaptic Activity and the Action Potential," *Neuroscience* 153 (2008): 556–570.

48. See John R. Searle, *Minds, Brains and Science* (Cambridge, MA: Harvard University Press, 1986), John R. Searle, *The Mystery of Consciousness* (New York: The New York Review of Books, 1990), and John R. Searle, *Mind: A Brief Introduction* (New York: Oxford University Press, 2005).

49. For a recent study of the effects of Tibetan Buddhist *tummo* (inner fire) meditation practice on core body temperature and EEG brain rhythms, see Maria Kozhevnikov et al., "Neurocognitive and Somatic Components of Temperature Increases During g-Tummo Meditation: Legend and Reality," *PLoS ONE* 8 (3): e58244. doi:10.1371/journal.pone.0058244.

50. Frans B. M. de Waal, "The Thief in the Mirror," *PLoS Biology* 6 (8): e201. doi:10.1371/journal.pbio.0060201.

51. See Michael Tomasello, *The Cultural Origins of Human* Cognition (Cambridge, MA: Harvard University Press, 1999), 89–90.

52. Frans B. M. de Waal, "Putting the Altruism Back Into Altruism: The Evolution of Empathy," *Annual Review of Psychology* 59 (2008): 279–300.

53. See Thompson, *Mind in Life*, chapter 13.

54. Tomasello, *The Cultural Origins of Human Cognition*, 89–90.

55. Ibid., 99–100.

56. Randy L. Buckner and Daniel C. Carrol, "Self-Projection and the Brain," *Trends in Cognitive Sciences* 11 (2007): 49–57.

57. David H. Ingvar, "'Memory of the Future': An Essay on the Temporal Organization of Conscious Awareness," *Human Neurobiology* 3 (1985): 126–136. See

also Daniel L. Schacter et al., "Remembering the Past to Imagine the Future: The Prospective Brain," *Nature Reviews Neuroscience* 8 (2007): 657–661.

58. Antonio Damasio, *The Feeling of What Happens: Body and Emotion in the Making of Consciousness* (New York: Harcourt Brace, 1999); Shaun Gallagher and Dan Zahavi, *The Phenomenological Mind*, 2nd ed. (London: Routledge, 2012).

59. Buckner and Carrol, "Self-Projection and the Brain."

60. Randy L. Buckner et al., "The Brain's Default Network: Anatomy, Function, and Relevance to Disease," *Annals of the New York Academy of Sciences* 1124 (2008): 1–38.

61. See Legrand and Ruby, "What Is Self-Specific?" and Christoff et al., "Specifying the Self for Cognitive Neuroscience."

62. Matthew A. Killingsworth and Daniel T. Gilbert, "A Wandering Mind Is an Unhappy Mind," *Science* 330 (2010): 932.

63. See Benjamin Baird et al., "Inspired by Distraction: Mind Wandering Facilitates Creative Incubation," *Psychological Science* 23 (2012): 1117–1122; Jonathan Smallwood et al., "Shifting Moods, Wandering Minds: Negative Moods Lead the Mind to Wander," *Emotion* 9 (2009): 271–276; and Killingsworth and Gilbert, "A Wandering Mind Is an Unhappy Mind."

64. Smallwood et al., "Shifting Moods, Wandering Minds."

65. Killingsworth and Gilbert, "A Wandering Mind Is an Unhappy Mind."

66. Ibid., 932.

67. Malia F. Mason et al., "Wandering Minds: The Default Network and Stimulus Independent Thoughts," *Science* 315 (2007): 393–395, and Kalina Christoff et al., "Experience Sampling During fMRI Reveals Default Network and Executive Systems Contributions to Mind Wandering," *Proceedings of the National Academy of Sciences USA* 106 (2009): 8179–8724.

68. Wendy Hasenkamp et al., "Mind Wandering and Attention During Focused Meditation: A Fine-Grained Temporal Analysis of Fluctuating Cognitive States," *Neuroimage* 59 (2012): 750–760.

69. See also Julie A. Brefczynski-Lewis et al., "Neural Correlates of Attentional Expertise in Long-Term Meditation Practitioners," *Proceedings of the National Academy of Sciences U.S.A.* 104 (2007): 11483–11488; Wendy Hasenkamp and Lawrence Barsalou, "Effects of Meditation Experience on Functional Connectivity of Distributed Brain Networks," *Frontiers in Human Neuroscience* 6 (2012): 1–13.

70. Judson A. Brewer et al., "Meditation Experience Is Associated with Differences in Default Mode Network Activity and Connectivity," *Proceedings of the National Academy of Sciences USA* 108 (2011): 20254–20259.

71. Norman A.S. Farb et al., "Attending to the Present: Mindfulness Meditation Reveals Distinct Neural Modes of Self-Reference," *Social Cognitive Affective Neuroscience* 2 (2007): 313–322.

72. See Jon Kabat-Zinn, *Wherever You Go, There You Are* (New York: Hyperion, 2004).

73. For contemporary philosophical reconstructions of the Yogācāra view, see Georges Dreyfus, "Self and Subjectivity: A Middle Way Approach," in Mark Siderits et al., eds., *Self, No Self? Perspectives from Analytical, Phenomenological, and Indian Traditions*, 114–156 (Clarendon: Oxford University Press, 2010), and Ganeri, *The Self*, chapter 8.

74. See Ganeri, *The Self*, 146–152.

75. Ibid., 148–152.

76. See Paul J. Griffiths, *On Being Mindless: Buddhist Meditation and the Mind-Body Problem* (LaSalle, IL: Open Court, 1986).

77. See Lambert Schmithausen, *Ālayavijñāna: On the Origin and Early Development of a Central Concept of Yogācāra Philosophy* (Tokyo: International Institute for Buddhist Studies, 1987); and William S. Waldron, *The Buddhist Unconscious: The Ālayavijnāna in the Context of Indian Buddhist Thought* (London and New York: RoutledgeCurzon, 2003).

78. See William S. Waldron, "Buddhist Steps to an Ecology of Mind: Thinking About 'Thoughts Without a Thinker,'" *Eastern Buddhist* 34 (2002): 1–52.

79. See Dan Lusthaus, *Buddhist Phenomenology: A Philosophical Investigation of Yogācāra Buddhism and the Ch'eng Wei-shih lun* (London and New York: RoutledgeCurzon 2002), 273–359.

80. See Dan Zahavi, "The Experiential Self: Objections and Clarifications," in Mark Siderits et al., eds., *Self, No Self? Perspectives from Analytical, Phenomenological, and Indian Traditions*, 56–78 (Clarendon: Oxford University Press, 2010).

81. See Ganeri, *The Self*, chapter 8.

82. For further discussion of this issue, see Miri Albahari, *Analytical Buddhism: The Two-Tiered Illusion of Self* (New York: Palgrave Macmillan, 2006). For a critical assessment of Albahari's views, see Aaron Henry and Evan Thompson, "Witnessing from Here: Self-Awareness from a Bodily Versus Embodied Perspective," in Shaun Gallagher, ed., *The Oxford Handbook of the Self* (Oxford and New York: Oxford University Press, 2011), 228–249.

83. See Sidney Shoemaker, "Self-Reference and Self-Awareness," in his *Identity, Cause, and Mind: Philosophical Essays* (Cambridge: Cambridge University Press, 1984), 6–18.

84. For further discussion, see Dan Zahavi, *Subjectivity and Selfhood: Investigating the First-Person Perspective* (Cambridge, MA: MIT Press, 2005).

85. See Ganeri, *The Self*, chapter 8.

86. See Metzinger, *Being No One* and *The Ego Tunnel*. See also Thomas Metzinger, "The No-Self Alternative," in Shaun Gallagher, ed., *The Oxford Handbook of the Self*, 297–315 (New York and Oxford: Oxford University Press, 2011).

87. Ganeri, *The Concealed Art of the Soul*, 200–203; *The Self*, 157, 161.

88. Ganeri, *The Concealed Art of the Soul*, 202.

89. See MacKenzie, "Enacting the Self."

90. Ganeri, *The Concealed Art of the Soul*, 204.

91. For Candrakīrti's view of the self, see James Duerlinger, "Candrakīrti's Denial of the Self," *Philosophy East and West* 34 (1984): 261–272; James Duerlinger, *The Refutation of Self in Indian Buddhism: Candrakīrti on the Selflessness of Persons* (London: Routledge, 2012); and Ganeri, *The Concealed Art of the Soul*, chapter 7.

92. See Duerlinger, "Candrakīrti's Denial of the Self," 263.

93. Ibid.

94. Batchelor, *Verses from the Center*, 115–116. These are the final lines of Batchelor's rendition of Nāgārjuna's chapter on the self.

BIBLIOGRAPHY

Abrams, M. H., et al., eds. *The Norton Anthology of English Literature, Volume I*, Sixth Edition. New York and London: Norton, 1993.

Albahari, Miri. *Analytical Buddhism: The Two-Tiered Illusion of Self.* New York: Palgrave Macmillan, 2006.

Alkire, Michael T., et al. "Consciousness and Anesthesia." *Science* 322 (2008): 876–880.

Anālayo. "*Nāma-rūpa.*" In A. Sharma, ed., *Encyclopedia of Indian Religions.* Berlin: Springer Science+Business Media, 2015.

——. *Satipaṭṭāna: The Direct Path to Realization.* Cambridge, England: Windhorse, 2003.

Angel, Leonard. "Reincarnation All Over Again: Evidence for Reincarnation Rests on Backward Reasoning." *The Skeptic* [USA] 9 (2002): 87–90.

Anton, Corey. "Dreamless Sleep and the Whole of Human Life: An Ontological Exposition." *Human Studies* 29 (2006): 181–202.

Āraṇya, Sāṃkhya-yogāchāra Swāmi Hariharānanda. *Yoga Philosophy of Patañjali.* Rendered into English by P. N. Mukerji. Albany: State University of New York Press, 1983.

Aristotle. "*De Somniis* (On Dreams)." In Richard McKeon, ed., *The Basic Works of Aristotle,* 618–625. New York: Random House, 1941.

Arnold, Dan. "Dharmakīrti's Dualism: Critical Reflections on a Buddhist Proof of Rebirth." *Philosophy Compass* 3 (2008): 1079–1096.

Arya, Pandit Usharbudh. *Yoga-Sūtras of Patañjali with the Exposition of Vyāsa. Volume I: Samādhi-pāda.* Honesdale, PA: The Himalayan International Institute, 1989.

Arzy, Shahar, et al. "Neural Basis of Embodiment: Distinct Contributions of Temporoparietal Junction and Extrastriate Body Area." *Journal of Neuroscience* 26 (2006): 8074–8081.

Aspell, Jane, and Olaf Blanke. "Understanding the Out-of-Body Experience from a Neuroscientific Perspective." In Craig D. Murray, ed., *Psychological Scientific Perspectives on Out of Body and Near Death Experiences*, 73–88. Hauppauge, NY: Nova Science Publishers, 2009.

Austin, James H. *Selfless Insight: Zen and the Meditative Transformations of Consciousness*. Cambridge, MA: MIT Press, 2009.

——. *Zen and the Brain: Toward an Understanding of Meditation and Consciousness*. Cambridge, MA: MIT Press, 1999.

Baird, Benjamin, et al. "Inspired by Distraction: Mind Wandering Facilitates Creative Incubation." *Psychological Science* 23 (2012): 1117–1122.

Batchelor, Stephen. *Buddhism Without Beliefs: A Contemporary Guide to Awakening*. New York: Riverhead, 1997.

——. *Verses from the Center: A Buddhist Vision of the Sublime*. New York: Riverhead, 2000.

Bateson, Gregory. *Steps to an Ecology of Mind*. New York: Ballantine, 1972.

Bauer, Roman. "In Search of a Neural Signature of Consciousness—Facts, Hypotheses, and Proposals." *Synthese* 141 (2004): 233–245.

Bayne, Tim. "Conscious States and Conscious Creatures: Explanation in the Scientific Study of Consciousness." *Philosophical Perspectives* 21 (2007): 1–22.

Begley, Sharon. *Train Your Mind, Change Your Brain*. New York: Ballantine, 2007.

Blackman, Sushila. *Graceful Exits: How Great Beings Die. Death Stories of Hindu, Tibetan Buddhist, and Zen Masters*. Boston: Shambhala, 1997.

Bitbol, Michel. "Death from the First-Person Standpoint." Unpublished article.

——. "Is Consciousness Primary?" *NeuroQuantology* 6 (2008): 53–71.

Bitbol, Michel, and Pier-Luigi Luisi. "Science and the Self-Referentiality of Consciousness." *Journal of Cosmology* 14 (2011): 207–223.

Blackmore, Susan J. *Beyond the Body: An Investigation of Out-of-the-Body Experiences*. Chicago: Academy Chicago, 1982, 1992.

Blake, Randolph and Nikos K. Logothetis. "Visual Competition." *Nature Reviews Neuroscience* 3 (2002): 13–21.

Blanke, Olaf. "Multisensory Brain Mechanisms of Bodily Self-Consciousness." *Nature Reviews Neuroscience* 13 (2012): 556–571.

Blanke, Olaf, and Shahar Arzy. "The Out-of-Body Experience: Disturbed Self-Processing at the Temporo-Parietal Junction." *The Neuroscientist* 11 (2005): 16–24.

Blanke, Olaf, and Sebastian Dieguez. "Leaving Body and Life Behind: Out-of-Body and Near-Death Experience." In Steven Laureys and Giulio Tononi, eds., *The Neurology of Consciousness*, 303–325. London: Academic Publishers, 2009.

Blanke, Olaf, and Christine Mohr. "Out-of-Body Experience, Heautoscopy, and Autoscopic Hallucination of Neurological Origin: Implications for Neurocognitive Mechanisms of Corporeal Awareness and Self-Consciousness." *Brain Research Reviews* 50 (2005): 184–199.

Blanke, Olaf, et al. "Linking Out-of-Body Experience and Self-Processing to Mental Own-Body Imagery at the Temporoparietal Junction." *Journal of Neuroscience* 25 (2005): 550–557.

——. "Out-of-Body Experience and Autoscopy of Neurological Origin." *Brain* 127 (2004): 243–258.

——. "Stimulating Illusory Own-Body Perceptions." *Nature* 419 (2002): 269–270.

Block, Ned. "Comparing the Major Theories of Consciousness." In Michael Gazzaniga, ed., *The Cognitive Neurosciences IV*, 1111–1122. Cambridge, MA: MIT Press, 2009.

——. "Consciousness, Accessibility, and the Mesh Between Psychology and Neuroscience." *Behavioral and Brain Sciences* 30 (2007): 481–548.

Block, Ned, and Cynthia MacDonald. "Consciousness and Cognitive Access." *Proceedings of the Aristotelian Society* CVIII, Part 3 (2008): 289–316.

Bodhi, Bhikkhu. *A Comprehensive Manual of Abhidhamma*. Onalaska, WA: Buddhist Publication Society, 1993, 1999.

——. *The Connected Discourses of the Buddha: A New Translation of the Saṃyutta Nikāya*. Somerville, MA: Wisdom, 2002.

——. *In the Buddha's Words: An Anthology of Discourses from the Pāli Canon*. Somerville, MA: Wisdom, 2005.

Borges, Jorge Luis. *Everything and Nothing*. Trans. Donald A. Yates, James E. Irby, John M. Fein, and Eliot Weinberger. New York: New Directions, 1999.

——. *Selected Poems*. Ed. Alexander Coleman. New York: Penguin, 1999.

Borjigin, Jimo, et al. "Surge of Neurophysiological Coherence and Connectivity in the Dying Brain." *Proceedings of the National Academy of Sciences U.S.A.* 110 (2013): 14432–14437.

Braithwaite, Jason J. "Towards a Cognitive Neuroscience of the Dying Brain." *The Skeptic* 21 (2008): 8–15.

Brefczynski-Lewis, Julie A., et al. "Neural Correlates of Attentional Expertise in Long-Term Meditation Practitioners." *Proceedings of the National Academy of Sciences U.S.A.* 104 (2007): 11483–11488.

Brewer, Judson A., et al. "Meditation Experience Is Associated with Differences in Default Mode Network Activity and Connectivity." *Proceedings of the National Academy of Sciences U.S.A.* 108 (2011): 20254–20259.

Britton, Willoughby B., and Richard R. Bootzin. "Near-Death Experiences and the Temporal Lobe." *Psychological Science* 15 (2004): 254–258.

Bromberg, Philip M. *Awakening the Dreamer: Clinical Journeys*. Mahwah, NJ: Analytic Press, 2006.

Brooks, Janice E., and Jay A. Vogelsong. *The Conscious Exploration of Dreaming: Discovering How We Create and Control Our Dreams*. Bloomington, IN: 1st Books Library, 2000.

Bryant, Edwin F. *The Yoga Sutras of Patañjali*. New York: North Point Press, 2009.

Buckner, Randy L., and Daniel C. Carrol. "Self-Projection and the Brain." *Trends in Cognitive Sciences* 11 (2007): 49–57.

Buckner, Randy L., et al.. "The Brain's Default Network: Anatomy, Function, and Relevance to Disease." *Annals of the New York Academy of Sciences* 1124 (2008): 1–38.

Busch, Niko A., and Ruffin VanRullen. "Spontaneous EEG Oscillations Reveal Periodic Sampling of Visual Attention." *Proceedings of the National Academy of Sciences* 107 (2010): 16048–16053.

Busch, Niko A., et al. "The Phase of Ongoing EEG Oscillations Predicts Visual Perception." *Journal of Neuroscience* 29 (2009): 7869–7876.

Buschman, Timothy J. and Earl K. Miller. "Shifting the Spotlight of Attention: Evidence for Discrete Computations in Cognition." *Frontiers in Human Neuroscience* 4 (2010): 1–9.

Buzsáki, György. *Rhythms of the Brain.* Oxford: Oxford University Press, 2006.

Cabezon, José Ignacio. "Buddhism and Science: On the Nature of the Dialogue." In B. Alan Wallace, ed., *Buddhism and Science: Breaking New Ground*, 35–68. New York: Columbia University Press, 2003.

Cahn, B. Rael, et al. "Occipital Gamma Activation During Vipassana Meditation." *Cognitive Processing* 11 (2010): 39–56.

Carter, Olivia, et al. "Meditation Alters Perceptual Rivalry in Tibetan Buddhist Monks." *Current Biology* 15 (2005): R412–R413.

Chalmers, David J. "Facing Up to the Problem of Consciousness." *Journal of Consciousness Studies* 2 (1995): 200–219.

——. "On the Search for a Neural Correlate of Consciousness." In Stuart R. Hameroff et al., eds., *Toward a Science of Consciousness II*, 219–229. Cambridge, MA: MIT Press, 1998.

Chapple, Christopher Key. *Yoga and the Luminous: Patañjali's Spiritual Path to Freedom.* Albany: State University of New York Press, 2008.

Christoff, Kalina, et al. "Experience Sampling During fMRI Reveals Default Network and Executive Systems Contributions to Mind Wandering." *Proceedings of the National Academy of Sciences U.S.A.* 106 (2009): 8179–8724.

——. "Specifying the Self for Cognitive Neuroscience." *Trends in Cognitive Sciences* 15 (2011): 104–112.

Clute, Holly L., and Warren J. Levy. "Electroencephalographic Changes During Brief Cardiac Arrest in Humans." *Anesthesiology* 73 (1990): 821–825.

Coleman, Graham, and Thupten Jinpa, eds. *The Tibetan Book of the Dead.* London: Penguin, 2007.

Colins, Steven. *Selfless Persons: Imagery and Thought in Theravāda Buddhism.* Cambridge: Cambridge University Press, 1982.

Comans, Michael. "The Self in Deep Sleep According to Advaita and Viśiṣṭādvaita." *Journal of Indian Philosophy* 18 (1990): 1–28.

Conze, Edward. *Buddhist Thought in India*. Ann Arbor: University of Michigan Press, 1962.

Cook, N. D. "The Neuron-Level Phenomena Underlying Cognition and Consciousness: Synaptic Activity and the Action Potential." *Neuroscience* 153 (2008): 556–570.

Cosmelli, Diego, et al. "Waves of Consciousness: Ongoing Cortical Patterns During Binocular Rivalry." *Neuroimage* 23 (2004): 128–140.

Cosmelli, Diego, and Evan Thompson. "Envatment Versus Embodiment: Reflections on the Bodily Basis of Consciousness." In John Stewart et al., eds., *Enaction: Towards A New Paradigm for Cognitive Science*, 361–386, Cambridge, MA: MIT Press, 2010.

——. "Mountains and Valleys: Binocular Rivalry and the Flow of Experience." *Consciousness and Cognition* 16 (2007): 623–641.

Couture, André. "The Problem of the Meaning of Yoganidrā's Name." *Journal of Indian Philosophy* 27 (1999): 35–47.

Craig, A. D. "How Do You Feel? Interoception: The Sense of the Physiological Condition of the Body." *Nature Reviews Neuroscience* 3 (2002): 655–666.

——. "How Do You Feel—Now? The Anterior Insula and Human Awareness." *Nature Reviews Neuroscience* 10 (2009): 59–70.

Crislip, Mark. "Near Death Experiences and the Medical Literature." *The Skeptic* [USA] 14 (2008): 14–15.

Cruse, Damian, et al. "Bedside Detection of Awareness in the Vegetative State: A Cohort Study." *The Lancet* 378 (2011): 2088–2094.

Dalai Lama. *Dzogchen: The Heart Essence of the Great Perfection*. Trans. Thupten Jinpa and Richard Barron (Chökyi Nyima). Ithaca, NY: Snow Lion, 2000.

——. "On the Luminosity of Being." *New Scientist* (May 24, 2003): 42.

——. *Sleeping, Dreaming, and Dying: An Exploration of Consciousness with the Dalai Lama*. Boston: Wisdom, 1996.

——. *The Universe in a Single Atom: The Convergence of Science and Spirituality*. New York: Morgan Road, 2005.

Damasio, Antonio. *The Feeling of What Happens: Body and Emotion in the Making of Consciousness*. New York: Harcourt Brace, 1999.

Damasio, Antonio, and Gil B. Carvalho. "The Nature of Feelings: Evolutionary and Neurobiological Origins." *Nature Reviews Neuroscience* 14 (2013): 143–152.

Dang-Vu, Thien Thanh, et al. "Neuroimaging of REM Sleep and Dreaming." In Patrick McNamara and Deirdre Barrett, eds., *The New Science of Dreaming, Volume I: The Biology of Dreaming*, 95–113. Westport, CT: Praeger, 2007.

Dasgupta, Surendranath. *A History of Indian Philosophy. Volume I*. Cambridge: Cambridge University Press, 1922.

Davis, Jake H., and Evan Thompson. "From the Five Aggregates to Phenomenal Consciousness: Towards a Cross-Cultural Cognitive Science." In Steven

Emmanuel, ed., *A Companion to Buddhist Philosophy*, 585–598. Hoboken, NJ: Wiley-Blackwell, 2013.

deCharms, R. Christopher, et al. "Control Over Brain Activation and Pain Learned by Using Real-Time Functional MRI." *Proceedings of the National Academy of Sciences U.S.A.* 102 (2005): 18626–18631.

Dement, William C., and Nathaniel Kleitman. "The Relation of Eye Movements During Sleep to Dream Activity: An Objective Method for the Study of Dreaming." *Journal of Experimental Psychology* 53 (1957): 89–97.

Dennett, Daniel C. "Are Dreams Experiences?" Reprinted in Daniel C. Dennett, *Brainstorms: Philosophical Essays on Mind and Psychology*, 129–148. Cambridge, MA: MIT Press/A Bradford Book, 1981.

——. *Consciousness Explained.* Boston: Little Brown, 1991.

——. "The Onus Re Experiences: A Reply to Emmett." *Philosophical Studies* 35 (1979): 315–318.

Descartes, René. *The Philosophical Writings of Descartes, Volume 2.* Trans. John Cottingham, Robert Stoothoff, and Dugald Murdoch. Cambridge: Cambridge University Press, 1985.

Destexhe, Alain, et al. "Are Corticothalamic 'Up' States Fragments of Wakefulness?" *Trends in Neurosciences* 30 (2007): 334–342.

Deutsch, Eliot. *Advaita Vedānta. A Philosophical Reconstruction.* Honolulu: University Press of Hawaii, 1969.

Deutsch, Eliot and Rohit Dalvi, eds. *The Essential Vedānta. A New Source Book of Advaita Vedānta.* Bloomington, IN: World Wisdom, 2004.

de Waal, Frans B. M. "Putting the Altruism Back Into Altruism: The Evolution of Empathy." *Annual Review of Psychology* 59 (2008): 279–300.

——. "The Thief in the Mirror." *PLoS Biology* 6(8): e201. doi:10.1371/journal.pbio.0060201

de Warren, Nicolas. "The Inner Night: Towards a Phenomenology of (Dreamless) Sleep." In Dieter Lohmar, ed., *On Time: New Contributions to the Husserlian Problem of Time-Consciousness.* Dordrecht: Springer Verlag, 2010.

Diekelmann, Susanne and Jan Born. "The Memory Function of Sleep." *Nature Reviews Neuroscience* 11 (2010): 114–126.

Di Paolo, Ezequiel. "Extended Life." *Topoi* 28 (2009): 9–21.

Doesburg, Sam M., et al. "Rhythms of Consciousness: Binocular Rivalry Reveals Large-Scale Oscillatory Network Dynamics Mediating Visual Perception." *PLoS ONE* 4 (7) (2009): e6142. doi:10.1371/journal.pone.0006142.

Doidge, Norman. *The Brain That Changes Itself. Stories of Personal Triumph from the Frontiers of Brain Science.* New York: Penguin, 2007.

Dresler, Martin, et al. "Dreamed Movement Elicits Activation in the Sensorimotor Cortex." *Current Biology* 21 (2011): 1–5.

——. "Neural Correlates of Dream Lucidity Obtained from Contrasting Lucid Versus Non-lucid REM Sleep: A Combined EEG/fMRI Study." *Sleep* 35 (2012): 1017–1020.

Dreyfus, Georges. *Recognizing Reality: Dharmakīrti's Philosophy and Its Tibetan Interpretations.* Albany: State University of New York Press, 1997.

——. "Self and Subjectivity: A Middle Way Approach." In Mark Siderits et al., eds., *Self, No Self? Perspectives from Analytical, Phenomenological, and Indian Traditions,* 114–156. Clarendon: Oxford University Press, 2010.

——. *The Sound of Two Hands Clapping: The Education of a Tibetan Buddhist Monk.* Berkeley: University of California Press, 2003.

Dreyfus, George and Evan Thompson. "Asian Perspectives: Indian Theories of Mind." In Philip David Zelazo et al., eds., *The Cambridge Handbook of Consciousness,* 89–114. New York and Cambridge: Cambridge University Press, 2007.

Duerlinger, James. "Candrakīrti's Denial of the Self." *Philosophy East and West* 34 (1984): 261–272

——. *The Refutation of Self in Indian Buddhism: Candrakīrti on the Selflessness of Persons.* London: Routledge, 2012.

Dunne, John D. *Foundations of Dharmakīrti's Philosophy.* Somerville, MA: Wisdom, 2004.

Edelmann, Jonathan, and William Bernet. "Setting Criteria for Ideal Reincarnation Research." *Journal of Consciousness Studies* 14 (2007): 92–101.

Ehrsson, Henrik H. "The Experimental Induction of Out-of-Body Experiences." *Science* 317 (2007): 1048.

Eich, Eric, et al. "Neural Systems Mediating Field and Observer Memories." *Neuropsychologia* 47 (2009): 2239–2251.

Emmett, Kathleen. "Oneiric Experiences." *Philosophical Studies* 34 (1978): 445–450.

Erlacher, Daniel, and Heather Chapin. "Lucid Dreaming: Neural Virtual Reality as a Mechanism for Performance Enhancement." *International Journal of Dream Research* 3 (2010): 7–10.

Erlacher, Daniel, and Michael Schredl. "Cardiovascular Responses to Dreamed Physical Exercise During REM Lucid Dreaming." *Dreaming* 18 (2008): 112–121.

——. "Do REM (Lucid) Dreamed and Executed Actions Share the Same Neural Substrate?" *International Journal of Dream Research* 1 (2008): 7–14.

——. "Time Required for Motor Activities in Lucid Dreams." *Perceptual and Motor Skills* 99 (2004): 1239–1242.

Fan, Jin, et al. "The Relation of Brain Oscillations to Attentional Networks." *Journal of Neuroscience* 27 (2007): 6197–6206.

Farb, Norman A. S., et al. "Attending to the Present: Mindfulness Meditation Reveals Distinct Neural Modes of Self-Reference." *Social Cognitive Affective Neuroscience* 2 (2007): 313–322.

Fell, Juergen et al. "From Alpha to Gamma: Electrophysiological Correlates of Meditation-Related States of Consciousness." *Medical Hypotheses* 75 (2010): 218–224.

Ferrarelli, Fabio, et al. "Experienced Mindfulness Meditators Exhibit Higher Parietal-Occipital EEG Gamma Activity During NREM Sleep." *PLoS ONE* 8 (8): e73417. doi:10.1371/journal.pone.0073417.

Flanagan, Owen. *The Bodhisattva's Brain: Buddhism Naturalized.* Cambridge, MA: MIT Press, 2011.

——. *Dreaming Souls: Sleep, Dreams, and the Evolution of the Conscious Mind.* New York: Oxford University Press, 2000.

——. *The Really Hard Problem: Meaning in a Material World.* Cambridge, MA: MIT Press, 2007.

Fort, Andrew. "Dreaming in Advaita Vedānta." *Philosophy East and West* 35 (1985): 377–386.

——. *The Self and Its States: A States of Consciousness Doctrine in Advaita Vedānta.* Delhi: Motilal Banarsidass, 1990.

Foulkes, David. *Children's Dreaming and the Development of Consciousness.* Cambridge, MA: Harvard University Press, 2002.

——. "Dream Reports from Different Stages of Sleep." *Journal of Abnormal and Social Psychology* 65 (1962): 14–25.

Foulkes, David W., and Gerald Vogel. "Mental Activity at Sleep Onset." *Journal of Abnormal Psychology* 70 (1965): 231–243.

Fox, Oliver. *Astral Projection.* New Hyde Park, NY: University Books, 1962.

French, Christopher C. "Near-Death Experiences in Cardiac Arrest." *Progress in Brain Research* 150 (2005): 351–367.

——. "Near-Death Experiences and the Brain." In Craig D. Murray, ed., *Psychological Scientific Perspectives on Out of Body and Near Death Experiences,* 187–204. Hauppauge, NY: Nova Science Publishers, 2009.

Freud, Sigmund. *The Intrepretation of Dreams.* Trans. Joyce Crick. New York: Oxford University Press, 1999.

——. *The Interpretation of Dreams.* Trans. James Strachey. New York: Basic Books, 1955.

Frost, Robert. *The Poetry of Robert Frost: The Collected Poems, Complete and Unabridged.* New York: Henry Holt, 1969.

Gackenbach, Jayne, and Matthew Rosie. "Presence in Video Game Play and Nighttime Dreams: An Empirical Inquiry." *International Journal of Dream Research* 4 (2011): 98–109.

Gaillard, Raphaël, et al. "Converging Intracranial Markers of Conscious Access." *PLoS Biology* 7 (3) (2009): e1000061. doi:10.1371/journal.pbio.1000061.

Gallagher, Shaun, and Dan Zahavi. *The Phenomenological Mind,* 2nd ed. London and New York: Routledge, 2012.

Gambhirananda, Swami, trans. *Eight Upanisads, Volume Two.* Kolkata, India: Advaita Ashrama, 1958.

Ganeri, Jonardon. *The Concealed Art of the Soul: Theories of Self and Practices of Truth in Indian Ethics and Epistemology.* Oxford: Oxford University Press, 2007.

——. "Emergentisms: Ancient and Modern." *Mind* 120 (2011): 671–703.

——. *The Self: Naturalism, Consciousness, and the First-Person Stance.* Oxford: Oxford University Press, 2012.

Ganesh, Shanti, et al. "How the Human Brain Goes Virtual: Distinct Cortical Regions of the Person-Processing Network Are Involved in Self-Identification with Virtual Agents." *Cerebral Cortex* 22 (2012): 1577–1585.

Garfield, Jay L. *The Fundamental Wisdom of the Middle Way: Nagarjuna's Mulamadhyamakakarika.* New York and Oxford: Oxford University Press, 1995.

Germain, Anne, and Tore A. Nielsen. "EEG Power Associated with Early Sleep Onset Images Differing in Sensory Content." *Sleep Research Online* 4 (2001): 83–90.

Gethin, Rupert. *The Foundations of Buddhism.* Oxford and New York: Oxford University Press, 1998.

Gho, Michel, and Francisco J. Varela. "A Quantitative Assessment of the Dependency of the Visual Temporal Frame Upon the Cortical Alpha Rhythm." *Journal of Physiology-Paris* 83 (1988–1989): 95–101.

Giddings, Gordon. *Dying in the Land of Enchantment: A Doctor's Journey.* Big Pine, CA: Lost Borders Press, 2012.

Gloor, Pierre. "Role of the Limbic System in Perception, Memory, and Affect: Lessons from Temporal Lobe Epilepsy." In Benjamin K. Doane and Kenneth E. Livingstone, eds., *The Limbic System: Functional Organization and Clinical Disorders,* 159–169. New York: Raven Press.

Goldstein, Joseph. *The Experience of Insight.* Boston: Shambhala, 1976.

Goleman, Daniel. *Destructive Emotions: A Scientific Dialogue with the Dalai Lama.* New York: Bantam, 2003.

Gombrich, Richard. *What the Buddha Thought.* London and Oakville, CT: Equinox, 2009.

Green, Celia, and Charles McCreery. *Lucid Dreaming: The Paradox of Consciousness During Sleep.* London: Routledge, 1994.

Greyson, Bruce. "Incidence and Correlates of Near-Death Experiences in a Cardiac Care Unit." *General Hospital Psychiatry* 25 (2003): 269–276.

——. "The Near-Death Experience Scale: Construction, Reliability, and Validity." *Journal of Nervous and Mental Disease* 171 (1983): 369–375.

——. "Near-Death Experiences." In V. S. Ramachandran, ed., *The Encyclopedia of Human Behavior,* Second Edition, vol. 2, 669–676. Academic Press, 2012.

Greyson, Bruce, et al. "'There Is Nothing Paranormal About Near-Death Experiences' Revisited: Comment on Mobbs and Watt." *Trends in Cognitive Sciences* 16 (2012): 446.

Griffiths, Paul J. *On Being Mindless: Buddhist Meditation and the Mind-Body Problem*. LaSalle, IL: Open Court, 1986.

Gupta, Bina. *Cit: Consciousness*. New Delhi: Oxford University Press, 2003.

———. *The Disinterested Witness: A Fragment of Advaita Vedānta Phenomenology*. Evanston, IL: Northwestern University Press, 1998.

Gyatrul Rinpoche. *Meditation, Transformation, and Dream Yoga*. Trans. B. Alan Wallace and Sangye Khandro. Ithaca, NY: Snow Lion, 1993 and 2002.

Hadot, Pierre. *Philosophy as a Way of Life: Spiritual Exercises from Socrates to Foucault*. Ed. and with an introduction by Arnold Davidson. Malden, MA: Blackwell, 1995.

Halifax, Joan. *Being with Dying: Cultivating Compassion and Fearlessness in the Presence of Death*. Boston: Shambhala Publications, 2008.

Hamilton, Sue. *Early Buddhism: A New Approach. The Eye of the Beholder*. Richmond, Surrey: Curzon Press, 2000.

Hanh, Thich Nhat. *Understanding Our Mind*. Berkeley, CA: Parallax Press, 2006.

Harrington, Anne, and Arthur Zajonc, eds. *The Dalai Lama at MIT*. Cambridge, MA: Harvard University Press, 2006.

Hasenkamp, Wendy, et al. "Mind Wandering and Attention During Focused Meditation: A Fine-Grained Temporal Analysis of Fluctuating Cognitive States." *Neuroimage* 59 (2012): 750–760.

Hayes, Richard. "Dharmakīrti on *Punarbhava*." In Egaku Mayeda, ed., *Studies in Original Buddhism and Mahayana Buddhism. Volume 1*, 111–129. Kyoto: Nagata Bunshodo, 1993.

Hayward, Jeremy W., and Francisco J. Varela, eds. *Gentle Bridges: Conversations with the Dalai Lama on the Sciences of Mind*. Boston: Shambhala Press, 1992.

Hearne, Keith M.T. "Lucid Dreams: An Electrophysiological and Psychological Study." Ph.D. diss. University of Liverpool, 1978.

Heidegger, Martin. *Being and Time*. Rrans. Joan Stambaugh. Albany: State University of New York Press, 1996.

Heller-Roazen, Daniel. *The Inner Touch: Archaeology of a Sensation*. New York: Zone Books, 2007.

Henry, Aaron, and Evan Thompson. "Witnessing from Here: Self-Awareness from a Bodily Versus Embodied Perspective." In Shaun Gallagher, ed., *The Oxford Handbook of the Self*, 228–249. Oxford and New York: Oxford University Press, 2011.

Hill, James. "The Philosophy of Sleep: The Views of Descartes, Locke, and Leibniz." *Richmond Journal of Philosophy* 6 (2004): 1–7.

Hirshfield, Jane. *Given Sugar, Given Salt*. New York: Harper Perennial, 2002.

Hobson, J. Allan. *Consciousness*. New York: Scientific American Library, 1999.

———. *Dreaming: An Introduction to the Science of Sleep*. Oxford: Oxford University Press, 2002.

——. "The Neurobiology of Consciousness: Lucid Dreaming Wakes Up." *International Journal of Dream Research* 2 (2009): 41–44.

——. "REM Sleep and Dreaming: Toward a Theory of Protoconsciousness." *Nature Reviews Neuroscience* 10 (2009): 803–813.

——. *13 Dreams Freud Never Had: The New Mind Science.* New York: Pi Press, 2005.

Hobson, J. Allan, and Robert McCarley. "The Brain as a Dream State Generator: An Activation-Synthesis Hypothesis of the Dream Process." *American Journal of Psychiatry* 134 (1977): 1335–1348.

Hobson, J. Allan, et al. "Dreaming and the Brain: Toward a Cognitive Neuroscience of Conscious States." *Behavioral and Brain Sciences* 23 (2000): 793–842.

Hoffman, Yoel. *Japanese Death Poems: Written by Zen Monks and Haiku Poets on the Verge of Death.* North Clarendon, VT: Charles E. Tuttle, 1986.

Holden, Janice Minder. "Veridical Perception in Near-Death Experiences." In Janice Minder Holden et al., eds., *The Handbook of Near-Death Experiences: Thirty Years of Investigation*, 185–211. Santa Barbara, CA: ABC-Clio, 2009.

Homer, *The Iliad.* Trans. Robert Fitzgerald. London: Everyman's Library, 1992.

Hopkins, Jeffrey. *Meditations on Emptiness.* London: Wisdom, 1983.

Hori, Tadao, et al. "Topographical EEG Changes and the Hypnagogic Experience." In *Anonymous Sleep Onset: Normal and Abnormal Processes*, 237–253. Washington, D.C.: American Psychological Association, 1994.

Horne, Jim. *Sleepfaring: A Journey Through the Science of Sleep.* Oxford: Oxford University Press, 2006.

Husserl, Edmund. *The Crisis of European Sciences and Transcendental Phenomenology.* Trans. David Carr. Evanston, IL: Northwestern University Press, 1970.

——. *On the Phenomenology of the Consciousness of Internal Time (1893–1917).* Trans. John Brough. Dordrecht: Kluwer Academic Publishers, 1991.

Hut, Piet, and Roger Shepard. "Turning the 'Hard Problem' Upside Down and Sideways." *Journal of Consciousness Studies* 3 (1996): 313–329.

Ichikawa, Jonathan. "Dreaming and Imagination." *Mind and Language* 24 (2009): 103–121.

Inada, Kenneth K. *Nagarjuna: A Translation of his Mulamadhyamakakarika with an Introductory Essay.* Tokyo: The Hokuseido Press, 1970.

Ingvar, David H. "'Memory of the Future': An Essay on the Temporal Organization of Conscious Awareness." *Human Neurobiology* 3 (1985): 126–136.

Iyengar, B.K.S. *Light on the Yoga Sutras of Patañjali.* London: Thorsons, 1996.

James, William. *The Principles of Psychology.* Cambridge, MA: Harvard University Press, 1981.

Jha, Amishi P., et al. "Mindfulness Training Modifies Subsystems of Attention." *Cognitive, Affective, and Behavioral Neuroscience* 7 (2007): 109–119.

Jinpa, Thupten. "Science as an Ally or a Rival Philosophy? Tibetan Buddhist Thinkers' Enagement with Modern Science." In B. Alan Wallace, ed., *Buddhism and Science: Breaking New Ground*, 71–85. New York: Columbia University Press, 2003.

Jun, Daoyun, and Matthew A. Wilson. "Coordinated Memory Replay in the Visual Cortex and Hippocampus During Sleep." *Nature Neuroscience* 10 (2007): 100–107.

Kabat-Zinn, Jon. *Wherever You Go, There You Are*. New York: Hyperion, 2004.

Kapleau, Philip Roshi. *The Three Pillars of Zen*. New York: Anchor Books, Doubleday, 1989.

Kelen, Betty. *Gautama Buddha in Life and Legend*. New York: Avon Books, 1967.

Kesarcordi-Watson, Ian. "An Ancient Indian Argument for What I Am." *Journal of Indian Philosophy* 9 (1981): 259–272.

Killingsworth, Matthew A., and Daniel T. Gilbert. "A Wandering Mind Is an Unhappy Mind." *Science* 330 (2010): 932.

Koch, Christof. *The Quest for Consciousness: A Neurobiological Approach*. Greenwood Village, CO: Roberts & Company, 2007.

Koyabashi, Eliane, et al. "Widespread and Intense BOLD Changes During Brief Focal Electrographic Seizures." *Neurology* 66 (2006): 1049–1055.

Kozhevnikov, Maria, et al., "Neurocognitive and Somatic Components of Temperature Increases During g-Tummo Meditation: Legend and Reality." *PLoS ONE* 8 (3): e58244. doi:10.1371/journal.pone.0058244.

Kroeger, Daniel, et al. "Human Brain Activity Patterns Beyond the Isoelectric Line of Extreme Deep Coma." *PLoS ONE* 8 (9): e75257. doi:10.1371/journal.pone.0075257.

Krueger, James M., et al. "Sleep as a Fundamental Property of Neuronal Assemblies." *Nature Reviews Neuroscience* 9 (2008): 910–919.

LaBerge, Stephen. "Lucid Dreaming." In Patrick McNamara and Deirdre Barrett, eds., *The New Science of Dreaming, Volume II: Content, Recall, and Personality Correlates of Dreams*, 307–328. Westport, CT: Praeger, 2007.

——. "Lucid Dreaming and the Yoga of the Dream State: A Psychophysiological Perspective." In B. Alan Wallace, ed., *Buddhism and Science: Breaking New Ground*, 233–258. New York: Columbia University Press, 2003.

——. "Signal-Verified Lucid Dreaming Proves That REM Sleep Can Support Reflective Consciousness." *International Journal of Dream Research* 3 (2010): 26–27.

Laberge, Stephen, and Howard Rheingold. *Exploring the World of Lucid Dreaming*. New York: Ballantine, 1990.

LaBerge, Stephen, et al. "Lucid Deaming Verified by Volitional Communication During REM Sleep." *Perceptual and Motor Skills* 52 (1981): 727–731.

——. "Psychophysiological Correlates of the Initiation of Lucid Dreaming." *Sleep Research* 10 (1981): 149.

Lati Rinpoche and Jeffrey Hopkins. *Death, Intermediate State and Rebirth in Tibetan Buddhism*. Ithaca, NY: Snow Lion, 1981.

Legrand, Dorothée. "Myself with No Body? Body, Bodily-Consciousness, and Self-Consciousness." In Daniel Schmicking and Shaun Gallagher, eds., *Handbook of Phenomenology and Cognitive Science*, 181–200. New York, Heidelberg, London: Springer, 2010.

——. "Pre-reflective Self-as-Subject from Experiential and Empirical Perspectives." *Consciousness and Cognition* 16 (2007): 583–599.

Legrand, Dorothée, and Perrine Ruby. "What Is Self-Specific? A Theoretical Investigation and Critical Review of Neuroimaging Results." *Psychological Review* 116 (2009): 252–282.

Lempert, Thomas, et al. "Syncope and Near-Death Experience." *The Lancet* 334 (1994): 829–830.

Lenggenhager, Bigna, et al. "Spatial Aspects of Bodily Self-Consciousness." *Consciousness and Cognition* 18 (2009): 110–117.

——. "Video Ergo Sum: Manipulating Bodily Self-Consciousness." *Science* 317 (2007): 1096–1099.

Le Van Quyen, Michel, et al. "Large-Scale Microelectrode Recordings of High-Frequency Gamma Oscillations in Human Cortex During Sleep." *Journal of Neuroscience* 30 (2010): 7770–7782.

Levitan, Lynne, and Stephen LaBerge. "Testing the Limits of Dream Control: The Light and Mirror Experiment." *Nightlight* 5 (2) (Summer 1993). http://www.lucidity.com/NL52.LightandMirror.html.

Levitan, Lynne, et al. "Out-of-Body Experiences, Dreams, and REM Sleep." *Sleep and Hypnosis* 1 (1999): 186–196.

Llinás, Rodolfo. *The I of the Vortex: From Neurons to Self*. Cambridge, MA: MIT Press, 2002.

Llinás, Rodolfo, and Urs Ribary. "Coherent 40-Hz Oscillation Characterizes Dream State in Humans." *Proceedings of the National Academy of Sciences U.S.A.* 90 (1993): 2078–2081.

——. "Perception as an Oneiric-Like State Modulated by the Senses." In Christof Koch and Joel L. Davis, eds., *Large-Scale Neuronal Theories of the Brain*, 111–124. Cambridge, MA: MIT Press, 1994.

Llinás, Rodolfo, et al. "The Neuronal Basis for Consciousness." *Philosophical Transactions of the Royal Society of London. Series B: Biological Sciences* 353 (1998): 1841–1849.

Logothetis, Nikos K. "Single Units and Conscious Vision." *Philosophical Transactions of the Royal Society of London B* 353 (1998): 1801–1818.

——. "Vision: A Window on Consciousness." *Scientific American* 281 (1999): 68–75.

Lopez, Donald, Jr. *Buddhism and Science: A Guide for the Perplexed*. Chicago: University of Chicago Press, 2008.

Lusthaus, Dan. *Buddhist Phenomenology: A Philosophical Investigation of Yogācāra Buddhism and the Ch'eng Wei-shih lun*. London and New York: RoutledgeCurzon, 2002.

Lutz, Antoine, and Evan Thompson. "Neurophenomenology: Integrating Subjective Experience and Brain Dynamics in the Neuroscience of Consciousness." *Journal of Consciousness Studies* 10 (2003): 31–52.

Lutz, Antoine, et al. "Attention Regulation and Monitoring in Meditation." *Trends in Cognitive Sciences* 12 (2008): 163–169.

——. "Changes in the Tonic High-Amplitude Gamma Oscillations During Meditation Correlate with Long-Term Practitioners' Verbal Reports." Association for the Scientific Study of Consciousness Annual Meeting, Poster Presentation, 2006.

——. "Guiding the Study of Brain Dynamics by Using First-Person Data: Synchrony Patterns Correlate with Ongoing Conscious States During a Simple Visual Task." *Proceedings of the National Academy of Sciences U.S.A.* 99 (2002): 1586–1591.

——. "Long-Term Meditators Self-Induce High Amplitude Gamma Synchrony During Mental Practice," *Proceedings of the National Academy of Sciences* 101 (2004): 16369–16373.

——. "Meditation and the Neuroscience of Consciousness: An Introduction." In Philip David Zelazo et al., eds., *The Cambridge Handbook of Consciousness*, 499–553. Cambridge: Cambridge University Press, 2007.

——. "Mental Training Enhances Attentional Stability: Neural and Behavioral Evidence." *Journal of Neuroscience* 29 (2009): 13418–13427.

MacKenzie, Matthew. "Enacting the Self: Buddhist and Enactivist Approaches to the Emergence of Self." In Mark Siderits et al., eds., *Self, No Self? Perspectives from Analytical, Phenomenological, and Indian Traditions*, 239–273. Oxford: Oxford University Press, 2011.

——. "The Illumination of Consciousness: Approaches to Self-Awareness in the Indian and Western Traditions." *Philosophy East and West* 57: 40–62.

Maclean, Katherine A., et al. "Intensive Meditation Training Improves Perceptual Discrimination and Sustained Attention." *Psychological Science* 21 (2010): 829–839.

Mair, Victor H. *Wandering on the Way: Early Taoist Tales and Parables of Chuang Tzu.* Honolulu: University of Hawaii Press, 1998.

Maquet, Pierre et al. "Human Cognition During REM Sleep and the Cortical Activity Profile within Frontal and Parietal Cortices: A Reappraisal of Functional Neuroimaging Data." *Progress in Brain Research* 150 (2005): 219–227.

Margulis, Lynn, and Dorion Sagan. *What is Life?* New York: Simon and Schuster, 1995.

Marsh, Michael N. *Out-of-Body and Near-Death Experiences: Brain-State Phenomena or Glimpses of Immortality?* Oxford: Oxford University Press, 2010.

Mason, L. I., et al. "Electrophysiological Correlates of Higher States of Consciousness During Sleep in Long-Term Practitioners of the Transcendental Meditation Program." *Sleep* 20 (1997): 102–110.

Mason, Malia F., et al. "Wandering Minds: The Default Network and Stimulus Independent Thoughts." *Science* 315 (2007): 393–395.

Massimini, Marcello, et al. "Breakdown of Cortical Effective Connectivity During Sleep." *Science* 309 (2005): 2228–2232.

Mathewson, Kyle E., et al. "To See or Not to See: Prestimulus α Phase Predicts Visual Awareness." *Journal of Neuroscience* 29 (2009): 2725–2732.

Maturana, Humberto R., and Francisco J. Varela. *Autopoiesis and Cognition: The Realization of the Living*. Boston Studies in the Philosophy of Science, vol. 42. Dordrecht: D. Reidel, 1980.

——. *The Tree of Knowledge: The Biological Roots of Human Understanding*. Boston: Shambhala/New Science Library, 1987.

Maury, Alfred. "Des hallucinations hypnagogiques ou des erreurs des sens dans l'etat intermédiare entre la veille et le sommeil" [Hypnagogic hallucinations or sensory errors in the intermediate state between wakefulness and sleep]. *Annales Medico-Psychologiques du système nerveux* 11 (1848): 26–40.

Mavromatis, Andreas. *Hypnagogia: The Unique State of Consciousness Between Wakefulness and Sleep*. London and New York: Routledge and Kegan Paul, 1987.

May, Todd. *Death*. Stocksfield: Acumen Publishing, 2009.

Mayer, Elizabeth Lloyd. *Extraordinary Knowing: Science, Skepticism, and the Inexplicable Powers of the Human Mind*. New York: Bantam, 2007.

McGinn, Colin. *Mindsight: Image, Dream, Meaning*. Cambridge, MA: Harvard University Press, 2004.

——. *The Mysterious Flame: Conscious Minds in a Material World*. New York: Basic Books, 2000.

McIsaac, Heather K., and Eric Eich. "Vantage Point in Episodic Memory." *Psychological Science* 9 (2002): 146–150.

McMahan, David L. *The Making of Buddhist Modernism*. New York: Oxford University Press, 2008.

McNamara, Patrick, et al. "'Theory of Mind' in REM and NREM Dreams." In Patrick McNamara and Deirdre Barrett, eds., *The New Science of Dreaming, Volume II: Content, Recall, and Personality Correlates*, 201–220. Westport, CT: Praeger, 2007.

Melloni, Lucia, et al. "Synchronization of Neural Activity Across Cortical Areas Correlates with Conscious Perception." *Journal of Neuroscience* 27 (2007): 2858–2865.

Metzinger, Thomas. *Being No One: The Self-Model Theory of Subjectivity*. Cambridge, MA: MIT Press/A Bradford Book, 2003.

——. *The Ego Tunnel: The Science of the Mind and the Myth of the Self*. New York: Basic Books, 2009.

——. "The No-Self Alternative." In Shaun Gallagher, ed., *The Oxford Handbook of the Self*, 297–315. New York and Oxford: Oxford University Press, 2011.

——. "Out-of-Body Experiences as the Origin of the Concept of a 'Soul.'" *Mind & Matter* 3 (2005): 57–84.

Mobbs, Dean, and Caroline Watt. "There Is Nothing Paranormal About Near-Death Experiences: How Neuroscience Can Explain Seeing Bright Lights, Meeting the Dead, or Being Convinced You are One of Them." *Trends in Cognitive Sciences* 15 (2011): 447–449.

Moeller, Hans-Georg. *Daoism Explained: From the Dream of the Butterfly to the Fishnet Allegory.* Chicago and LaSalle, IL: Open Court, 2004.

Monroe, Robert. *Journeys Out of the Body.* New York: Doubleday, 1971.

Moody, Raymond A., Jr. *Life After Life: The Investigation of a Phenomenon—Survival of Bodily Death.* San Francisco: Harper, 1975, 2001.

Muldoon, Sylvan, and Hereward Carrington. *The Projection of the Astral Body.* New York: Samuel Weiser, 1969.

Murphy, Nancey, et al., eds. *Downward Causation and the Neurobiology of Free Will.* Berlin: Springer, 2009.

Muzur, Amir, et al. "The Prefrontal Cortex in Sleep." *Trends in Cognitive Sciences* 6 (2002): 475–481.

Myers, Frederick H.M. *Human Personality and Its Survival of Bodily Death.* London: Longmans, Green, 1903.

Nabokov, Vladimir. *Speak, Memory. An Autobiography Revisited.* New York: Vintage International, 1989.

Nagendra, Ravindra P., et al. "Meditation and Its Regulatory Role on Sleep." *Frontiers in Neurology* 3 (2012) Article 54: 1–3.

Naiman, Rubin R. *Healing Night: The Science and Spirit of Sleeping, Dreaming, and Awakening.* Minneapolis: Syren Book Company, 2006.

Ñaṇamoli, Bhikkhu, and Bodhi, Bhikkku, trans. *The Middle Length Discourses of the Buddha. A Translation of the Majjhima Nikāya.* Somerville, MA: Wisdom, 1995.

Nielsen, Tore A. "Describing and Modeling Hypnagogic Imagery Using a Systematic Self-Observation Procedure." *Dreaming* 5 (1995): 75–94.

——. "A Review of Mentation in REM and NREM Sleep: 'Covert' REM Sleep as a Possible Reconciliation of Two Opposing Models." *Behavioral and Brain Sciences* 23 (2000): 851–866.

Nietzsche, Friedrich. *The Birth of Tragedy and Other Writings.* Ed. Raymond Geuss and Ronald Speirs. Cambridge: Cambridge University Presss, 1999.

Nigro, Georgia, and Ulric Neisser. "Point of View in Personal Memories." *Cognitive Psychology* 15 (1983): 467–482.

Nir, Yuval, and Guilio Tononi. "Dreaming and the Brain: From Phenomenology to Neurophysiology." *Trends in Cognitive Sciences* 14 (2009): 88–100.

Norbu, Chogyal Namkhai. *Dream Yoga and the Practice of the Natural Light.* Ithaca, NY: Snow Lion, 1992.

Obeyesekere, Gananath. *Imagining Karma: Ethical Transformation in Amerindian, Buddhist, and Greek Rebirth*. Berkeley: University of California Press, 2002.

O'Flaherty, Wendy Doniger. *Dreams, Illusion and Other Realities*. Chicago: University of Chicago Press, 1984.

Olivelle, Patrick, trans. *Upanisads*. Oxford: Oxford University Press, 1996.

Owen, Adrian M., and Martin R. Coleman. "Functional Neuroimaging of the Vegetative State." *Nature Reviews Neuroscience* 9 (2008): 235–243.

Pace-Schott, Edward F. "The Frontal Lobes and Dreaming." In Patrick McNamara and Deirdre Barrett, eds., *The New Science of Dreaming, Volume I: The Biology of Dreaming*, 115–154. Westport, CT: Praeger, 2007.

Padmasambhava. *Natural Liberation: Padmasambhava's Teachings on the Six Bardos*. Trans. B. Alan Wallace. Boston: Wisdom, 1998.

Pagel, J. F. *The Limits of Dream. A Scientific Exploration of the Mind/Brain Interface*. Oxford: Elsevier/Academic Press, 2008.

Palva, Satu, et al. "Early Neural Correlates of Conscious Somatosensory Perception." *Journal of Neuroscience* 25 (2005): 5248–5258.

Parnia, Sam, and Peter Fenwick. "Near Death Experiences in Cardiac Arrest: Visions of a Dying Brain or Visions of a New Science of Consciousness." *Resuscitation* 52 (2002): 5–11.

Peigneux, Philippe, et al. "Are Spatial Memories Strengthened in the Human Hippocampus During Slow-Wave Sleep?" *Neuron* 44 (2004): 535–545.

Pessoa, Luiz, et al. "Target Visibility and Visual Awareness Modulate Amygdala Responses to Fearful Faces." *Cerebral Cortex* 16 (2006): 366–375.

Petitmengin, Claire. "Describing One's Subjective Experience in the Second Person: An Interview Method for the Science of Consciousness." *Phenomenology and the Cognitive Sciences* 5 (2009): 229–269.

Petitmengin, Claire, ed. *Ten Years of Viewing from Within: The Legacy of F. J. Varela*. Special issue of the *Journal of Consciousness* 16 (10–12) (October–December 2009).

Petitmengin, Claire, et al. "Anticipating Seizure: Pre-Reflective Experience at the Center of Neuro-Phenomenology." *Consciousness and Cognition* 16 (2007): 746–764.

——. "Seizure Anticipation: Are Neurophenomenological Approaches Able to Detect Preictal Symptoms?" *Epilepsy and Behavior* 9 (2006): 298–306.

Petkova, Valeria I., and H. Henrik Ehrsson. "If I Were You: Perceptual Illusion of Body Swapping." *PLoS ONE* 3 (12) (2008): e3832. doi:10.1371/journal.pone.0003832.

Phillips, Stephen. *Yoga, Karma, and Rebirth: A Brief History and Philosophy*. New York: Columbia University Press, 2009.

Ponlop, Dzogchen. *Mind Beyond Death*. Ithaca, NY: Snow Lion, 2006.

Proust, Marcel. *The Way by Swann's*. Trans. Lydia Davis. London: Penguin, 2003.

Pruden, Leo M., trans. *Abhidharmakosabhasyam, by Louis de la Vallée Poussin. Vols 1–4*. Berkeley, CA: Asian Humanities Press, 1991.

Rabten, Geshe. *The Mind and Its Functions*. Rrans. Stephen Batchelor. Mt. Pelerin, Switzerland: Tharpa Choeling, 1981.

Rāmānuja. *The Vedānta-Sūtras with the Commentary by Rāmānuja. Sacred Books of the East, Volume 48*. Trans. George Thibaut, 1904. http://www.sacred-texts.com/hin/sbe48/index.htm.

Ram-Prasad, Chakravarthi. "Dreams and Reality: The Śaṅkarite Critique of Vijñānavāda." *Philosophy East and West* 43 (1993): 405–455.

——. *Indian Philosophy and the Consequences of Knowledge: Themes in Ethics, Metaphysics, and Soteriology*. Hampshire, England and Burlington, VT: Ashgate, 2007.

Rasch, Björn, et al. "Odor Cues During Slow-Wave Sleep Prompt Declarative Memory Consolidation." *Science* 315 (2007): 1426–1429.

Raveh, Daniel. "*Ayam aham asmīti:* Self-Consciousness and Identity in the Eighth Chapter of the *Chāndogya Upaniṣad* vs Śankara's *Bhāṣya.*" *Journal of Indian Philosophy* 36 (2008): 319–333.

Revonsuo, Antti. *Inner Presence: Consciousness as a Biological Phenomenon*. Cambridge, MA: MIT Press, 2006.

Ring, Kenneth. *Life at Death: A Scientific Investigation of Near-Death Experience*. New York: Conward, McCann & Geoghegan, 1980.

Rock, Andrea. *The Mind at Night: The New Science of How and Why We Dream*. New York: Basic Books, 2004.

Rodriguez, Eugenio, et al. "Perception's Shadow: Long-Distance Synchronization of Human Brain Activity." *Nature* 397 (1999): 430–433.

Roebuck, Valerie J., trans. *The Upanisads*. London: Penguin, 2003.

Rooksby, Bob, and Sybe Terwee. "Freud, Van Eeden and Lucid Dreaming." http://www.spiritwatch.ca/LL%209(2)%20web/Rooksby_Terwee%20paper.htm.

Rushton, Cynda Hylton, et al. "Impact of a Contemplative End-of-Life Training Program: Being with Dying." *Palliative and Supportive Care* 7 (2009): 405–414.

Sabom, Michael. *Light and Death*. Grand Rapids, MI: Zondervan, 1998.

Saint-Denys, d'Hervey de. *Dreams and How to Guide Them*. London: Duckworth, 1982.

Sandars, N. K., trans. *The Epic of Gilgamesh*. London: Penguin, 1972.

Saraswati, Swami Satyananda. *Yoga Nidra*. 6th ed. Munger, Bihar, India: Yoga Publications Trust, 1998.

Sartre, Jean-Paul. *The Imaginary: A Phenomenological Psychology of the Imagination*. Trans. Jonathan Webber. London and New York: Routledge, 2004.

Schacter, Daniel L. "The Hypnagogic State: A Critical Review of the Literature." *Psychological Bulletin* 83 (1976): 452–481.

——. *Searching for Memory: The Brain, the Mind, and the Past*. New York: Basic Books, 1996.

Schacter, Daniel L., et al. "Remembering the Past to Imagine the Future: The Prospective Brain." *Nature Reviews Neuroscience* 8 (2007): 657–661.

Scharfstein, Ben-Ami. *A Comparative History of World Philosophy: From the Upanishads to Kant.* Albany: State University of New York Press, 1998.

Schmithausen, Lambert. *Ālayavijñāna: On the Origin and Early Development of a Central Concept of Yogācāra Philosophy.* Tokyo: International Institute for Buddhist Studies, 1987.

Schwartz, Sophie, and Pierre Maquet. "Sleep Imaging and the Neuropsychological Assessment of Dreams." *Trends in Cognitive Sciences* 6 (2002): 23–30.

Searle, John R. "Consciousness." *Annual Review of Neuroscience* 23 (2000): 557–578.

——. *Mind: A Brief Introduction.* New York: Oxford University Press, 2005.

——. *Minds, Brains and Science.* Cambridge, MA: Harvard University Press, 1986.

——. *The Mystery of Consciousness.* New York: The New York Review of Books, 1990.

Sebel, Peter S., et al. "The Incidence of Awareness During Anesthesia: A Multicenter United States Study." *Anesthesia and Analgesia* 99 (2004): 833–839.

Sharf, Robert H. "Buddhist Modernism and the Rhetoric of Meditative Experience." *Numen* 42 (1995): 228–283.

——. "The Rhetoric of Experience and the Study of Religion." *Journal of Consciousness Studies* 7 (2000): 267–287.

Sharma, Arvind. *Sleep as a State of Consciousness in Advaita Vedānta.* Albany: State University of New York Press, 2004.

Sharma, Ramesh Kumar. "Dreamless Sleep and Some Related Philosophical Issues." *Philosophy East and West* 51 (2001): 210–231.

Shoemaker, Sidney. "Self-Reference and Self-Awareness." In *Identity, Cause, and Mind: Philosophical Essays*, 6–18. Cambridge: Cambridge University Press, 1984.

Siderits, Mark. *Buddhism as Philosophy.* Indianaoplis, IN: Hackett, 2007.

Siderits, Mark, and Shōryū Katsura. *Nāgārjuna's Middle Way: Mūlamadhyamakakārikā.* Somerville, MA: Wisdom, 2013.

Siderits, Mark, et al., eds. *Self, No Self? Perspectives from Analytical, Phenomenological, and Indian Traditions.* Clarendon: Oxford University Press, 2010.

Silberer, Herbert. "Report on a Method of Eliciting and Observing Certain Symbolic Hallucination Phenomena." In D. Rapaport, ed., *Organization and Pathology of Thought.* New York: Columbia University Press, 1951.

Slagter, Heleen, et al. "Mental Training Affects Distribution of Limited Brain Resources," *PLoS Biology* 5 (6) (2007): e138. doi:10.1371/journal.pbio.0050138.

——. "Theta Phase Synchrony and Conscious Target Perception: Impact of Intensive Mental Training." *Journal of Cognitive Neuroscience* 21 (2009): 1536–1549.

Slater, Mel, et al. "First Person Experience of Body Transfer in Virtual Reality." *PLoS ONE* 5 (5) (2010): e10564. doi:10.1371/journal.pone.0010564.

Smallwood, Jonathan, et al. "Shifting Moods, Wandering Minds: Negative Moods Lead the Mind to Wander." *Emotion* 9 (2009): 271–276.

Smilek, Daniel, et al. "When '3' is a Jerk and 'E' is a King: Personifying Inanimate Objects in Synesthesia." *Journal of Cognitive Neuroscience* 19 (2007): 981–992.

Sogyal Rinpoche. *The Tibetan Book of Living and Dying*. San Francisco: HarperSanFrancisco, 1993.

Sokolowski, Robert. *An Introduction to Phenomenology*. New York: Cambridge University Press, 2000.

Solms, Mark. "Dreaming and REM Sleep Are Controlled by Different Brain Mechanisms." *Behavioral and Brain Sciences* 23 (2000): 843–850.

Solms, Mark, and Oliver Turnbull. *The Brain and the Inner World: An Introduction to the Neuroscience of Subjective Experience*. New York: Other Press, 2002.

Sosa, Ernest. "Dreams and Philosophy." *Proceedings and Addresses of the American Philosophical Association* 79 (2) (2005): 7–18.

Spitzer, H., et al. "Increased Attention Enhances Both Behavioral and Neuronal Performance." *Science* 15 (1988): 338–340.

Stenstrom, Philippe, et al. "Mentation During Sleep Onset Theta Bursts in a Trained Participant: A Role for NREM Stage 1 Sleep in Memory Processing?" *International Journal of Dream Research* 5 (2012): 37–46.

Stevenson, Ian. *Children Who Remember Previous Lives: A Question of Reincarnation*, rev. ed. Jefferson, NC and London: McFarland and Company, 2000.

——. *European Cases of the Reincarnation Type*. Jefferson, NC and London: McFarland and Company, 2003.

Stickgold, Robert, et al. "Replaying the Game: Hypnagogic Images in Normals and Amnesics." *Science* 290 (2000): 350–353.

Strawson, Galen. *Real Materialism and Other Essays*. Oxford: Clarendon Press, 2008.

——. "Realistic Monism: Why Physicalism Entails Panpsychism." *Journal of Consciousness Studies* 13 (2006): 3–31.

Stryk, Lucien, et al. *Zen Poems of China and Japan*. New York: Grove Press, 1973.

Stumbrys, Tadas, et al. "Induction of Lucid Dreams: A Systematic Review of the Evidence." *Consciousness and Cognition* 21 (2012): 1456–1475.

Szczepanowski, Remigiusz, and Luiz Pessoa. "Fear Perception: Can Objective and Subjective Awarenes Measures Be Dissociated?" *Journal of Vision* 7 (2007): 1–17.

Tanaka, Hideki, et al. "Topographical Characteristics and Principal Component Structure of the Hypnagogic EEG." *Sleep* 20 (1997): 523–534.

Tang, Yi-Yuan, et al. "Central and Autonomic Nervous System Interaction Is Altered by Short-Term Meditation." *Proceedings of the National Academy of Sciences U.S.A.* 106 (2009): 8865–8870.

——. "Short-Term Meditation Training Improves Attention and Self-Regulation." *Proceedings of the National Academy of Sciences U.S.A.* 104 (2007): 17152–17156.

Tart, Charles. *The End of Materialism: How Evidence of the Paranormal Is Bringing Science and Spirit Together*. Oakland, CA: New Harbinger Publications, 2009.

Tart, Charles, ed. *Altered States of Consciousness*. New York: Anchor Books, Double-day, 1972.

Theise, Neil. "Now You See It, Now You Don't." *Nature* 435 (2005): 1165.

Tholey, Paul. "Techniques for Inducing and Manipulating Lucid Dreams." *Perceptual and Motor Skills* 57 (1983): 79–90.

Thompson, Evan. *Colour Vision: A Study in Cognitive Science and the Philosophy of Perception*. London: Routledge, 1995.

——. "Living Ways of Sense-Making." *Philosophy Today* SPEP Supplement 2011: 114–123.

——. *Mind in Life: Biology, Phenomenology, and the Sciences of Mind*. Cambridge, MA: Harvard University Press, 2007.

——. "Neurophenomenology and Francisco Varela." In Anne Harrington and Arthur Zajonc, eds., *The Dalai Lama at MIT*, 19–24. Cambridge, MA: Harvard University Press, 2003, 2006.

Thompson, Evan, and Diego Cosmelli. "Brain in a Vat or Body in a World? Brainbound Versus Enactive Views of Experience/" *Philosophical Topics* 39 (2011): 163–180.

Thompson, Evan, and Francisco J. Varela. "Radical Embodiment: Neural Dynamics and Consciousness." *Trends in Cognitive Sciences* 5 (2001): 418–425.

Thompson, William Irwin. *Coming Into Being: Artifacts and Texts in the Evolution of Consciousness*. New York: St. Martin's Press, 1996, 1998.

——. *Nightwatch and Dayshift*. Princeton, NJ: Wild River Review Books, 2014.

Tomasello, Michael. *The Cultural Origins of Human Cognition*. Cambridge, MA: Harvard University Press, 1999.

Tononi, Giulio. "Consciousness as Integrated Information: A Provisional Manifesto." *Biological Bulletin* 215 (2008): 216–242.

Tononi, Giulio, and Christof Koch. "The Neural Correlates of Consciousness: An Update." *Annals of the New York Academy of Sciences* 1124 (2008): 239–261.

Tononi, Giulio, and Marcello Massimini. "Why Does Consciousness Fade in Early Sleep?" *Annals of the New York Academy of Sciences* 1129 (2008): 330–334.

Tranströmer, Tomas. *Selected Poems 1954–1986*. Hopewell, NJ: Ecco Press, 1987.

Valberg, J. J. *Dream, Death, and the Self*. Princeton, NJ: Princeton University Press, 2007.

van der Kloet, Dalena, et al. "Fragmented Sleep, Fragmented Mind: The Role of Sleep in Dissociative Symptoms." *Perspectives on Psychological Science* 7 (2012): 159–175.

Van Eeden, Frederik. "A Study of Dreams." In Charles Tart, ed., *Altered States of Consciousness*, 147–160. New York: Anchor Books, 1969. Also available at: http://www.lucidity.com/vanEeden.html.

Vanhaudenhuyse, Audrey, Marie Thonnard, and Steven Laureys. "Towards a Neuro-Scientific Explanation of Near-Death Experiences?" In Jean-Louis Vincent, ed.,

Yearbook of Intensive Care and Emergency Medicine, 961–968. Berlin: Springer-Verlag, 2009.

van Lommel, Pim. *Consciousness Beyond Life: The Science of Near-Death Experience.* New York: Harper One, 2010.

van Lommel, Pim, et al. "Near-Death Experience in Survivors of Cardiac Arrest: A Prospective Study in the Netherlands." *The Lancet* 358 (2001): 2039–2045.

VanRullen, Ruffin, and Christof Koch. "Is Perception Discrete or Continuous?" *Trends in Cognitive Sciences* 7 (2003): 207–213.

VanRullen, Ruffin, et al. "The Blinking Spotlight of Attention." *Proceedings of the National Academy of Sciences U.S.A.* 104 (2007): 19204–19209.

——. "Ongoing EEG Phase as a Trial-by-Trial Predictor of Perceptual and Attentional Variability." *Frontiers in Psychology* 2 (2011): 1–9.

Vass, Arpad A. "Dust to Dust: The Brief, Eventful Afterlife of a Human Corpse." *Scientific American* (September 2010): 56–58.

Vass, Arpad A., et al. "Beyond the Grave: Understanding Human Decomposition." *Microbiology Today* 28 (2001): 190–192.

Varela, Francisco J. "The Creative Circle: Sketches on the Natural History of Circularity." In Paul Watzlavick, ed., *The Invented Reality: How Do We Know What We Believe We Know?*, 309–323. New York: Norton, 1984.

——. "Neurophenomenology: A Methodological Remedy for the Hard Problem." *Journal of Consciousness Studies* 3 (1996): 330–350.

Varela, Francisco J., and Jonathan Shear, eds. *The View from Within: First-Person Approaches to the Study of Consciousness.* Thorverton: Imprint Academic, 1991.

Varela, Francisco J., et al. "The Brainweb: Phase Synchronization and Large-Scale Integration." *Nature Reviews Neuroscience* 2 (2001): 229–239.

——. *The Embodied Mind: Cognitive Science and Human Experience.* Cambridge, MA: MIT Press, 1991; expanded ed., 2015.

——. "Perceptual Framing and Cortical Alpha Rhythm." *Neuropsychologia* 19 (1981): 675–686.

Vincent, Justin L., et al., "Evidence for a Frontoparietal Control System Revealed by Intrinsic Connectivity." *Journal of Neurophysiology* 100 (2008): 3328–3342.

Vladimirov, Nikita, and Victor Sourjik. "Chemotaxis: How Bacteria Use Memory." *Biological Chemistry* 390 (2009): 1097–1104.

Vogel, Gerald, et al. "Ego Functions and Dreaming During Sleep Onset." In Charles Tart, ed., *Altered States of Consciousness,* 77–94. New York: Anchor Books, Doubleday, 1972.

Voss, Ursula, et al. "Lucid Dreaming: A State of Consciousness with Features of Both Waking Consciousness and Non-lucid Dreaming." *Sleep* 32 (2009): 1191–1200.

Wackerman, Jiri, et al. "Brain Electrical Activity and Subjective Experience During Altered States of Consciousness: Ganzfeld and Hypnagogic States." *International Journal of Psychophysiology* 46 (2002): 123–146.

Waldron, William S. "Buddhist Steps to an Ecology of Mind: Thinking About 'Thoughts Without a Thinker.'" *Eastern Buddhist* 34 (2002): 1–52.

———. *The Buddhist Unconscious: The Ālayavijñāna in the Context of Indian Buddhist Thought.* London and New York: RoutledgeCurzon, 2003.

Walker, Matthew P. "The Role of Sleep in Cognition and Emotion." *Annals of the New York Academy of Sciences* 1156 (2009): 168–197.

Wallace, B. Alan. *Contemplative Science: Where Buddhism and Neuroscience Converge.* New York: Columbia University Press, 2007.

———. *Genuine Happiness: Meditation as a Path to Fulfillment.* Hoboken, NJ: Wiley, 2005.

———. *The Taboo of Subjectivity: Toward a New Science of Consciousness.* New York: Oxford University Press, 2000.

Wallace, B. Alan, and Brian Hodel. *Dreaming Yourself Awake: Lucid Dreaming and Tibetan Dream Yoga for Insight and Transformation.* Boston: Shambhala, 2012.

Wangyal, Tenzin Rinpoche. *The Tibetan Yogas of Dream and Sleep.* Ithaca, NY: Snow Lion, 1998.

Warren, Jeff. *Head Trip: Adventures on the Wheel of Consciousness.* Toronto: Random House Canada, 2007.

Westerhoff, Jan. *Nagarjuna's Madhyamaka: A Philosophical Introduction.* New York and Oxford: Oxford University Press, 2009.

———. *Twelve Examples of Illusion.* New York: Oxford University Press, 2010.

Willson, Martin. *Rebirth and the Western Buddhist.* London: Wisdom, 1987.

Wilson, Matthew A., and Bruce L. McNaughton. "Reactivation of Hippocampal Ensemble Memories During Sleep." *Science* 265 (1994): 676–679.

Windt, Jennifer M. "The Immersive Spatiotemporal Hallucination Model of Dreaming." *Phenomenology and the Cognitive Sciences* 9 (2010): 295–316.

Windt, Jennifer Michelle, and Thomas Metzinger. "The Philosophy of Dreaming and Self-Consciousness: What Happens to the Experiential Subject During the Dream State?" In Patrick McNamara and Deirdre Barrett, eds., *The New Science of Dreaming, Volume III: Cultural and Theoretical Perspectives on Dreaming,* 193–247. Westport, CT: Praeger, 2007.

Woerlee, G. M. "An Anaesthesiologist Examines the Pam Reynolds Story, Part I: Background Considerations." *The Skeptic* [UK] 18 (1) (2005): 14–17.

———. "An Anaesthesiologist Examines the Pam Reynolds Story. Part II: The Experience." *The Skeptic* [UK] 18 (2) (2005): 16–20.

———. "The Man with the Dentures." http://www.neardeath.woerlee.org/man-with-the-dentures.php.

———. *Mortal Minds: The Biology of Near-Death Experiences.* Amherst, NY: Prometheus Books, 2003.

———. "Pam Reynolds Near Death Experience," http://www.neardeath.woerlee.org/pam-reynolds-near-death-experience.php.

——. "Setting the Record Straight. Commentary on an Article by Pim van Lommel," http://www.neardeath.woerlee.org/setting-the-record-straight.php.

Yogananda, Paramhansa. *Autobiography of a Yogi.* Reprint of the Philosophical Library 1946 First Edition. Nevada City, CA: Crystal Clarity Publishers, 2005.

Zahavi, Dan. "The Experiential Self: Objections and Clarifications." in Mark Siderits et al., eds., *Self, No Self? Perspectives from Analytical, Phenomenological, and Indian Traditions,* 56–78. Clarendon: Oxford University Press, 2010.

——. *Subjectivity and Selfhood: Investigating the First-Person Perspective.* Cambridge, MA: MIT Press, 2005.

Zhuangzi. *Zhuangzi: The Essential Writings.* Trans. Brook Ziporyn. Indianapolis, IN: Hackett, 2009.

INDEX

Abhidharma philosophical tradition:
duration of moments of consciousness,
40, 45–46; and intentionality
("aboutness") of the mind, 36; mental
factors, 37–39, 48, 58; and phenomenal
vs. access consciousness, 49; and
stream of consciousness/discrete
moments of awareness, 34–39, 56–59
access consciousness, 7–8, 48–49, 248,
257, 341
acetylcholine, 175
Advaita Vedānta school of Indian
philosophy: and distinguishing
waking from dreaming, 195; and
dreamless sleep state, 248–50; and
memory vs. inference of sleep,
240–46; and "seed sleep," 260–61;
and witnessing not knowing, 246–48
"After Apple-Picking" (Frost), 111–12
agency, 109, 185, 208, 211
Alkire, Michael, 252–53
alpha waves, 31, 42, 43, 118, 120, 270
amygdala, 131, 137, 305, 313
Anālayo, 25, 320
Anderson, Adam, 353–55
anesthesia, 252–53, 306, 309
animals: mirror test for self-recognition, 345;
and perception, 14, 27; and sensorimotor
selfhood, 332–34; sentience/feeling
in nonhuman organisms, 335–36,
339–44. *See also* bacteria; cells

annihilationism, 322
anoxia, 311–12
anterior cingulate cortex, 137, 172, 352, 353
Aristotle, 160
astral body, 204, 218, 220–21
Atiśa, 276
ātman, 2–3, 5, 9–11, 23, 195
attention: attentional blink experiment,
49–56; attentional perspective and
out-of-body experiences, xxxvii (*see
also* out-of-body experiences); contrast
to concentration and mindfulness,
38–39; different aspects of, 44, 45;
and discrete moments of awareness,
44–50; and dream state, 136–37, 152,
181–82; focused attention meditation,
xxxiii, 51–52, 54–56, 76, 351, 352; and
imagination, 181–82; and infants and
social self-making, 346; and lucid
dreaming, 151–53, 158; as mental
factor (Abhidharma tradition), 37,
38, 48; and metacognition, 135; and
"name-and-form," 25; and nonlucid
dreaming, 136–37; and REM sleep, 186;
spellbound attention in hypnagogic
state, xxxv, xxxvi, 122–27; "spotlight"
metaphor, 44; and Vipassanā
meditation, 51–57; and waking state,
xxxiv, 9, 351; and wandering mind, 351.
See also meditation
Austin, James, 121–23